INTELLECTUAL PROPERTY AND ENTREPRENEURSHIP

ADVANCES IN THE STUDY OF ENTREPRENEURSHIP, INNOVATION AND ECONOMIC GROWTH

Series Editor: Gary D. Libecap

Volume 10: Legal, Regulatory and Policy Changes that Affect Entrepreneurial Midsize Firms, 1998

Volume 11: The Sources of Entrepreneurial Activity, 1999

Volume 12: Entrepreneurship and Economic Growth in the American Economy, 2000

Volume 13: Entrepreneurial Inputs and Outcomes: New Studies of Entrepreneurship in the United States, 2001

Volume 14: Issues in Entrepreneurship: Contracts, Corporate Characteristics and Country Differences, 2002

ADVANCES IN THE STUDY OF ENTREPRENEURSHIP,
INNOVATION AND ECONOMIC GROWTH VOLUME 15

INTELLECTUAL PROPERTY AND ENTREPRENEURSHIP

EDITED BY

GARY D. LIBECAP

The University of Arizona, USA

2004

ELSEVIER
JAI

Amsterdam – Boston – Heidelberg – London – New York – Oxford – Paris
San Diego – San Francisco – Singapore – Sydney – Tokyo

ELSEVIER B.V.
Sara Burgerhartstraat 25
P.O. Box 211,
1000 AE Amsterdam
The Netherlands

ELSEVIER Inc.
525 B Street, Suite 1900
San Diego,
CA 92101-4495
USA

ELSEVIER Ltd
The Boulevard, Langford
Lane Kidlington,
Oxford OX5 1GB
UK

ELSEVIER Ltd
84 Theobalds Road
London
WC1X 8RR
UK

First edition 2004

Library of Congress Cataloging in Publication Data
A catalog record is available from the Library of Congress.

British Library Cataloguing in Publication Data
A catalogue record is available from the British Library.

ISBN: 0-7623-1102-9
ISSN: 1048-4736 (Series)

♾ The paper used in this publication meets the requirements of ANSI/NISO Z39.48-1992 (Permanence of Paper). Printed in The Netherlands.

CONTENTS

LIST OF CONTRIBUTORS *vii*

INTRODUCTION
Gary D. Libecap *ix*

PROCURING KNOWLEDGE
Stephen M. Maurer and Suzanne Scotchmer *1*

ECONOMIC PERSPECTIVES ON OPEN SOURCE
Josh Lerner and Jean Tirole *33*

SUBMARINES AND TECHNOLOGICAL INNOVATION: U.S. CONTINUATION PATENTING IN SOFTWARE AND BIOTECHNOLOGY TECHNOLOGIES IN THE 1980s AND 1990s
Stuart J. H. Graham and David C. Mowery *71*

FIRM STRATEGIES AND TRENDS IN PATENT LITIGATION IN THE UNITED STATES
Deepak Somaya *103*

CROSSING THE GREAT DIVIDE: USING ADVERSE POSSESSION TO RESOLVE CONFLICTS BETWEEN THE ANTITRUST AND INTELLECTUAL PROPERTY REGIMES
Constance E. Bagley and Gavin Clarkson *149*

INCOMPLETE CONTRACTING AND THE STRUCTURE OF R&D JOINT VENTURE CONTRACTS
Suzanne E. Majewski and Dean V. Williamson *201*

WILL MP3 DOWNLOADS ANNIHILATE THE RECORD INDUSTRY? THE EVIDENCE SO FAR
Stan J. Liebowitz 229

STRATEGIES THAT WORK WHEN PROPERTY RIGHTS DON'T
Bharat Anand and Alexander Galetovic 261

LIST OF CONTRIBUTORS

Bharat Anand	Harvard Business School, Boston, Massachusetts, USA
Constance E. Bagley	Harvard Business School, Boston, Massachusetts, USA
Gavin Clarkson	School of Information, University of Michigan, Ann Arbor, Michigan, USA
Alexander Galetovic	Center of Applied Economics, Universidad de Chile, Chile
Stuart J. H. Graham	DuPree College of Management, Georgia Institute of Technology, Atlanta, Georgia, USA
Josh Lerner	Harvard Business School, Boston, Massachusetts, USA
Gary D. Libecap	Karl Eller Center, The University of Arizona, Tucson, USA
Stan J. Liebowitz	School of Management, University of Texas at Dallas, Texas, USA
Suzanne E. Majewski	Antitrust Division, U.S. Department of Justice, Economic Analysis Group, Washington, DC, USA
Stephen M. Maurer	Goldman School of Public Policy, University of California, Berkeley, California, USA
David C. Mowery	Walter A. Haas School of Business, University of California Berkeley and National Bureau of Economic Research, Haas School of Business, Berkeley, California, USA

Suzanne Scotchmer	Department of Economics and Goldman School of Public Policy, University of California, Berkeley, California, USA
Deepak Somaya	Robert H. Smith School of Business, University of Maryland, College Park, Maryland, USA
Jean Tirole	Institut d'Economie Industrielle, Toulouse, France
Dean V. Williamson	Antitrust Division, U.S. Department of Justice, Economic Analysis Group, Washington, DC, USA

INTRODUCTION

Although issues of intellectual property rights would seem to be ones of interest only to obscure groups of academics and lawyers, they have become topics of everyday discussion among the regular population. Alleged copyright infringements by teenagers downloading music from the internet and accompanying threats of prosecution as well as charges of strategic patenting to harm competitors in recent high profile antitrust cases have placed intellectual property into public and political debate.

Property rights to more tangible assets, such as land, have long been the source of economic and legal investigation as well as of public policy. Ownership of real property is relatively uncontroversial in an economy and society based upon private property. Indeed, much of the early legislative and legal structure of the U.S. was focused on the definition and support of property rights to natural resources and other real assets.[1] Intellectual property, especially patents, was recognized as important in the U.S. Constitution as critical for stimulating technological progress. Nevertheless, intellectual property did not become prominent until the mid-20th century as technology assumed a commanding role in productivity growth. As the search for new technology and its adoption became more important for lowering costs and providing new and refined products and services, the provision and protection of intellectual property became critical. The incentive effects of secure property rights for promoting research and development, investment, production, and exchange are well known. And these are the major arguments for patents, copyrights and other forms of intellectual property. Concerns about infringement on patents and copyrights from unauthorized production in other countries and downloading from the internet within the U.S. reflect, at least in part, worries about whether such activities will erode property rights sufficiently to reduce investment and production. Recently, there has been another concern, however, about the strategic use of patenting by companies as a means of blocking entry into markets, and here the overall social welfare effects are less apparent.

The papers in this volume represent some of the leading work on intellectual property. They address the question of how to create incentives to develop new technologies and how to protect those technologies once developed from theft. They also ask when valuable property might be developed even under weak ownership conditions. Other papers address how firms balance the tradeoffs in considering costly patent litigation and they examine the antitrust implications.

Chapter one, "Procuring Knowledge" by Stephen Maurer and Suzanne Scotchmer begins the volume with an examination of the mechanisms used to stimulate creativity, especially for basic scientific research, when there is a hybrid of public and private sponsorship. Contests, prizes, and government grants increasingly characterize the process, and Maurer and Scotchmer explore why these are used rather than more traditional forms of intellectual property. The public goods aspects of basic research and discovery are important justifications for government subsidies. The authors catalogue a range of incentive schemes that are in use for knowledge creation, and many are appropriate only within a specific environment. Maurer and Scotchmer model the creative process and the necessary incentive mechanisms that are suggested by it. They then turn to the efficacy of prize systems that are used to stimulate innovation. There are special problems of identifying and rewarding those with the best ideas when the science is very basic and its uses still to be determined. The authors provide an economic rationale for the federal grant process where ideas for funding originate with the grantees and there is no enforcement mechanism to ensure that the proposal is feasible nor any method to reclaim payment if the objective is not met. Maurer and Scotchmer explore the complicated problem of sponsoring innovation when it is unknown what the best approach might be and how to evaluate it against competing processes. Finally, they turn to controversies regarding the public subsidization of innovations that also receive private property protection. The issues raised by Maurer and Scotchmer show that this is indeed a fitting chapter to start the volume.

In Chapter two, "Economic Perspectives on Open Source," Josh Lerner and Jean Tirole examine why individual programmers and commercial companies engage in open source software development when intellectual property protection is absent. Open source software is a public good, open for free riding and other sorts of abuses, with little opportunity for developers to directly capture the returns as "residual claimants." Yet, open source software is growing and attracts both programmers and companies. Indeed, some very well known software products, such as the Apache web server and the Linux operating system are at least in part, open source software. In this chapter, Lerner and Tirole use the insights of industrial organization and labor economics to provide some tentative answers as to why open source software would be used. In particular, they focus on the "career concerns" of programmers to explain their participation and use industrial organization insights to explain why private firms participate. They examine the histories of four projects, Apache, Linux, Pearl, and Sendmail, to demonstrate the usefulness of open source in product development. Lerner and Tirole note that open source encouraged cooperation in creating new software and that issues of documentation, distribution, licensing, leadership, and identification of contributors were successfully addressed. Typically, an elite group of programmers initially participated in open source software development, and having this "core group" status provided both valuable peer recognition and motivation for collaboration. Within companies, the authors found that collaborative software development served as a signaling device to demonstrate the accomplishments and skills of the programmers involved. They also could improve the performance of their companies through the introduction of open source products and through the relationships that were established among programmers and organizations. Lerner and Tirole point out that commercial software companies have been challenged by open source software and have employed various strategies to address it.

Patenting issues are addressed in the next group of chapters. In Chapter three, "Submarines and Technological Innovation: U.S. Continuation Patenting in Software and Biotechnology Technologies in the 1980s and 1990s," Stuart Graham and David Mowery examine the interaction between patent systems in novel areas of technology, such as biotechnology and software, and firms' patent strategies. A particular problem for assigning property rights based on novelty in new areas is that there is little well-defined "prior art" so that there potentially is a wide range of overlap with existing patents. Accordingly, inventors and entrepreneurs operate in an environment of high uncertainty where they may find that aspects of their innovations may prove to be patented, exposing them to litigation and the invalidation of their patents. As part of the growth of the biotech and software industries there has been increased use of "defensive patenting" by firms to retain some "freedom to innovate" and to acquire a bargaining position for cross-licensing negotiations. Graham and Mowery focus on the strategic use of patent "continuations," which are reapplications and revisions of patent applications. These continuations maintain the priority of the earliest filing date, but delay the final issue of a patent and therefore remain relatively hidden from patent searches. This activity enables an invention to remain hidden for a long period of time, as a "submarine." If technologies covered by patents issued after the priority date held by the submarine patent became widely used, the holder of the submarine patent can pursue infringement actions against other patent holders. The authors argue that this strategic action added uncertainty to R&D and the legitimacy of new product introductions, and revisions were made to U.S. patent law to reduce the use of such devices. Graham and Mowery illustrate the use of submarine patents with a case study of Rambus, Inc., which used a continuation process as a hold up strategy in semiconductors. To illustrate the use of submarine patents, the authors provide a case study of Rambus Inc., which used a continuation process as a hold up strategy in semiconductors. The company succeeded in patent infringement litigation against other semiconductor manufacturers. Graham and Mowery also provide examples from biotechnology for comparison with software. They provide a list of hypotheses for examination of continuation patents and use a data set of all such U.S. patents issued from 1987 to 1999. They note that after 1995 and legislative intervention the use of continuations in software declined, as it did for biotechnology. Nevertheless, the continuation rate among biotech patents still remained well above that for overall patents.

In Chapter four, "Firm Strategies and Trends in Patent Litigation in the United States," Deepak Somaya points out that patent litigation has been rising rapidly in the United States since the mid-1980s, and especially in high-technology industries. He examines the motivation for firms to engage in litigation since it is costly both financially, in diverting management's attention from other activities, and in requiring additional expertise from lawyers, engineers, and scientists. He examines firms in the computer and research medicines (pharmaceuticals plus medical biotechnology) industries and focuses on both the strategic uses of patents by firms and corresponding decisions to litigate. Somaya argues that the strategies pursued by firms with their patents have a significant influence on their decisions to file suit, and on the outcomes of litigation. Patent strategy is an issue for technology firms that rely on intellectual property rights for their competitive advantage. In his empirical analysis, the author investigates reasons for non-settlement of patent disputes. He tests five hypotheses using an extensive data set for both industries. The empirical analysis includes 709 patent suits filed between 1983 and 1993 – 449 in computers and 260 in research medicines. Somaya finds that patent suits are less likely to be settled if patentees, or their opponents, have strong stakes in the litigated patent, especially in the computer industry. The evidence for mutual hold up threats to insure access to technologies is less strong. Nevertheless, it appears that when firms use multiple technologies in final products, they are more likely to settle to avoid hold up of their commercial activities. Somaya's analysis demonstrates how patents function as strategic assets, serving as a source of competitive advantage. Accordingly, firms must make tradeoffs in determining their patent litigation and licensing strategies.

Constance Bagley and Gavin Clarkson in Chapter five, "Crossing the Great Divide: Using Adverse Possession to Resolve Conflicts Between the Antitrust and Intellectual Property Regimes," examine a similar problem on the control of intellectual property. They focus on two questions: First, under what circumstances must the holder of a patent or a copyright or the owner of a trade secret allow others to use that intellectual property? Second, under what circumstances can the holder of an intellectual property right use that right to make it difficult for another party to succeed in a related market? These issues were raised in *United States v. Microsoft* and continue to cloud the environment for technology commercialization. Intellectual property laws grant exclusive rights to the holders of patents, copyrights, and trade secrets, but these privileges can also result in monopoly like powers that counter antitrust laws. Bagley and Clarkson trace the jurisprudential path to describe the problem. They briefly survey relevant parts of the Sherman Act and then point out that refusals to deal in intellectual property are most likely to run afoul of antitrust when they are used to "punish" others for doing business with a competitor, such as when Microsoft charged OEMS higher prices for Microsoft's Windows operating system if those OEMs installed Netscape Navigator. Two areas are particularly problematic, according to the authors: A monopolist has a duty to deal with its rivals when it controls an essential facility, a resource necessary to its rivals' survival that they cannot feasibly duplicate. The second case occurs when the monopolist uses its intellectual property power to change pre-existing distributions, business practices, or to gain monopoly powers in other markets through tying agreements or similar practices. The authors illustrate the problems with litigation involving Kodak, Xerox, and Microsoft. They discuss some of the intellectual property and antitrust principles that should guide the development of any potential solution to situations where holders of intellectual property rights refuse to deal with other firms. They then propose a solution that uses a variation on the real property concept of adverse possession for intellectual property and include an extension of the Essential Facilities Doctrine for industries that exhibit network effects. Bagley and Clarkson argue that intellectual property owners should be able to exclude other firms from developing aftermarkets within the scope of the original patent. If other firms are not excluded and aftermarkets develop, then they propose that after a *sufficient period of time*, adverse possession would preclude the original intellectual property owners from excluding other firms. Bagley and Clarkson illustrate how their proposal might work in both the U.S. and in Europe. Their proposed solution allows cases to be resolved quickly and efficiently via summary judgment.

Chapter six, "Incomplete Contracting and the Structure of R&D Joint Venture Contracts," is by Suzanne Majewski and Dean Williamson. In this chapter, Majewski and Williamson empirically examine how firms structure joint R&D agreements to protect and assign the intellectual property rights. The contracts they examine involve very uncertain contingencies and important governance issues for the control and sharing of intellectual property. They survey the literature on incomplete contracts, describe the problem facing firms in joint R&D activities, and develop hypotheses for testing. They use a data set of 96 contracts drawn from data from filings made to the U.S. Department of Justice under the National Cooperative Research Act of 1984 (NCRA) and National Cooperative Research and Production Act of 1992 (NCRPA). The contracts describe the arrangements under which the partners form, operate, and share from the collaboration. Weak property rights limit appropriability of the returns and hence there are benefits from strengthening them through a variety of disclosure restrictions and limits on the subsequent use of intellectual property. Majewski and Williamson, however, note that there is a risk involved from such restrictions. Some of the parties may use the restrictions to strategically block or hold up the others well after the venture has terminated. Costly litigation, negotiations for licensing agreements, or halting of production are potential outcomes. Accordingly, the collaborating parties must balance the benefits of restricting access and use of intellectual property with the costs of potential hold ups. The authors argue that to mitigate such hold up risks, the duration of the restrictions might be limited. They also examine other options and examine their data to see which is used. The chapter illustrates the complicated settings in which research and development contracts are written when there is uncertainty about the outcomes, potential problems in securing property rights, and the threat of subsequent strategic hold ups.

The final two chapters examine the problems of illegal downloads of intellectual property from the internet, the impact, and firm strategies to address them. Chapter seven, "Will MP3 Downloads Annihilate the Record Industry? The Evidence So Far," is by Stan Liebowitz. He examines one of the most controversial issues regarding intellectual property, the extent of damage that occurs through unauthorized electronic computer downloading of copyrighted music from the internet. Napster and MP3 downloads and their possible impact on the recording industry, as well as the industry's attempt to thwart such downloads are the most notorious examples of the problem. With MP3 downloads, recording companies are less able to protect their property rights, and companies have sued to block such practices. Legislation has been introduced to allow copyright owners to introduce actions to block the unauthorized distribution of copyrighted items. At center to the debate is whether or not such infringement of property rights results in serious economic damage to the industry. The recording industry was successful in legal action against Napster in making such arguments, but Liebowitz cautions that the evidence on the impact is slim indeed. In this chapter he provides some additional information on the likely effect of MP3 downloads on the recording industry. He concludes that such downloads likely have harmed the industry, but not by much. Measurement of the impact is not easy both because of missing data and because of the existence of multiple factors that could impinge upon music sales. Liebowitz focuses on album sales and analyzes the effects of changes in relative prices, income, recording formats and technologies, music quality, changes in the size and distribution of the population of MP3 users, and distribution channels. Based on these data, Liebowitz concludes that MP3 downloads are hurting the recording industry, but the effect does not appear to be as large as the industry claims. He is cautious because of the limited nature of the data. He notes that individuals may, as Napster claimed, be sampling music via downloads and buying CDs. He does not believe, however, that the existing evidence supports strong, draconian measures to protect the record industry. He supports reliance on market based solutions, such as adoption of digital right management technologies.

The final chapter, "Strategies that Work When Property Rights Don't," by Bharat Anand and Alexander Galetovic also address the problem of weak property rights and the specific challenge facing recording studios with Napster. The authors note that to combat piracy, the studios first turned to the courts, while also looking for increasingly sophisticated technologies that might make it difficult to pirate digitized content. But they had limited success and the industry increasingly searched for "new business models" that might allow them to capture returns from their investments in the financing and marketing of music. Anand and Galetovic argue that the challenge facing the music, and, increasingly, the movie studios, is not unique to the media and entertainment industry, but rather is faced in a broad array of industries where it is difficult to establish secure property rights over investments or products, especially ideas. The authors argue that firms do continue to invest in assets or products that they cannot own, and they examine various strategies available to capture value from assets over which property rights are difficult to establish. These strategies encompass a variety of mechanisms, including adapting internal organization and governance of projects in ways that allow them to overcome the problem of weak property rights. In other cases, firms alter their organizational boundaries and use the ownership of related assets where property rights are stronger to mitigate the problem. Anand and Galetovic develop a model to provide a stylized setting for examining the problem of coping with weak property rights and then draw five basic principles: Diversity, Limits of Law, Penalties to Prevent Holdup, Rewards to Prevent Holdup, and Strategies to Prevent Holdup. They apply these principles to examine the actions of firms in the music and software industries in confronting piracy and to develop some decision traps for managers to avoid when tackling the difficulties posed by insecure or non-existent property rights over core assets.

NOTE

1. See Libecap (1989), Barzel (1989), and Anderson and McChesney (2003) for instance.

REFERENCES

Anderson, T. L., & McChesney, F. S. (2003). *Property rights: Cooperation, conflict, and law*. Princeton: Princeton University Press.
Barzel, Y. (1989). *Economic analysis of property rights*. New York: Cambridge University Press.
Libecap, G. D. (1989). *Contracting for property rights*. New York: Cambridge University Press.

Gary D. Libecap
Editor

PROCURING KNOWLEDGE

Stephen M. Maurer and Suzanne Scotchmer

ABSTRACT

There is growing public interest in alternatives to intellectual property including, but not limited to, prizes and government grants. We collect various historical and contemporary examples of alternative incentives, and show when they are superior to intellectual property. We also give an explanation for why federally funded R&D has moved from an intramural activity to largely a grant process. Finally, we observe that much research is supported by a hybrid system of public and private sponsorship, and explain why this makes sense in some circumstances.

1. INTRODUCTION

Patents and other intellectual property rights (IPRs) became stronger during the 1980s and 1990s than at any point in the preceding century. Nevertheless, other mechanisms for eliciting and investing in new ideas have remained important. This is particularly true in the public sector, where federal spending accounts for almost a third of the nation's total R&D. (The fraction is higher in most other countries.) Funding mechanisms used by the public sector routinely include in-house development, procurement through competitive bidding, and research grants to universities and promising scientists.

The public sector also uses hybrids that mix sponsorship with intellectual property. The 1980 Bayh-Dole and Stevenson-Wydler Acts authorized the patenting and licensing of federally funded research outputs, and the Federal Technology

Intellectual Property and Entrepreneurship
Advances in the Study of Entrepreneurship, Innovation and Economic Growth,
Volume 15, 1–31

ISSN: 1048-4736/doi:10.1016/S1048-4736(04)01501-2

Transfer Act of 1986 authorized the formation of Cooperative Research and Development Agreements (CRADA's) between national laboratories and industry, granting intellectual property to the industrial partners.

Despite the prominence of intellectual property incentives, prize-based and other incentives remain strong, and may even be undergoing a renaissance. Beginning in the Clinton Administration, NASA and Congress have debated a prize-based approach to new space technology (Space Projects, 2003). The Defense Advanced Research Projects Agency (DARPA) has offered a $1 million prize designed to elict a 40-fold improvement in robotic off-road vehicle design (DARPA, 2003; Holden, 2003). Even the private sector has begun to experiment with prizes, and in at least one offering, the prize mechanism has become institutionalized. Eli Lilly Corporation has established a company called Innocentive to elicit new ideas for solving problems in biology, chemistry, biochemistry, and materials science. "Seeker companies" post problems on Innocentive's web site and offer rewards ranging from $5,000 to $100,000. Scientists are invited to submit solutions, and the seeker companies pick the best ones (Innocentive, 2003).

In this paper we catalog some of the many incentive schemes that have been suggested and used. We argue that the sensibleness of each scheme – indeed, the attractiveness of intellectual property – cannot be debated without reference to a specific environment of knowledge creation. The diversity of models and incentive schemes discussed here suggest that no single incentive mechanism dominates, but that different models of knowledge creation call for different incentive schemes. There are also creative environments where the social value of an innovation is not appropriable by private firms or intellectual property rights are insufficient to cover costs.[1]

The main objective of the incentive schemes we discuss is to elicit investments in new knowledge. Another objective is to do this at the lowest possible cost. The cost of investing in knowledge has two components: resource costs and the diminution of value that may arise through proprietary pricing and deadweight loss. The latter particularly afflicts intellectual property. Although raising funds for general revenue is not costless, the associated inefficiencies are usually thought to be less onerous than taxing a single market, as proprietary pricing does. This observation leads to the conjecture that, in many creative environments, some form of public procurement may be preferred.

We begin in Section 2 with a model of the creative process, and then discuss the incentive mechanisms that naturally flow from it. The efficacy of prize systems, discussed in Section 3, will depend largely on whether prizes can be made to reflect the values of delivered innovations. But even if that is possible, prize incentives, like patent incentives, are a crude mechanism for discriminating among ideas for R&D investment. In Section 4 we discuss contests that are targeted specifically to the problem of finding the innovators with the best ideas. In Section 5 we give an economic rationale for the federal grant process, in which ideas for funding originate with grantees, there is no enforcement mechanism to ensure that the grantee's proposal is feasible, and no means to reclaim payment if the objective is not accomplished. In Section 6 we turn to the virtues of sponsorship for innovations such as nuclear fusion for energy, where it is *ex ante* unknown what the best approach is. In Section 7 we address what is probably the most controversial development in the R&D establishment, the fact that many innovations are both subsidized by public sponsors, and also receive intellectual property protection. We show that, provided industry is required to give matching funds, this system can solve the dual problems of generating enough funding for "big science," while also tapping into industry's expertise, to avoid ill-conceived ideas.

2. IDEAS AND INNOVATIONS

To compare incentive schemes from a normative point of view, we must have some notion of an efficient investment plan. How much money should be devoted to R&D? Where should R&D dollars be channeled? Which incentive system is mostly likely to achieve the desired goals?

From a business point of view, an "efficient" investment plan is the one that maximizes profit – investing in knowledge is only "efficient" if it leads to profit. However, the business point of view does not help society as a whole choose among incentive mechanisms. Profit is only a partial view of benefits, since it does not account for benefits that accrue directly to users. These are especially important for innovations funded by the public sector and put in the public domain.

To compare the efficiency of incentive mechanisms, we must first have a notion of which R&D investments are efficient. The notion that underlies this paper is a standard economics notion, namely that an R&D investment is efficient if it provides higher (discounted) consumers' surplus net of costs than any other feasible investment or investment path, at least in expectation. This is a definition that takes account of how investments in knowledge are funded. If investments are funded by intellectual property, then they generate less consumers' surplus (or more deadweight loss) than if they are funded by public sponsors and put in the public domain.

Just as importantly, efficiency of an R&D investment, and therefore efficiency of incentive mechanisms, can only be understood within a model of the creative process. We will use the "ideas" model of O'Donoghue, Scotchmer and Thisse

(OST, 1998; see also Scotchmer, 2004), which makes a clean separation between an exogenous process that generates ideas for innovations, and the decisions whether to invest in them. An innovation requires both an idea and an investment in it.

Not all economic models of technological advance are set up to evaluate efficiency in investment, or to compare incentive mechanisms. For example, the evolutionary model of Nelson and Winter (1982) (see also Mokyr, 1990, Chap. 11) is a positive model of technical change, rather than a normative model. Regardless of the incentive mechanisms in place, firms are assumed to invest in R&D when profit falls below a certain level. There is no notion of rationality that drives investment decisions.

Another class of models relies on an exogenous production function for knowledge that is commonly known to everyone (Scherer, 1984; Shavell & van Ypserele, 2001; Wright, 1983, and earlier models using the Poisson process, summarized by Reinganum, 1989). These models do not recognize that ideas can be scarce. Indeed, an important intuition for why intellectual property may be a better mechanism than any form of government sponsorship is that it can harness the creativity that is widely dispersed among humans. Intellectual property can encourage investment in an idea that only a single person thinks of. But as we shall see, other mechanisms also have merit in such environments.

We describe an idea by a pair (v, c) (per-period social value and cost).[2] The variable v represents the per-period consumers' surplus with competitive supply. If the social value lasts forever, then, if the invention is in the public domain, the discounted social value is $(1/r)v$. If the invention is marketed by a proprietary firm, we will assume that the per-period profit is πv, where π is a fraction less than one. Then the proprietary profit available under a patent that lasts for discounted length T is πvT.[3] We assume that the associated deadweight loss is dv per period, so that the associated deadweight loss is dvT.

Following the economics literature on contracting, we will make a distinction between the case that the per-period social value, v, or the cost of achieving it, c, is *observable* to a third party, and the case that it is *verifiable* by an enforcing party such as a court. If it is not verifiable, then it is impossible to enforce a contract in which payments depend on that variable. This limits the kinds of incentive contracts that are available.

We begin with a preliminary comment that relates patents and prizes. If the social value of an invention is verifiable, then prizes clearly dominate intellectual property as an incentive system provided the efficiency concern is aggregate deadweight loss. Whatever the duration, T, of the proposed protection, a prize equal to $\rho(v) = \pi vT$ entails the same incentives to invent as the IP system has, but with less deadweight loss, provided the prize giver puts the innovation in the public domain. This observation remains true in other models of knowledge creation, such as where there is a knowledge production function.

If the value v is observable, an invention authority can generally do even better, by offering a prize $\rho(v)$ that is tailored to the distribution of costs. In fact, it might seem natural to base rewards directly on the R&D cost c, since that would enable a sponsor to minimize the amount of money that must be raised for rewards. In practice, however, this would not work. First, a sponsor cannot observe cost by consulting the researcher's accounting data. The economic definition of cost is the minimum cost required to achieve the result. Wasteful or inefficient efforts should not count. If the sponsor simply reimbursed the accounting costs, he might, for example, give the researcher an incentive to go to the beach under the guise of attending research conferences. Second, in most research endeavors, the laboratories and researchers' time are spread among many research projects. Overhead costs must be apportioned among the projects. No one except the researcher, and possibly not even him or her, can know how to apportion the costs.

Finally, and perhaps most importantly, some research efforts do not pay off with certainty. PhRMA, the trade organization of the pharmaceutical firms, estimates that fewer than one in five drug development efforts results in a successful drug. The failures obviously cannot be identified in advance, else the drug companies would avoid them. If only the successful drugs result in profitable IP rights or prizes, the cost being covered by the IP rights or prizes would have to include the cost of failures as well as successes. If only the successful drugs are rewarded, they must be rewarded at least five times more than the average cost per drug discovery effort.

Conditional on v, the sponsor should have some subjective notion of the distribution of costs. An optimally chosen prize should reflect this conditional distribution. If costs for ideas with value v are lower on average for some types of innovations, then presumably the prize $\rho(v)$ should also be lower. For other types of innovations, the expected costs might be higher on average, and the prize should be higher. In making these judgments, the sponsor recognizes that, since the cost c is unobservable, a prize $\rho(v)$ smaller than the social value v/r may not cover cost, and a prize larger than the patent value πvT may be larger than cost, and attract too many competitors. The benefit of giving lower prizes must be balanced against the possibility of discouraging some innovations.

A prize based on the value created will not be credible if the sponsor has opportunity to renege, or is thought likely to renege. This is where verifiability by a court may be important. For a prize system to work, something about the merits of the delivered invention must be observable to an enforcer, even if it is only a noisy proxy for v.[4] It also helps if the prize giver can make himself credible in a repeat game, as was true to a large extent in 18th century and 19th century France.

The examples of incentive mechanisms that we give below address two types of creative environments: those where the creation of knowledge addresses a known need, and those where the need had not been identified, or at least articulated by a sponsor, prior to someone thinking of the idea. In the latter context, it is natural to call the idea *scarce*, in the sense that there are no substitute ideas that would address the same economic need.

The distinction between ideas that address well-known needs and ideas that are scarce has implications for what types of mechanisms can be used. If the need is known in advance, then it makes sense for a sponsor to post a prize to address it, or otherwise solicit solutions *ex ante*. The prizes would typically be contingent on some performance standard dictated by the need. In contrast, an incentive system to reward scarce ideas could not very easily be established in advance, at least with a performance standard. The terms of the reward must be established *ex post*.

The distinction between well known needs and scarce ideas is, of course, fuzzy. A need can have varying degrees of specificity, from "demonstrate that Maxwell's equations are correct" to "improve the efficiency of harnessing water power," both discussed below.

We close this section with three examples that illustrate the scarcity of ideas, and also illustrate the growing realization that, before good R&D investments can be made, the individuals with good ideas must be identified.

The first example is NASA's Institute for Advanced Concepts, which spends $4 million per year funding ideas for space propulsion. The funding is in two phases. Initial grants are up to $75,000, which is entirely for exploring an idea. The agency then funds Phase II grants of up to $500,000 to validate them. The two phases arguably correspond to the idea and the investment, although the second stage is still rather far from a usable innovation. Speculation is encouraged: the director urges audiences "Don't let your preoccupation with reality stifle your imagination." Not surprisingly, results have been mixed. Some grants (antimatter rockets, solar sails, space elevators) re-work ideas that have been widely discussed for decades. Other grants have elicited risky but physically plausible schemes. In fact, NASA has already picked up one idea – creating magnetic "sails" to surf the solar wind – for further development. Critics charge that another ("hydrino" propulsion) rests on pseudoscience ruled out by the known laws of physics (Reichhardt, 2002).

The second example concerns jet fighters. The U.S. Air Force has always recognized that an essential part of advancing aircraft capabilities is attracting new ideas. The current F-22 design effort began in 1981 with a formal request to industry for ideas. Suggestions ranged from an ultra-lightweight fighter to a 120,000 lb. "battle cruiser." The Air Force then took the best ideas from each proposal and prepared detailed specifications. Even then, the Air Force resisted prototyping because that would reduce the number of participants and ideas. It was not until 1986 that Lockheed and Northrup were selected for the remaining competition (Sweetman, 1991).

The third example comes from the private sector. In 2002, Google announced its first annual prize for a scalable computer program that "does something interesting" to the company's internal library of pre-parsed web pages. The winning entry merges Google's data with census bureau address data to create a database that can limit searches by geographic area. (Contestants granted Google a worldwide, non-exclusive license to make, sell, or use any technologies they developed (Google, 2002).) Although the winning contestant was required to submit working code, the main benefit to Google resided more in the idea than in the implementation. Any programmer could have turned the idea into an innovation; it was the idea itself that had value, and that is what Google was looking for.

3. PRIZES: LINKING PAYMENTS TO VALUE

An important class of mechanisms are prizes. These come in two types that correspond to the two innovative environments that arise from well-known needs and scarce ideas. *Targeted* prizes, which are posted *ex ante* and have clear performance standards, reward solutions to needs that originate with sponsors. The inventor's idea is a solution to the sponsor's stated need. *Blue-sky* prizes are given when no need has been stated in advance, but judges are allowed to "know it when they see it." (In blue-sky prizes, "the sky is the limit.") If the reward is specified in advance, it cannot be tailored to the idea. For example, Nobel prizes are a fixed award, regardless of the nature of the accomplishment. More often, as we shall see, blue-sky prizes are tailored *ex post* to the value of the innovation, but this is a difficult problem.

The problem with blue-sky prizes is how to tailor them to the value v. This same problem carries over to contests, discussed below. If a prize cannot depend on the value of the innovation that is delivered, then the sponsor cannot avoid rewards to useless innovations or rewards to the wrong innovators.

We illustrate two methods that have been used to ensure that a prize reflects the value v, namely, making the prize conditional on a verifiable performance standard, and giving the inventor the option to choose intellectual property protection instead. In the latter case, the prize effectively becomes a patent *buy-out*. The inventor will not agree unless the prize is at least as large as the patent value πvT, and in this sense, the prize is constrained to reflect the value.

An example of how the buy-out works occurred when the French inventors of photography, Daguerre and Nahin, sold their rights in an *ex post* negotiation in 1839. The inventors received pensions totaling 10,000 francs per annum in exchange for revealing the secrets of the process at a joint meeting of the French Academies of Art and Science. Afterwards, the process was put in the public domain (Newhouse, 1988). We can only presume that the French inventors received value commensurate with their invention, since they would not have accepted a prize less than the patent value.[5]

In contrast, John Hyatt chose patent protection over a prize for inventing celluloid in the 19th century (Porter, 1994). Hyatt had originally invented celluloid in order to claim a prize posted by a manufacturer of billiard balls who wanted to replace ivory. However, when Hyatt realized that his invention had wider applicability, he chose patent protection instead, apparently judging the value of the patent to be greater than the prize.

Michael Kremer (1998) suggests a direct way to buy out the patent at a price that reflects value. Since the value of the invention is likely to be observable by rival firms after the invention has been made, the patent authority can turn the patent into a prize as follows. The invention authority appropriates the patent, and auctions it to rival firms. He promises that there is a small probability that he will actually deliver the patent to the highest bidder in return for the bid price. The rivals will bid the same as if the patent would be transferred with probability one, thus revealing their valuations. The winning bid should be (close to) the private value $\pi v T$. The sponsor can divide the winning bid by πT to get an estimate of the value v. Kremer suggests that the sponsor pay the inventor a prize equal to his estimate of the social value, v/r, from general revenue. With low probability he transfers the patent to the winning bidder in return for the bid price. With high probability, he puts the innovation in the public domain, thus turning the patent system into a prize system. This scheme is enforceable, provided the parameters πT are specified in advance, so that a court could enforce it.

The greatest virtue of patent buy-outs is that they allow the government to confiscate inventions with high public consequences, such as pharmaceuticals, without hurting the innovator. If anything, the Kremer scheme increases the incentive to invest, since the innovator's reward v/r is larger than the value of the patent $\pi v T$. However, such a high prize also leads to social waste if it incites rival firms to duplicate each other's costs in a wasteful race. Under these circumstances, it may be better to choose a prize $\rho(v)$ below the social value, as discussed in the previous section.

An example of linking the prize to verifiable performance standards occurred in the Lyonnaise silk-weaving industry. Members of the Fabrique Lyonnaise could make improvements to weaving on their own initiative, and then petition a prize committee for remuneration. It is unclear whether most of these improvements would alternatively have been patentable, but the terms of the *ex post* reward evidently did not rely on that alternative. Instead, the prize committee set the terms of the rewards based on performance criteria, such as the number of weavers who adopted it (Foray & Hilaire-Perez, 2000). Such terms allowed the committee to ensure that the prize was given only in return for value.

Targeted prizes were common in France along with blue-sky prizes. In 1795 a prize was offered for a means to preserve food to feed Napoleon's vast armies and navy. The prize was awarded in 1810 to Nicolas Appert on condition that he publish the technique and put it in the public domain (Porter, 1994). His technique, which involved heat-sterilization of food packed in bottles, is still in use. (If the solution seems obvious, we should remember that the causes of spoilage were not understood.) Other targeted prizes led to improvements of the steam engine (Porter, 1994) and also water power, which led to the first water turbine (Strandh, 1979).

One of the most famous targeted prizes was for discovering how to determine longitude (Sobel, 1995). By the time the English prize of £20,000 was posted in 1714, similar prizes in other countries had gone unclaimed. The failure to determine longitude was the principal cause of a large and growing number of naval and commercial catastrophes. No one disputed the high value of a solution, but mariners despaired of finding one. In the course of the English competition, two very different ideas for solving the problem emerged. John Harrison had the idea to engineer seaworthy clocks that would allow the mariner to keep London time, and thus compare local time to London time. Such a comparison would tell him how far East or West he had sailed. The other idea was to use lunar observations in comparison with the stars to ascertain longitude. Both were developed in parallel, but the prize eventually went to Harrison.

In the modern renaissance of prizes, the X Prize Foundation was established in 1996 with a $10 million prize for the first private firm to carry three passengers to a suborbital height of 100 km twice within a single two week period.[6] To date, 23 teams representing seven countries have entered (X Prize Foundation, 2003). Observers report that at least three teams are likely to produce actual hardware. One effort has reported that it plans to spend $5–7 million in the effort. Some commentators believe that several teams have a reasonable chance of claiming the prize before it expires in 2005 (Hoffman, 2003).

Another modern targeted prize was $30 m announced in 1992 to develop a new, "Super Efficient Refrigerator" (Penn, 1993; Zuckerman, 2003). In order to receive payment the winner had to sell 250,000 copies of the new refrigerator by 1997. Although Whirlpool won the contest, they did not meet the performance standard, selling only 200,000 units, and the prize was not paid.

Finally, we mention two targeted prizes with performance standards that have been offered for inventions that have great value to the sponsor or to society, but no commercial value. The first was a German prize for demonstrating that Maxwell's equations of electromagnetism were correct. The prize was offered by the Berlin Academy in 1879, and the feat was accomplished by Henrich Hertz, albeit long after the prize expired (Bryant, 1988).

A modern prize with similar character is the RSA Factoring Challenge, which offers a $250,000 prize for the first team to factor a specified 2048-bit integer. Rewards for smaller specified numbers start at $10,000 (RSA Factoring Challenge, 2003). RSA sells a commercial implementation of public key encryption whose security rests on the computational difficulty of factoring large numbers. If the prize goes unclaimed, RSA has more confidence in the security of its system. This is a case where the sponsor is better off if it does not get what it is looking for. The successful claimant cannot profit from the discovery in any way other than the prize.

These examples show that prizes can be linked to value by: (i) administering prizes in a legal environment where intellectual property protection is also available; and (ii) making the prize contingent on a performance standard. These techniques can apply to either targeted prizes or blue-sky prizes, although patent protection is probably a more important backup for blue-sky prizes, as a way to avoid *ex post* hold-up.

So far these arguments have been entirely about the feasibility of linking prizes to the value of the innovation. They do not inform us about the optimal *size* of the prize. If costs are "expected" to be much lower than value, then it is sensible to set a prize much lower than the social value, even if the social value can be observed. The appropriate size of a prize should also be tailored to the number of participants that the sponsor wishes to attract.

In the simple ideas model given above, it is desirable to avoid duplication of costs. When ideas are scarce, this is not a problem – scarcity means that only one inventor can fill the market niche defined by his idea. In that case, the size of the prize should simply balance the possibility of giving up the innovation (when the prize turns out to be less than cost) against the social cost of over-rewarding the innovator. Such a balancing must be done by reference to a subjective distribution of costs, conditional on value.

For targeted prizes, ideas are typically not scarce. There are typically several ideas that could fill the targeted need. In such an environment, simple prizes share a defect of intellectual property, even if the prize can depend on the value. There is no way to ensure that only the agent with the best idea invests, or even that a single agent invests. Those problems can be partly addressed by the contests we discuss in the next section.

4. RESEARCH CONTESTS

A difficult problem for a public sponsor is when the prize cannot be made to reflect the value of the delivered invention. The sponsor is then in jeopardy of giving a prize for no value. Just as bad, potential inventors may refuse to take the bait, fearing that the sponsor will renege on promises. In this environment, intellectual property starts to look attractive, since the value of an intellectual property right is automatically linked to the social value of the invention.

However, we should not give up on prizes too soon. For one thing, there are many inventions for which intellectual property will be ineffectual because the social value cannot be appropriated. Examples that spring to mind are military wares and pure scientific knowledge. Fortunately, the sponsor still has options.

We discuss how the sponsor can elicit investments in innovations of high value even if the value is not verifiable *ex post* to an enforcement body. He will do this by setting up contests that involve enforceable contracts. In two of the contests we discuss, the contract terms will not involve the enforcement of any payment conditional on the observation of value. These two contests have the downside that the best contest may nevertheless lead to duplicated efforts of the contestants. The inability to make enforceable contracts in which payments depend on the delivered quality is costly.

4.1. A Simple Commitment to Pay

The first commitment device is extremely simple, and also very common. The sponsor sets a prize, and commits only to give it away. He does not make the payment conditional on any specified performance, but rather announces his objectives, and lets the contestants choose their strategies. *Ex post* he will choose the contestant who comes closest to meeting his objective. There is no reason for an enforcement body to be involved, except to make sure that the prize money is actually paid to someone. The sponsor cannot renege on paying the prize, and has no reason to pay it to any contestant other than the one who comes closest to his objective.

Much of the development of airplanes was motivated by such contests, especially before World War II. The Gordon Bennett, Schneider, Thompson and Bendix prizes supported air races, typically offering purses between $5,000 to $10,000.[7] This promoted fierce competition from 1911 until 1939. While it is impossible to measure the technological impact of these competitions with any certainty, prize winners routinely outperformed contemporary military aircraft in speed and endurance until the late 1930s. More specifically, the participation of two companies in pre-war racing – Rolls-Royce and Supermarine – pioneered technologies for advanced aircraft and engines that were eventually incorporated in World War II designs like the British Spitfire. Although much of this effort was subsidized by government loans of pilots and specially-built aircraft, other racing machines were built by private investors in search of prize money. The most famous example was the lethal (but very fast) "GeeBee" design that dominated racing in the early 1930s (Porter, 1994; Vorderman, 1991).

The next two contest-like mechanisms that we discuss are more complicated than a simple commitment to give away the money, but they have additional efficiency properties. As we have already mentioned, a simple commitment to pay a fixed prize shares a defect with patent incentives, namely, that it provides no control over which contestants enter, or even how many. Suppose that a prize or patent is worth $100. Suppose there are two potential contestants, one of which can enter for a cost of $70, and the other of whom can enter for a cost of $60. Only one will enter, since entry by both means that the expected winning of $50 will not cover the cost of either. But if the higher-cost firm enters, then the lower-cost firm will not. There is no guarantee that it is the lower-cost firm that enters in equilibrium. Similarly, if the two firms have different prospects for success, there is no guarantee that it is the better firm that enters. A further problem is that, depending on costs as compared to the prize, both firms may enter even if investment by a single firm would suffice. The incentive system then leads to a wasteful duplication of costs. A good incentive mechanism should avoid these forms of waste.

We now discuss two more refined contests that to some extent address these problems. Suppose that two firms have substitute ideas (v_i, c_i), $i = 1, 2$, for how to serve the sponsor's need. We assume that, if the sponsor could observe these values, then he would choose the idea that provides the greatest social surplus, that is, the sponsor would choose firm 1's idea if $(1/r)v_1 - c_1 \geq (1/r)v_2 - c_2$, and would otherwise choose firm 2's idea. This is the firm with the lower cost if $v_1 = v_2$. The problem is that the sponsor cannot observe (v_1, c_1), (v_2, c_2). The firms may have an incentive to lie about them if they think that lying will garner a profitable government contract.

4.2. A Vickrey Auction

The Vickrey auction goes some distance toward eliciting efficient investment, but only works if the value can be verified *ex post*.[8] So far as we know, this has not been used in practice. In the second-price (Vickrey) auction, the sponsor asks each prospective innovator i to report the social surplus $s_i = (1/r)v_i - c_i$ that it could provide. The firms report some values s_1, s_2. Of course the sponsor cannot verify that either firm is reporting s_i honestly. In addition, he does not know whether a high value of the reported surplus s_i means that v_i is high or c_i is low. However this will not matter, as the promised payments will give the firms an incentive to be honest.

The sponsor chooses the firm that reports the highest net surplus. Suppose that this is firm 1. He promises a payment to the winning firm, firm 1, that is equal to

$(1/r)v_1 - s_2$. Notice that the payment depends on the *other* firm's bid, s_2, not on the winning bidder's bid s_1. The sponsor then asks firm 1 to invest. After paying its own costs c_1, firm 1 ends up with the payment $[(1/r)v_1 - s_2 - c_1] = s_1 - s_2$.

The second-price auction has the following properties: Assuming that the firms report their respective surpluses honestly, then: (1) the payment to the winning firm, say firm 1, will be close to (but no smaller than) the cost c_1 if s_1 is close to s_2; and (2) the winning firm makes non-negative profit by delivering the innovation in return for the specified payments.

We can now ask whether the premise is valid: Does either firm have anything to gain by misrepresenting its net surplus to the sponsor? In the case that $s_1 < s_2$, does firm 1 want to overstate s_1? Unless its lie causes firm 1 to win the bid instead of losing it, the lie has no effect, since firm 1's payment does not depend on its own bid s_1. If the lie is large enough to change the outcome, then firm 1's profit is $s_1 - s_2$, which would be negative. Firm 1 is better off losing the bid and making zero profit.

In the case that $s_1 > s_2$, does firm 1 want to understate the surplus s_1? Unless its lie causes firm 1 to lose the bid instead of winning it, the lie again has no effect. If the lie is large enough to change the outcome, then firm 1 makes zero profit by losing the bid instead of earning $s_1 - s_2$, which is positive. Thus, firm 1 has no incentive to lie about s_1. The same argument applies to firm 2.

Thus, the most important feature of the second-price auction is that each firm has an incentive to report faithfully on the net surplus it can deliver, and the sponsor can safely pick the firm that claims the highest surplus.

The second-price auction is particularly attractive if the surplus available from the two rivals is expected to be similar. In that case, the payment to the winning bidder will be close to the winning bidder's cost, and equal if $s_1 = s_2$. In general, however, a sponsor would care about the size of the transfer that he must pay to the winning bidder. The social cost of raising funds for general revenue is smaller than the social cost of taxing a single market, but is still not zero. The second-price auction yields an efficient outcome in the sense that the high-surplus firm is asked to invest, and there is no duplication of cost.

The second-price (Vickrey) auction assumes that the *ex post* payment can depend on the delivered value. But that is the difficulty we would like to avoid. The contest we now discuss, following Che and Gale (forthcoming), is a hybrid between the commitment prize and the Vickrey auction.

4.3. A Prototype Contest

The *ex ante* problem of the sponsor is to elicit investment in the best idea, given that he cannot condition his payments on the value of the delivered innovation.

But in order to elicit investment, he needs a mechanism that commits him to pay off *ex post*, instead of trying to negotiate a low price once the prototypes are delivered. If the sponsor cannot commit to pay a price above production cost, the innovators will not invest.

The prototype contest solves this problem by allowing the firms and the sponsor to make contingent contracts before any investments are made. The contracts specify what price the sponsor will pay to each innovator, contingent only on buying that firm's innovation. The sponsor's observation of quality is reflected only in the decision to buy. The only enforcement problem is to make sure that *if* the inventor is chosen, the price will be as specified in the contract written before the innovator invested. Each firm would like to get a high price if its innovation is chosen, but a high price increases the chance of not being chosen *ex post*. This constrains the prices demanded by the firms in negotiating the contracts *ex ante*. On the other hand, since the contingent contracts are negotiated before costs are sunk, the firms will not offer prices so low that they do not cover costs.

To see this mechanism more explicitly, suppose the two potential innovators have ideas (v_1, c_1), (v_2, c_2).[9] Suppose, for concreteness, that $v_1 > v_2$. If v_1 is very close to v_2, then in an *ex post* auction, the winning bid would be close to zero, and neither bidder would be likely to cover its cost in expectation. This is remedied in the contest we now describe by letting the firms bid *ex ante*, before they sink their costs. For simplicity, assume that the qualities v_1, v_2 are known to all parties, even if they cannot be verified in court. Thus, all the parties know what qualities of innovations will be delivered *ex post*. The firms will announce *ex ante* the prices (ρ_1, ρ_2) at which they are willing to sell their innovations *ex post*. The firm will receive this price, but only if chosen by the sponsor *ex post*.

Our objective is now to characterize the firms' equilibrium bids. These are bids such that neither firm has an incentive to revise its bid, assuming that the other firm's bid is fixed.

We first argue that the contingent bid prices cannot be deterministic. To see this, suppose that $v_1 = v_2$ and $c_1 = c_2 = c$, and consider what prices the firms will demand. A natural guess is $(\rho_1, \rho_2) = (2c, 2c)$. These are the minimum prices the firms could demand and still cover their costs in expectation, assuming that the tie-breaking rule is to randomize evenly between the firms. Since each firm would win the bid with probability 1/2, the revenues would be $(1/2)\rho_1 = (1/2)\rho_2 = c$.

However, these prices are not an equilibrium. If firm 2 demands price $2c$ in the event that the sponsor chooses that firm, then firm 1 can improve profit by reducing its own demand to $\rho_1 = 2c - \times$, where $\times$ is a small number. With prices $(\rho_1, \rho_2) = (2c - \times, 2c)$, the sponsor will choose firm 1, and firm 1 will make profit $\rho_1 - c = c - \times$ instead of 0. This shows that the zero-profit prices $(\rho_1, \rho_2) = (2c, 2c)$ are not an equilibrium. Of course there are no lower prices that are an equilibrium either, since at least one firm would then not cover cost in expectation.

The solution to this problem is mixed strategies. Instead of choosing a deterministic price, each firm chooses a probability distribution over prices, and the price actually offered to the sponsor is a random draw from this distribution.

An equilibrium will have the property that the firms randomize on whether they develop the innovation, as well as on the price. With some probability each firm will drop out, which means that it does not innovate and demands a zero price. If both firms drop out, the sponsor will fail in his objective of procuring the innovation by setting up a contest. We will denote the cumulative distributions on price by F_1, F_2: $[0, v/r] \to [0, 1]$. For each $\rho \in [0, v/r]$, the probability that the firm chooses a price no larger than ρ is $F_i(\rho)$, for $i = 1, 2$. The firm will never choose a price larger than (v/r) because the sponsor would never pay a price greater than the value of the innovation. The price $\rho = 0$ will imply that the firm does not innovate, and a positive price will imply that the firm does innovate. Thus $F_1(0)$, $F_2(0)$ are the probabilities that the two firms do not innovate.

The firms' decisions whether to innovate and what price to demand will be made independently of each other, not observing each other's choices. Their choices are commitments in the sense that they cannot change the contract *ex post* if, for example, the other firm's price turns out to be lower.

For the example with symmetric ideas $(v_1, c_1) = (v_2, c_2) = (v, c)$, there is an equilibrium in which the firms drop out with probabilities $F_1(0) = F_2(0) = c/(v/r)$, and the distributions on prices are $F_1 = F_2 = F$, defined as follows:

$$F(\rho) = \begin{cases} \dfrac{c}{v/r} & \text{if } 0 \leq \rho \leq c \\[2ex] 1 - \dfrac{c}{\rho} + \dfrac{c}{v/r} & \text{if } c \leq \rho \leq \dfrac{v}{r} \end{cases}$$

This implies that with probability $c/(v/r)$ each firm drops out and asks for price $\rho = 0$, that the firm never chooses any other price between 0 and c, and that the probability distribution has density function c/ρ^2 for prices between c and v/r. That is, conditional on innovating, the firm puts most of its probability weight on a price near the cost c, with the probability weight declining to the maximum (v/r).

This is an equilibrium because each price in the support of the distribution yields the same expected profit as any other price, namely zero, and no other price would yield greater profit. If firm 1 drops out (chooses price $\rho = 0$), it makes zero profit. If firm 1 develops the innovation and demands any price ρ in the interval $[c, v/r]$, firm 1's expected profit is $\rho[F_2(0) + 1 - F_2(\rho)] - c = \rho[c/\rho] - c = 0$. The term $[F_2(0) + 1 - F_2(\rho)]$ represents the probability that firm 1 wins the bid *ex post*. With probability $F_2(0)$ firm 2 drops out, and with probability $1 - F_2(\rho)$ firm 2 innovates but demands a higher price than firm 1's price.

These strategies hold the firms to the lowest possible expected profit that will induce them to invest, given that two of them are asked to innovate. However, it is important to notice that, aside from the oddity of random prices, the investment decisions are inefficient. With some probability, the sponsor does not get the innovation, and even if he gets it, there is a large probability that the costs will be duplicated. Allowing duplication is how the sponsor induces rivalry to keep the procurement price down.

The reader can work out how this mechanism must be modified if the firms' ideas (v_1, c_1), (v_2, c_2) are different. In that case, the random prices, conditional on both firms innovating, may have the consequence that the sponsor does not always choose the highest-value innovation. Instead, the innovator chooses the innovation that generates the highest surplus, which will be the lower-value innovation if the innovator also demands a very low price.

The main example of prototype competition has occurred in military procurement. In the 1960s, the U.S. Air Force usually acquired prototypes from a single vendor after a contest to choose the best written proposal, and then decided whether to accept it. That procedure led to widely criticized cost overruns in the 1960s. In the 1970s the Air Force moved to a system where two rival companies received contracts to build prototypes followed by a flight competition to demonstrate quality. This process led to the F-16 and F-18 fighter jets (Sweetman, 1991; Sweetman et al., 1987).

While these contracts do not mirror the exact mechanism discussed above, they show that the sponsor can use prototype competition to keep quality high and costs low, at least relatively. Rogerson (1994) points out that the payoff to winning the prototype competition is a lucrative production contract. According to Rogerson, such procurement can be viewed as a three stage process consisting of: (a) a design phase in which multiple firms pursue competing designs; (b) a selection phase in which a limited number of firms compete to produce prototypes and/or a final design; and (c) a production phase, typically involving sole source production by a single firm. He argues that the production phases allows the DoD to "award larger prizes to more important innovations, at least in a rough sense" as well as providing ongoing incentives to improve the product after initial adoption. Firms that reach the production phase typically enjoy economic profits (above the normal return to capital) amounting to 4.4%. The super-normal rate of return can be viewed as a prize.

5. GOVERNMENT GRANTS

In this section we turn from the grand challenge prizes, such as those for longitude and food preservation, and focus on the more modest stuff of academic and industrial life, the small innovations that move knowledge forward in increments. Our premise here is that ideas are scarce, and that the agent with the idea should make an informed choice whether to invest in it. Under intellectual property incentives, or a prize system such as that in Lyon, the informed choice will follow the profit motive. But for most ordinary science, neither prizes nor patents have been the main mechanism for funding research. Instead, sponsors have employed scientists directly, or given them up-front grants in return for some unenforceable promise to implement a self-generated idea. How could that possibly work?

The grant process is a relatively modern invention. For most of history, publicly sponsored research was in-house. This dates back at least to ancient Egypt, where the engineer Imhotep was hired to build bigger and better pyramids (DeCamp, 1980). European monarchs routinely promoted knowledge creation by attaching gifted individuals to their courts, for example, Archimedes, Kepler, Brahe, Euler and Lagrange. Later examples of government employees who became famous innovators include Charles Darwin, science officer on the Beagle, and Herman Hollerith, the census bureau inventor of punch card technology (Porter, 1994). In modern versions of intramural research, the emphasis has been on teams. During WWII, the U.S. government hired large groups of academics to produce weapon prototypes, e.g. radar at the MIT Radiation Laboratory, nuclear weapons at Los Alamos, torpedos at the Harvard Underwater Sound Laboratory, and rockets at the Jet Propulsion Laboratory. Additional advances were generated within the government's own laboratories, notably the invention of an advanced "laminar flow" wing that was used in the "Mustang" fighter plane of WWII (Sweetman, 1984). This arrangement is continued by NASA and the national laboratories funded by the Department of Energy.

The paid researchers who are most familiar to readers of this article are university faculty members. This arrangement also has a long history. The preeminent example in the ancient world was Egypt's Library of Alexandria, which supported domestic and foreign scholars who were free to work on their own ideas. Resident scholars included Archimedes, Hipparchos, Eratosthenes, Euclid, and Hero of Alexandria. These researchers made fundamental advances in astronomy and mathematics, as well as a variety of clever mechanical and clockwork devices. The mechanic Ketsibios, also employed there, was widely acclaimed as "The Edison of the Ancient World" (Casson, 2001; DeCamp, 1980). Later, the famous inventors Galileo, Newton, Dalton, Volta, Ampere and Maxwell (Porter, 1994) all worked as employees of universities, rather than in response to project-specific incentives.

The industrial laboratories that came into existence in the 20th century mimic the university model, especially with the establishment of campus-like settings for research in electronics and biotechnology. Probably the most famous of these is Xerox Parc in Palo Alto, which originated many of the computer technologies that can probably be found on the reader's desktop.

Before WWII, R&D funding by the U.S. federal government was a small percentage of total R&D, and what existed was mostly intramural. The nature of public funding for R&D changed precipitously after the war.[10] The federal government became a primary source of R&D funds, of which 70% is currently given out as extramural funding. The percentage of total R&D that is paid for by the federal government has fallen steadily between 1953 and the present, from about 2/3 to 1/3, but it is still much larger than the percentage in the 1930s, which was between 12 and 20% (Mowery & Rosenberg, 1998, p. 27).

To explain the virtues of the grant process, we shall retain our model of innovation in which agents have ideas for innovations, and must be given incentives to invest in those ideas. Even though the grant giver cannot monitor the grantee or withdraw payment in case of failure, and has limited ability to monitor the past record, the system can do a very good job of selecting the best researchers and making sure that they perform. The grantor's main instrument of coercion is that he can cut off funding for future projects if a grantee fails to deliver on a past grant.

As we show, the system works by self-selection. The funding agency evaluates ideas (proposals), but will only fund them, even if good, if the researcher has not failed to deliver in the past. In order to stay in the system, the researcher will have to be honest about his ideas for grants – else he won't succeed – and must actually bear the cost of implementing his idea. It is only the researchers with relatively fertile minds who will have an incentive to do this, and this supports the funding agency's objective. However, the granting agency cannot actually observe the fertility of the researcher's mind or keep track of how many innovations the researcher has made. We show that, nevertheless, the grant system works well to motivate effort and select the best researchers. (If even the best and most honest researchers can fail by mistake rather than malfeasance, the argument we give must be modified so that the grantor can also use information on the history of successes.)

Since the fertility of a researcher's mind can be captured either in the rate at which he has ideas or in the value of the ideas, we will assume for simplicity that all ideas have the same value and cost (v, c), but that researchers receive ideas at different rates $\lambda \geq 0$ per year, where λ represents the fertility of the researcher's mind. The objective of the grant agency is to reward the high-λ researchers.

When the researcher receives an idea, he can file a grant proposal with the sponsor, and the sponsor will decide whether or not to fund him. The researcher does not need to worry about getting paid, because he gets the money up front. However, whether or not the sponsor can observe the merits of the idea (v, c), he has to worry that the researcher either cannot execute the idea or prefers to pocket the award and go to the beach.

In a given time period, a researcher with parameter λ has an average of λ ideas. (The parameter can be less than one, in which case the researcher will not, on average, have an idea in every period.) We can think of the parameter λ as measuring how creative he is. For a researcher with creativity λ, the expected present discounted value of investing in all the ideas he has in a given period at date t is the following, when r is the discount rate and the size of the grant per idea is ρ:

$$\frac{\lambda}{(1-r)^t}(\rho - c)$$

Suppose now that the researcher has received a grant of size ρ. He must decide whether to perform the research or go to the beach. His net gain if he fails to perform is the saved cost c. His net loss from lost future grants is

$$\sum_{t=0}^{\infty}\frac{\lambda}{(1-r)^t}(\rho - c) = \lambda(\rho - c)\sum_{t=0}^{\infty}\frac{1}{(1-r)^t} = \frac{\lambda}{r}(\rho - c)$$

Thus, he will perform instead of pocketing the money if

$$c \leq \frac{\lambda}{r}(\rho - c)$$

or

$$c \leq \frac{\lambda}{\lambda + r}\rho \tag{1}$$

We can see that, for fixed awards ρ, only researchers who expect to have lots of ideas (high λ) will perform in return for future options on grants. If the inequality (1) holds for any λ, it also holds for any researcher with a higher value of λ. We can also see that, for a fixed rate of idea formation λ, researchers will only perform if the rewards ρ are high enough. And, of course, they will never perform if the award ρ is smaller than the cost c.

Let H be the distribution of λ in the population of researchers, so that $1 - H(\lambda)$ is the fraction of researchers with parameter greater than λ. For a fixed award size ρ, let $\hat{\lambda}(\rho)$ be the minimum λ for which (1) holds, that is, the value for which (1) holds with equality. The function $\hat{\lambda}$ is decreasing in ρ: For higher prospective awards, even less creative researchers are willing to perform in return for future options on grants. Then the number of funded researchers $1 - H(\hat{\lambda}(\rho))$ increases with the size of the award, ρ. The total budget of the sponsor per unit time is

$$B(\rho) = \int_{\hat{\lambda}(\rho)}^{\infty} \lambda\rho \, \mathrm{d}H(\lambda)$$

In this system, if the sponsor wants to ensure that only the more creative researchers are funded, he must give awards without rationing. The only researchers who continue to apply for grants are those with creativity parameters larger than $\hat{\lambda}(\rho)$, and all of them are funded. The only way to increase the amount of research (number of researchers) is to increase ρ for all researchers, which means that the budget increases by more than the payments added for new researchers.

It is worth understanding what is lost by the grant giver if he cannot condition his awards on past success. Conditioning the size of awards on past success will reduce the grant giver's total budget, conditional on a fixed amount of innovation. Because the sponsor cannot observe λ, or condition grants on λ, sponsorship is more costly than it would be otherwise.

If the sponsor could observe the fertility λ of the researcher's mind, he could make different payments for different researchers; the grant per idea, ρ, would be a function of λ. For each λ, he would choose a prize ρ(λ) for which (1) holds as an equality. The function ρ would then be a decreasing function of λ; more productive researchers would receive less money per funded idea. However, one can also see from (1) that, since $c(\lambda + r) = \rho(\lambda)\lambda$, the more creative researchers (those with higher λ) would receive more grant funding per unit time, $\lambda\rho(\lambda)$.

The fertility of each researcher's mind will depend, among other things, on whether he is operating in an environment of "open science" (David, 2003). If open science means that each discovery is published, hence shared, each researcher will be stimulated to have ideas at a higher rate. We can thus interpret open science to imply that each researcher's λ is higher than otherwise. With higher λ's, a given total grant budget will support a higher rate of progress.

6. PUBLIC SPONSORS AND THE DIRECTION OF RESEARCH

For some research objectives, there may be several paths to the result, each one risky in its own idiosyncratic way. That is, one of the things that is unknown is whether a particular approach will succeed. In this environment, an idea is a triple (v, c, p), where v is the value of the specified objective, c is the expected cost of a particular research approach, and p is the probability that the approach will fail.

It is natural to assume that the successes and failures of different approaches are independent. Thus, if there are several ideas $\{(v_i, c_i, p_i)\}_{i \in S}$ for how to achieve the objective, and the sponsor elicits investment in all of them, the probability of success is $1 - \Pi_{i \in S} p_i$. (This model follows Wright, 1983, except that in Wright's model all competitors have the same idea. Only the sponsor is uninformed.)

In this environment, the main efficiency question is how many approaches will be taken, and which ones. One of the difficulties of patent incentives is that they can yield either too much or too little entry (too many or too few approaches). To see this, consider the case that $v_i = v$ for each i (the potential knowledge is the same for each firm), $c_i = c$ for each i (the approaches are equally costly) and $p_i = p$ for each firm (they have the same likelihood of failure), but the approaches are nevertheless different, reflected in the fact that the probabilities of failure are independent. Suppose that the value of the property right will be πvT, as above. If n approaches are taken, the probability of success is $1 - \Pi_i p_i = 1 - p^n$. Then the total expected revenue of the firms collectively is given by a function f defined as

$$f(n) = (1 - p^n)\pi vT$$

The social value of the innovation can be written $kf(n)$ for some number $k > 1$. This is because the social value is $[(m + \pi)vT + ((1/r) - T)v](1 - p^n) > (1 - p^n)\pi vT$, where mv is the per-period consumers' surplus that is available with proprietary pricing.

The number of approaches that maximizes social welfare, namely, the n^* that maximizes $kf(n) - cn$, will not be the same as the number of approaches $\hat{n}$ that are taken by competitive firms in a patent race. Firms will enter with different approaches up to the point where $f(\hat{n})/\hat{n} = c$; that is, they dissipate the rent. To see that the optimal number n^* can be either larger or smaller than the equilibrium of the race, $\hat{n}$, notice first that, since f is a concave function originating at 0, it holds for all n that $f'(n) < (f(n)/n)$. The optimal n^* satisfies $kf'(n^*) = c$, which implies that $kf(n^*)/(n^*) > c$. If k is close to 1, then, since $f(n)/n$ is decreasing, $\hat{n} > n^*$. But as k becomes large, n^* must also become large, with the consequence that it may hold that $n^* > \hat{n}$.

It is difficult to remedy the incorrect patent incentives. If too few firms would enter, it is hard to entice more, since the additional entrants would make negative profit in expectation. If too many firms would enter, it would be hard to exclude them, especially if the ideas were common knowledge. If some competitor dropped out, another competitor would enter with the unused approach.

Thus, if the approaches to solving the problem are common knowledge, the public sector can do a better job of coordinating the different approaches, by adjusting their relative funding or funding the appropriate selection from them.

The advantages of public sponsorship become even more pronounced if we augment the model to account for learning. We will not give a formal treatment, but will summarize the idea. An R&D effort typically takes place over time. As portions of the cost *c* are invested, more is learned about the probability that each approach will be successful. If one approach seems particularly promising, then it is rational to shift resources away from the less promising toward the more promising. At some point the promising venture may play itself out, and then the resources should again be reorganized.

This rational shifting of research priorities can best take place if the intermediate knowledge learned is aggregated in a systematic way. A problem with private IP incentives is that firms have no incentive to share knowledge in a way that is socially efficient. Their incentives to share knowledge are polluted by a desire to win the race, and hence to get rivals to drop out, even if that is inefficient. The rivals know that each of them has such an incentive, and therefore the firms' signalling about their private states of knowledge will be suspect.[11] An example of different ideas embodying different (uncertain) approaches can be found in the attempt to develop nuclear fusion as an energy source. Since 1945, governments have understood that current fossil-fuel and nuclear-energy technologies cannot meet society's demand for electricity indefinitely. By contrast, a successful fusion technology would generate essentially limitless energy from an isotope, deuterium, found in sea water. Researchers in the U.S., Europe, Russia, and Japan have all been interested in this problem, using different approaches, at considerable expense. Over the past 25 years, the U.S. expenditure on fusion-related R&D has consistently ranged (in constant dollars) from $100 million to $500 million per year (Kulcinski & Santarisu, 1998). The current budget stands at $257 million (Seife, 2003).

There are at least twenty approaches to the problem of generating energy through nuclear fusion, depending on how one counts.[12] Almost all of these approaches were understood at the very beginning, in 1951–1952 (Herman, 1990; Lindl, 1998). Despite years of public funding and experimentation, none of the approaches has been proved to achieve goals – and none has been definitively ruled out by experiment. The technical approach favored by the Russians – so-called tokamaks – leaped ahead in trials for a brief period in the 1960s, and worldwide funding shifted toward that approach. By 1972, at least seventeen new tokamaks were completed or under construction outside the USSR (Herman, 1990). Despite the disappointing record, partisans in the global partnership remain confident, and in 1986 they hatched a plan to spend $10 billion commercial-scale fusion reactor. The plan has been controversial due to the unproven nature of the technology. The commitment of funds was cut in half after the U.S. briefly dropped out in 1998 when detractors were advocating other approaches and more basic science (Seife, 2003).

It is unclear whether private ventures would ever invest in an innovation as costly as nuclear fusion, regardless of the potential payoff, and that may be the single most compelling reason that it should be funded publicly, if at all. But the other important lesson embedded in the example is that researchers can and do learn from each other. The tokamak may or may not end up the survivor, but it is significant that researchers responded to experimental evidence in its favor. Private competitors might have suppressed such evidence in order to maintain a competitive advantage.

7. MIXED PRIVATE/PUBLIC INCENTIVES

In contrast to many of the examples given above, and also in contrast to how the public/private question is usually posed by academic economists, incentives are generally not provided entirely through intellectual property or entirely through public sponsorship. The modern economy has evolved such that public and private funds are often mixed.

There are at least three pieces of legislation that underlie that phenomenon. The most important are probably the Bayh-Dole and Stevenson-Wydler Acts of 1980, which authorize the patenting of innovations developed with federal funds. They affect the national laboratories operated by the Department of Energy, such as Lawrence Berkeley Laboratory, Livermore Laboratory, Los Alamos, and many others, which collectively account for about 3% of total R&D spending in the U.S. They also affect most research done in universities, which collectively account for about 11% of total R&D spending in the U.S., but receive about 60% of their research funds from the federal government.[13] Subsequent to the Bayh-Dole Act, most research universities have instituted Offices of Technology Licensing. Most such offices prefer exclusive licensing (Jensen & Thursby, 2001), suggesting that they intend to exercise their rights as profit maximizers, just as a private entity would.

The other legal foundation is the Federal Technology Transfer Act of 1986. This is a tool that lets national laboratories (and other federal enterprises) create partnerships with private entities, called CRADAs, and thus spin off their innovations into the private sector. By the mid-1990s, about 1000 CRADA's were being formed each year, almost half in partnership with the Department of Energy.[14]

Finally, there are occasional examples of public/private incentives, resulting in intellectual property rights for the private sector, that fall outside these two broad categories. DARPA has offered a $1 million Grand Challenge prize for winning a race among robotic ground vehicles (DARPA, 2003). Entrants not only compete for this rather substantial prize, but will own any intellectual property that results.

Many commentators question the rationale for funding R&D directly with public funds, but at the same time offering intellectual property rights. The fear is that inventors get funded twice – once with grants and once with IP – thus compounding the burden to users and taxpayers. In addition, such lucrative funding can lead to wasteful duplication of efforts.

Many of these public/private ventures involve "big science," where one might argue that neither the private sector nor public sector alone can "afford" to fund it. That argument is not very convincing, since large benefits should attract large investments, even if the funds have to be borrowed. In any case, no organization has better ability to raise money than the federal government.

Nevertheless, the public/private partnership resulting in intellectual property rights may have merit. Our argument is not focussed on the difficulty of raising money, but rather on how the system of dual funding can overcome other problems that would otherwise afflict both sponsors.

For big science, industry has the problem that, although an innovation may have some commercial value, the commercial value under existing intellectual property laws may not be sufficient to cover costs. This is especially true for investments that have unappropriable social benefits, or where intellectual property rights are so narrow that the benefits will be eroded by competitors. The public sector could simply fund such projects, but then it faces the problem of choosing the ones that are likely to be fruitful, or making sure the funds are used as intended, especially when the expertise resides mostly in the private sector.

We have already shown how a grantor can overcome these problems for small, frequent innovations, simply by threatening to cut the innovator out of the grant process if he or she does not deliver enough innovations, or innovations of high enough value. But "big science" will not produce frequent deliveries of small innovations, regardless of how meritorious it is. In fact, for many big science projects such as energy through nuclear fusion, the real quality indicator is the likelihood of success, and even the best project may fail. The public sponsor therefore needs some mechanism to screen for the right investments, and to make sure that the researchers invest as directed.

We suggest that a system of matching funds, where it is industry matching the government subsidy rather than vice versa, together with partial payment to industry through intellectual property rights, can solve the dual problems of ensuring that industry covers its cost, and avoiding wrong investments by the public sector.

Each point in Fig. 1 represents an idea (v, c). We will suppose that industry is the repository of the best information about investments, so that the value and cost (v, c) is known to the researcher, but not to a government sponsor. The objective of the government sponsor is to invest in those ideas for which $(1/r)v - c > 0$. The value of the intellectual property available to a private firm if it invests on its own is πvT, and without a government subsidy, it will invest if $\pi vT - c > 0$. In Fig. 1, the area under the lower diagonal line, labeled πvT, represents the ideas for which the private incentives in intellectual property would be sufficient.

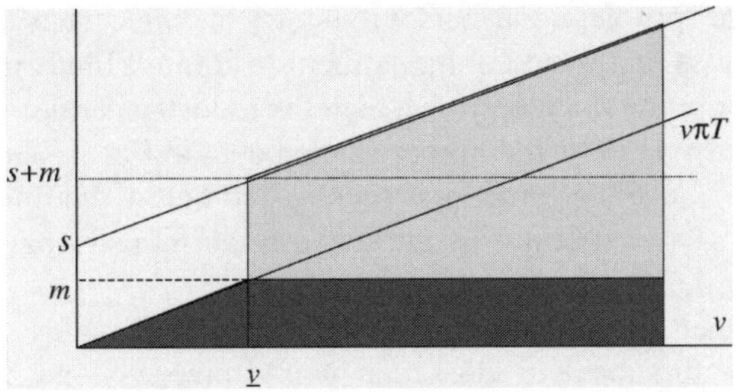

Fig. 1. Hybrid Incentives: Intellectual Property with Public Subsidies.

Suppose, however, that there are high-value ideas that are more costly than the value of the intellectual property right (above the line πvT), but still worth doing ($(1/r)v > c > \pi vT$). The government might like to sponsor these.

Suppose that the government simply offers a subsidy s. Then all the ideas (v, c) under the higher diagonal line $\pi vT + s$ will be undertaken. The subsidy will increase research, but indiscriminately. In particular, there are likely to many low-v ideas, toward the origin in Fig. 1, for which the subsidy of s is a waste of money. It is easy to imagine an endless series of subsidy claims for worthless innovations.

The government can solve this problem by insisting that the claimant make a matching commitment of funds in some amount, say m. Then in order to claim the subsidy, the sponsor and the claimant will contribute s and m respectively to the research budget, and invest in ideas suggested by the claimant. If the claimant suggests an idea (v, c) such that $c > s + m$, then he must provide the required supplement. In that case, the industry partner pays $c - s$ rather than m. If the claimant suggests an idea (v, c) such that $c < s + m$, then the surplus goes to supporting graduate students or other research enterprises. For simplicity, we assume in Fig. 1 that the claimant receives intellectual property in amount πTv on the subsidized innovations, although in practice, the sponsor may require a reduction in intellectual property rights for some broader social purpose. For example, the NIH imposes guidelines under which research tools must be made freely available to other academic researchers, with an intent to protect other grantees of the NIH.

We can now see what happens under this incentive mechanism.

First, ideas (v, c) that satisfy $c < m$ will not be subsidized. The industry partner gets the intellectual property rights whether or not he is subsidized, and it is more profitable to get these property rights for the lower cost c rather than m. Thus, the dark shaded area that is below both the diagonal line πvT and below the horizontal line m represents ideas that industry will invest in without claiming any subsidy.

It is only in ideas (v, c) that satisfy $\pi vT > m$ that the industry partner would claim a subsidy. These ideas are all to the right of v in Fig. 1. This is what solves the "moral hazard" problem. Partners will not try to collect subsidies on worthless innovations because that would obligate them to commit funds in amount m.

We can further refine the ideas for which the industry partner would be willing to accept a subsidy, namely, the higher shaded area in Fig. 1. These are the ideas to the right of v, above the horizontal line m, and below the line $\pi v T + s$. Ideas such that $c > \pi v T + s$ will not be attractive to an industry partner because the partner must pay the surfeit $c - s$.

We thus see that the partnership with mandatory matching funds will allow the public sector to subsidize ideas that would otherwise not elicit investment (those between the lines $\pi v T + s$ and $\pi v T$, to the right of v), without causing the sponsor to hemorrhage money in subsidies to worthless innovations.

In recent years, industry matching requirements have become increasingly important in government aircraft procurement. During the 1980s, industry contributed roughly one-half the cost of developing prototypes for the F-22 and F-23 fighters (Sweetman, 1991). During the 1990s, NASA agreed to spend $912 million on the X-33 spacecraft, provided that Lockheed contributed $212 million. Lockheed, which had agreed to pay for any cost overruns, eventually spent $357 million before the project was cancelled in 2001 (NASA, 2001a, b).

The rationale contained in the Bayh-Dole Act is not the one given here. The rationale is not about *ex ante* incentives to create innovations, but focuses instead on the *ex post* incentive of universities and national laboratories to disseminate them. The basic assertion is that licenses on university patent rights, especially exclusive licenses, give industry licensees enough protection to make the collateral investments required to commercialize those innovations.

But there is an equally compelling argument that publicly sponsored research should be put in the public domain. This would arguably maximize the number of users, and therefore maximize the number of follow-on innovations. Whether the *ex post* rationale for the Bayh-Dole Act is convincing depends on whether one believes that intellectual property rights are constructed sensibly to begin with. If the intellectual property for follow-on innovations is sufficient, then exclusive licenses on the underlying publicly sponsored inventions are not necessary. And if insufficient, then the Bayh-Dole Act remedies a defect of intellectual property law. Wouldn't it be better to fix the thing that's broken?

8. CONCLUSION

Public procurement provides the option to pay for knowledge out of general revenue, which is generally thought to impose less deadweight loss than taxing a single market, and then to put the knowledge in the public domain. Thus, public sponsorship of knowledge that is destined for the public domain has a natural advantage over intellectual property: it avoids the inefficient exclusion of users whose willingness to pay is smaller than the proprietary price, but larger than the marginal cost of supply. For such exclusion, we use the term "deadweight loss."

However, intellectual property also has a natural advantage over public sponsorship: it ensures that only the users pay the costs of inventions, and since these payments are voluntary, no citizen can be made worse off by an invention. The latter advantage is particularly persuasive for knowledge that is narrowly targeted to a special audience, but it is not determinative even there. Users of knowledge in two disjoint targeted audiences could all be better off by making reciprocal subsidies, each contributing their tax dollars toward the knowledge that is useful to the other targeted audience, under the condition that no one is excluded from use.

On grounds that citizens can provide reciprocal subsidies, we have implicitly taken the view in this paper that it is best to avoid deadweight loss (proprietary pricing). This pre-judges the choice between public and private sponsorship, and in doing so, allows us to focus on other important aspects of the creative environment, in particular, which schemes are best at harnessing the creative genius that is widely dispersed among humans.

It is worth noting that sometimes there is no alternative to public sponsorship funded through taxation, since it may be impossible to exclude beneficiaries from using new knowledge, and therefore impossible to charge each user a price.

Conversely, invention can sometimes create its own reward without any formal incentive mechanism. In the early days of airplane development, manufacturers often subsidized the aircraft that were used in competitions for long-distance flights, since winning could be expected to create commercial advantages.[15] Benefits presumably included publicity and gains to reputation, the opportunity to develop trade secrets and tacit knowledge, and the firm's normal share of externalities to aviation as a whole. In the case of air racing, prizes were also supported by gate receipts (Vorderman, 1991). The X-Prize competitors have similar business plans. Contestants hope to tap a small but lucrative market for passenger flights into space (Hoffman, 2003).

But in addition, many racing aircraft received massive subsidies from governments and, in the 1930s, also from charitable or patriotic individuals (Sweetman, 1984; Vorderman, 1991). The pattern again continues in the X-Prize competition, where at least two competitors are reportedly bankrolled by dot.com millionaires. Most of the X-Prize itself was raised from such donors (Hoffman, 2003). Motivations include civic pride, patriotism, and an ideological commitment to spaceflight as an end in itself (X Prize Foundation, 2003).

NOTES

1. One important examples is what Mokyr (2002) calls "propositional knowledge." Even if the knowledge that "DNA is a double helix" were patentable, it would be hard to collect a royalty on every use of that knowledge.

2. The interpretation of an idea will depend on how ideas are related to each other. In OST, ideas follow each other in a progression up a quality ladder; a later idea must be preceded by the earlier ones. That context is not discussed here.

3. For tidier notation, we always use discounted time instead of "real time." If the real length is τ, then the discounted length is $T = \int_0^{\tau} e^{-rt}\, dt$. There is a one-to-one map between real time and discounted time, and the maximum discounted time is $1/r$.

4. Gallini and Scotchmer (2002) point out that the same problem of verifiability arises with patents, but only when they are litigated.

5. Contemporary observers contrasted this procedure with what happened in England, where a better process was stifled by patent litigation (Newhouse, 1988).

6. Long distance flight has traditionally been supported by prizes. Famous examples include the first flight across the English channel in 1909 ($5,000), the first flight across the North Atlantic in 1919 (£10,000), and the first non-stop flight from New York to Paris in 1927 ($50,000). The 1927 contest, which resulted in Lindbergh's flight is particularly interesting. The prize was originally offered in 1919 for the first flier to accomplish the feat before 1924. This incentive may have been insufficient given the technologies of the time; in any case, there were no takers. The prize was then renewed for an additional five year period drawing a total of nine teams into the race, although at least one of these efforts failed to obtain adequate financing. The remaining teams spent a total of $400,000 in the effort. Eight of these aircraft either crashed on take off or disappeared. Lindbergh's successful effort cost $25,000 (Jablonski, 1972; X Prize Foundation, 2003).

7. These amounts tend to understate the incentive, since a single aircraft could compete in multiple races and win multiple purses. In a few cases, aircraft were also able to compete in successive years (Vorderman, 1991).

8. It is named after economist William S. Vickrey, who first exposited its remarkable properties in his (1961) paper. The simplest example is where an auctioneer wants to transfer an object to the agent with the highest valuation, but cannot observe the valuations. He asks each one to state his valuation, then gives the object to the highest bidder, but only charges a price equal to the second highest bid. In that situation, there is only one unobservable for each bidder, and nothing needs to be observed *ex post*. In the version given here, each agent has two unknown variables, cost and value. The value must be observed *ex post*, but not the cost.

9. The model presented here is simpler than that of Che and Gale. In their model, the firms can choose what qualities to develop, using a more standard model in which there is a known function that yields quality in return for investment effort. The firms announce their qualities in advance, coupled with a random distribution on prices as below.

10. For a concise account, see Mowery and Rosenberg (1998, Chap. 2). For the basic data on R&D funding, see NSF, Science Indicators 2002.

11. See Scotchmer and Green (1990) for a model that is similar in that firms may induce shakeouts by revealing their intermediate progress in pursuit of a goal.

12. Approaches are conventionally divided between "magnetic confinement schemes" that use powerful magnets and/or electric fields to contain ultra-hot gas and "inertial

confinement" schemes that extract energy from fuel pellets before they explode. Technologically distinct methods for magnetic confinement include pinch effect machines, stellarators, tokamaks, spherical tokamaks, spherical torus machines, spheromaks, electric tokamaks, colliding beam machines, magnetic mirrors, floating multipole machines, reversed field pinch machines, magnetic targets, inertial electrostatic confinement machines, and Z pinch machines. Inertial confinement schemes include direct drive by lasers, direct drive by ion beams, and indirect drive by lasers. The foregoing categories are not exclusive. Dark horse candidates have included cold fusion, sonoluminescence, and dense plasma focus machines. For a detailed survey, see National Fusion Science Energy Web Site (1998).

13. NSF, Science Indicators 2002.

14. See Maurer (2002) for a more complete discussion.

15. In one striking example, Bellanca told the Spirit of St. Louis team that it was willing to cut the price of its aircraft from $25,000 to $15,000 – but only if they let the company name its own pilot. Lindbergh declined (Jablonski, 1972).

REFERENCES

Bryant, J. H. (1988). *Henrich Hertz: The beginning of microwaves*. Piscataway, NJ: Institute of Electrical and Electronics Engineers.

Casson, L. (2001). *Libraries in the ancient world*. New Haven, CT: Yale University Press.

Che, Y.-K., & Gale, I. (forthcoming). Optimal design of research contests. *American Economic Review*.

David, P. (2003). The economic logic of "open science" and the balance between private property rights and the public domain in scientific data and information: A primer. In: National Research Council (Eds), *The Role of the Public Domain in Scientific and Technical Data and Information*. Washington, DC: National Academies Press.

DeCamp, L. S. (1980). *The ancient engineers*. New York: Ballentine Press.

Defense Advanced Research Projects Agency (2003). DARPA Grand Challenge, http://www.darpa.mil/grandchallenge/.

Foray, D., & Hilaire-Perez, L. (2000). The economics of open technology: Collective organization and individual claims in the "Fabrique Lyonnaise" during the old regime. Mimeo.

Gallini, N., & Scotchmer, S. (2002). Intellectual property: When is it the best incentive mechanism? *Innovation Policy and the Economy*, *2*, 51–78. Cambridge, MA: MIT Press.

Google (2002). Google programming contest. http://www.google.com/programming-contest/.

Herman, R. (1990). *Fusion: The search for endless energy*. New York: Cambridge University Press.

Hoffman, C. (2003). The right stuff. *Wired* (July), 134–145.

Holden, C. (2003). DARPA's desert dash. *Science*, *300*, 733.

Innocentive (2003). About Innocentive. http://www.innocentive.com/about/.

Jablonski, E. (1972). *Atlantic fever*. New York: Macmillan.

Jensen, R., & Thursby, M. (2001). Proofs and prototypes for sale: The licensing of university inventions. *American Economic Review*, *91*, 240–259.

Kremer, M. (1998). Patent buyouts: A mechanism for encouraging innovation. *Quarterly Journal of Economics*, *CXIII*, 1137–1168.

Kulcinski, G., & Santarisu, J. (1998). Reducing the barriers to fusion electric power. *Journal of Fusion Energy*, *17*, 17–23.

Lindl, J. D. (1998). *Inertial confinement fusion*. New York: American Institute of Physics Press.
Maurer, S. M. (2002). Promoting and disseminating knowledge: The public/private interface. National Academies of Sciences, http://www7.nationalacademies.org/biso/Maurer_background_paper.html.
Mokyr, J. (1990). *The lever of riches*. Oxford: Oxford University Press.
Mokyr, J. (2002). *The gifts of Athena*. Princeton, NJ: Princeton University Press.
Mowery, D., & Rosenberg, N. (1998). *Paths of innovation*. New York: Cambridge University Press.
NASA (2001a). Press release (March 1). ftp://ftp.hq.nasa.gov/pub/pao/pressrel/2001/01-031.txt.
NASA (2001b). Ames X-33 public web site. http://asm.arc.nasa.gov/x33/x33home.shtml.
National Fusion Science Energy Web Site (1998). Fusion innovations: A report from ICC98. http://www.fusionscience.org/ICC98/ICC98_Sum.pdf.
National Science Foundation (2002). Science indicators 2000. www.nsf.gov/sbe/srs/seind02/pdf_v2.htm.
Nelson, R. R., & Winter, S. D. (1982). *An evolutionary theory of economic change*. Cambridge, MA: Belknap Press of Harvard University Press.
Newhouse, E. L. (Ed.) (1988). *Inventors and discoverers: Changing our world*. Washington, DC: National Geographic Society.
O'Donoghue, T., Scotchmer, S., & Thisse, J.-F. (1998). Patent breadth, patent life and the pace of technological progress. *Journal of Economics and Management Strategy*, *7*, 1–32.
Penn, C. (1993). Super efficient refrigerator finalists. *Home energy magazine*. http://hem.dis.anl.gov/eehem/93/930305.html.
Porter, R. (1994). *The biographical dictionary of scientists*. New York: Oxford University Press.
Reichhardt, T. (2002). Out of this world. *Nature*, *420*, 10–11.
Reinganum, J. (1989). The timing of innovation: Research, development and diffusion. In: R. Schmalensee & R. D. Willig (Eds), *Handbook of Industrial Organization* (pp. 849–908). Amsterdam: Elsevier.
Rogerson, W. (1994). Economic incentives and the defense procurement process. *Journal of Economic Perspectives*, *8*, 65–90.
RSA Security (2003). Cryptographic challenges. http://www.rsasecurity.com/rsalabs/challenges/index.html.
Scherer, F. M. (1984). *Innovation and growth*. Cambridge, MA: MIT Press.
Scotchmer, S. (2004, forthcoming). *Innovation and incentives*. Cambridge, MA: MIT Press.
Scotchmer, S., & Green, J. (1990). Novelty and disclosure in patent law. *The RAND Journal of Economics*, *21*, 131–146.
Seife, C. (2003). United States rejoins international fusion-research project. *Science*, *299*, 801.
Shavell, S., & van Ypserele, T. (2001). Rewards versus intellectual property rights. *Journal of Law and Economics*, *44*, 525.
Sobel, D. (1995). *Longitude: The true story of a lone genius who solved the greatest scientific problem of his time*. New York: Penguin Books.
SpaceProjects.com (2003). Competitive prices regarding space as a waste-reducing means of NASA procurement. http://www.spaceprojects.com/prizes/.
Strandh, O. (1979). *A history of the machine*. New York: A&W Publishers.
Sweetman, W. (1984). *High speed flight*. London, UK: Janes.
Sweetman, W. (1991). *YF-22 and YF-23 advanced tactical fighters: Stealth, speed and agility for air superiority*. Osceola, WI: Motorbooks.
Sweetman, W., Gething, M., Richardson, D., Spick, M., & Gunston, W. (1987). *The great book of modern warplanes*. New York: Salamander.

Vickrey, W. (1961). Counterspeculation, auctions, and competitive sealed tenders. *Journal of Finance*, *16*, 8–37.

Vorderman, D. (1991). *The great air races*. New York: Bantam Books.

Wright, B. (1983). The economics of invention incentives: Patents, prizes and research contracts. *American Economic Review*, *73*, 691–707.

X Prize Foundation Web Site (2003). http://www.xprize.org/.

Zuckerman, S. (2003). The new wave of energy efficient refrigerators. http://www.ecomall.com/greenshopping/icebox2.htm.\newpage.

ECONOMIC PERSPECTIVES ON OPEN SOURCE

Josh Lerner and Jean Tirole

ABSTRACT

There has been a recent surge of interest in open source software development, which involves developers at many different locations and organizations sharing code to develop and refine programs. To an economist, the behavior of individual programmers and commercial companies engaged in open source projects is initially startling. This paper makes a preliminary exploration of the economics of open source software. We highlight the extent to which labor economics, especially the literature on career concerns', can explain many of these projects' features. Aspects of the future of open source development process, however, remain somewhat difficult to predict with off-the-shelf' economic models.

1. INTRODUCTION

In recent years, there has been a surge of interest in open source software development. Interest in this process, which involves software developers at many different locations and organizations sharing code to develop and refine software programs, has been spurred by three factors:

- *The rapid diffusion of open source software.* A number of open source products, such as the Apache web server, dominate their product categories. In the personal

Intellectual Property and Entrepreneurship
Advances in the Study of Entrepreneurship, Innovation and Economic Growth,
Volume 15, 33–69

ISSN: 1048-4736/doi:10.1016/S1048-4736(04)01502-4

computer operating system market, International Data Corporation estimates that the open source program Linux has between seven to twenty-one million users worldwide, with a 200% annual growth rate. Many observers believe it represents a leading potential challenger to Microsoft Windows in this important market segment.

- *The significant capital investments in open source projects*. Over the past two years, numerous major corporations, including Hewlett Packard, IBM, and Sun, have launched projects to develop and use open source software. Meanwhile, a number of companies specializing in commercializing Linux, such as Red Hat and VA Linux, have completed initial public offerings, and other open source companies such as Cobalt Networks, Collab.Net, Scriptics, and Sendmail have received venture capital financing.
- *The new organization structure*. The collaborative nature of open source software development has been hailed in the business and technical press as an important organizational innovation.

Yet to an economist, the behavior of individual programmers and commercial companies engaged in open source processes is startling. Consider these quotations by two leaders of the open source community:

> The idea that the proprietary software social system – the system that says you are not allowed to share or change software – is unsocial, that it is unethical, that it is simply wrong may come as a surprise to some people. But what else can we say about a system based on dividing the public and keeping users helpless? (Stallman, 1999).

> The "utility function" Linux hackers is maximizing is not classically economic, but is the intangible of their own ego satisfaction and reputation among other hackers. [Parenthetical comment deleted] Voluntary cultures that work this way are actually not uncommon; one other in which I have long participated is science fiction fandom, which unlike hackerdom explicitly recognizes "egoboo" (the enhancement of one's reputation among other fans) (Raymond, 1999b).

It is not initially clear how these claims relate to the traditional view of the innovative process in the economics literature. Why should thousands of top-notch programmers contribute freely to the provision of a public good? Any explanation based on altruism[1] only goes so far. While users in less developed countries undoubtedly benefit from access to free software, many beneficiaries are well-to-do individuals or Fortune 500 companies. Furthermore, altruism has not played a major role in other industries, so it would have to be explained why individuals in the software industry are more altruistic than others.

This paper seeks to make a preliminary exploration of the economics of open source software. Reflecting the early stage of the field's development, we do not seek to develop new theoretical frameworks or to statistically analyze large samples. Rather, we focus on four "mini-cases" of particular projects: Apache,

Linux, Perl, and Sendmail.[2] We seek to draw some initial conclusions about the key economic patterns that underlie the open source development of software. We find that much can be explained by reference to economic frameworks. We highlight the extent to which labor economics, in particular the literature on "career concerns," and industrial organization theory can explain many of the features of open source projects.

At the same time, we acknowledge that aspects of the future of open source development process remain somewhat difficult to predict with "off-the-shelf" economic models. In the final section of this paper, we highlight a number of puzzles that the movement poses. It is our hope that this paper will have itself an "open source" nature: that it will stimulate research by other economic researchers as well.

Finally, it is important to acknowledge the relationship with the earlier literature on technological innovation and scientific discovery. The open source development process is somewhat reminiscent of the type of "user-driven innovation" seen in many other industries. Among other examples, Rosenberg's (1976) studies of the machine tool industry and von Hippel's (1988) of scientific instruments have highlighted the role that sophisticated users can play in accelerating technological progress. In many instances, solutions developed by particular users for individual problems have become more general solutions for wide classes of users. Similarly, user groups have played an important role in stimulating innovation in other settings: certainly, this has been the case from the earliest days of the computer industry (e.g. Caminer et al., 1996).

A second strand of related literature examines the adoption of the scientific institutions ("open science," in Dasgupta & David's, 1994 terminology) within for-profit organizations. Henderson and Cockburn (1994) and Gambardella (1995) have highlighted that the explosion of knowledge in biology and biochemistry in the 1970s triggered changes in the management of R&D in major pharmaceutical firms. In particular, a number of firms encouraged researchers to pursue basic research, in addition to the applied projects that typically characterized these organizations. These firms that did so enjoyed substantially higher R&D productivity than their peers, apparently because the research scientists allowed them to more accurately identify promising scientific developments (in other words, their "absorptive capacity" was enhanced) and because the interaction with cutting-edge research made these firms more attractive to top scientists. At the same time, the encouragement of "open science" processes has not been painless. Cockburn, Henderson and Stern (1999) highlight the extent to which encouraging employees to pursue both basic and applied research led to substantial challenges in designing incentive schemes, because of the very different outputs of each activity and means through which performance is measured.[3]

But as we shall argue below, certain aspects of the open source process – especially the extent to which contributors' work is recognized and rewarded – are quite distinct from earlier settings. This study focuses on understanding this contemporaneous phenomenon, rather than seeking to make a general evaluation of the various cooperative schemes employed over time.

2. THE NATURE OF OPEN SOURCE SOFTWARE

While media attention to the phenomenon of open source software has been recent, the basic behaviors are much older in their origins. There has long been a tradition of sharing and cooperation in software development. But in recent years, both the scale and formalization of the activity have expanded dramatically with the widespread diffusion of the Internet.[4] In the discussion below, we will highlight three distinct eras of cooperative software development.

2.1. The First Era: Early 1960s to the Early 1980s

Many of the key aspects of the computer operating systems and the Internet were developed in academic settings such as Berkeley and MIT during the 1960s and 1970s, as well as in central corporate research facilities where researchers had a great deal of autonomy (such as Bell Labs and Xerox's Palo Alto Research Center). In these years, the sharing by programmers in different organizations of basic operating code of computer programs – the source code – was commonplace.[5]

Many of the cooperative development efforts in the 1970s focused on the development of an operating system that could run on multiple computer platforms. The most successful examples, such as Unix and the C language used for developing Unix applications, were originally developed at AT&T's Bell Laboratories. The software was then installed across institutions, being transferred freely or for a nominal charge. Many of the sites where the software was installed made further innovations, which were in turn shared with others. The process of sharing code was greatly accelerated with the diffusion of Usenet, a computer network begun in 1979 to link together the Unix programming community. As the number of sites grew rapidly (e.g. from 3 in 1979 to 400 in 1982), the ability of programmers in university and corporate settings to rapidly share technologies was considerably enhanced.

These cooperative software development projects were undertaken on a highly informal basis. Typically no effort to delineate property rights or to restrict reuse of the software were made. This informality proved to be problematic in the early

1980s, when AT&T began enforcing its (purported) intellectual property rights related to Unix.

2.2. The Second Era: Early 1980s to the Early 1990s

In response to these threats of litigation, the first efforts to formalize the ground rules behind the cooperative software development process emerged. This ushered in the second era of cooperative software development. The critical institution during this period was the Free Software Foundation, begun by Richard Stallman of the MIT Artificial Intelligence Laboratory in 1983. The foundation sought to develop and disseminate a wide variety of software without cost.

One important innovation introduced by the Free Software Foundation was a formal licensing procedure that aimed to preclude the assertion of patent rights concerning cooperatively developed software (as many believed that AT&T had done in the case of Unix). In exchange for being able to modify and distribute the GNU software (as it was known), software developers had to agree to make the source code freely available (or at a nominal cost). As part of the General Public License (GPL, also known as "copylefting"), the user had to also agree not to impose licensing restrictions on others. Furthermore, all enhancements to the code – and even code that intermingled the cooperatively developed software with that developed separately – had to be licensed on the same terms. It is these contractual terms that distinguish open source software from shareware (where the binary files but not the underlying source code are made freely available, possibly for a trial period only) and public-domain software (where no restrictions are placed on subsequent users of the source code).[6]

This project, as well as contemporaneous efforts, also developed a number of important organizational features. In particular, these projects employed a model where contributions from many developers were accepted (and frequently publicly disseminated or posted). The official version of the program, however, was managed or controlled by a smaller subset of individuals closely involved with the project, or in some cases, an individual leader. In some cases, the project's founder (or his designated successor) served as the leader; in others, leadership rotated between various key contributors.

2.3. The Third Era: Early 1990s to Today

The widespread diffusion of Internet access in the early 1990s led to a dramatic acceleration of open source activity. The volume of contributions and diversity of contributors expanded sharply, and numerous new open source projects emerged, most notably Linux (a UNIX operating system developed by Linus Torvalds in 1991). As discussed in detail below, interactions between commercial companies and the open source community also became commonplace in the 1990s.

Another innovation during this period was the proliferation of alternative approaches to licensing cooperatively developed software. During the 1980s, the GPL was the dominant licensing arrangement for cooperatively developed software. This changed considerably during the 1990s. In particular, Debian, an organization set up to disseminate Linux, developed the "Debian Free Software Guidelines" in 1995. These guidelines allowed licensees greater flexibility in using the program, including the right to bundle the cooperatively developed software with proprietary code. These provisions were adopted in early 1997 by a number of individuals involved in cooperative software development, and were subsequently known as the "Open Source Definition." As the authors explained:

> *License Must Not Contaminate Other Software*. The license must not place restrictions on other software that is distributed along with the licensed software. For example, the license must not insist that all other programs distributed on the same medium must be open-source software. Rationale: Distributors of open-source software have the right to make their own choices about their own software (Open Source Initiative, 1999).

These new guidelines did not require open source projects to be "viral": they need not "infect" all code that was compiled with the software with the requirement that it be covered under the license agreement as well. At the same time, they also accommodated more restrictive licenses, such as the General Public License.

The past few years have seen unprecedented growth of open source software. At the same time, the movement has faced a number of challenges. We will highlight two of these here: the "forking" of projects (the development of competing variations) and the development of products for high-end users.

One issue that has emerged in a number of open source projects is the potential for programs splintering into various variants. In some cases, passionate disputes over product design have led to the splintering of open source projects into different variants. Examples of such splintering are the Berkeley Unix program and Sendmail during the late 1980s.

Another challenge has been the apparently lesser emphasis on documentation and support, user interfaces,[7] and backward compatibility found in at least some open source projects. The relative technological features of software developed in open source and traditional environments are a matter of passionate discussion. Some members of the community believe that this production method dominates traditional software development in all respects. But many open source advocates argue that open source software tends to be geared to the more sophisticated users.[8] This point is made colorfully by one open source developer:

> [I]n every release cycle Microsoft always listens to its *most ignorant customers*. This is the key to dumbing down each release cycle of software for further assaulting the non personal-computing population. Linux and OS/2 developers, on the other hand, tend to listen to their *smartest* customers . . . The good that Microsoft does in bringing computers to non-users is outdone by the curse that they bring on experienced users (Nadeau, 1999).

Certainly, the greatest diffusion of open source projects appears to be in settings where the end users are sophisticated, such as the Apache server installed by systems administrators. In these cases, users are apparently more willing to tolerate the lack of detailed documentation or easy-to-understand user interfaces in exchange for the cost savings and the possibility of modifying the source code themselves. In several projects, such as Sendmail, project administrators chose to abandon backward compatibility in the interests of preserving program simplicity.[9] One of the rationales for this decision was that administrators using the Sendmail system were responsive to announcements that these changes would be taking place, and rapidly upgraded their systems. In a number of commercial software projects, it has been noted, these types of rapid responses are not as common. Once again, this reflects the greater sophistication and awareness of the users of open source software.

The debate about the ability of open source software to accommodate high-end users' needs has direct implications for the choice of license. The recent popularity of more liberal licenses and the concomitant decline of the GNU license are related to the rise in the "pragmatists'" influence. These individuals believe that allowing proprietary code and for-profit activities in segments that would otherwise be poorly served by the open-source community will provide the movement with its best chance for success.

2.4. Who Contributes?

Computer system administrators, database administrators, computer programmers, and other computer scientists and engineers represented about 2.1 million jobs in the United States in 1998. (Unless otherwise noted, the information in this paragraph is from U.S. Department of Labor, 2000.) A large number of these workers – estimated at between 5 and 10% – are either self-employed or retained on a project-by-project basis by employers. Computer-related positions are projected by the federal government to be among the fastest-growing professions in the next decade.

The distribution of contributors to open source projects appears to be quite skewed. This is highlighted by an analysis of 25 million lines of open source code,

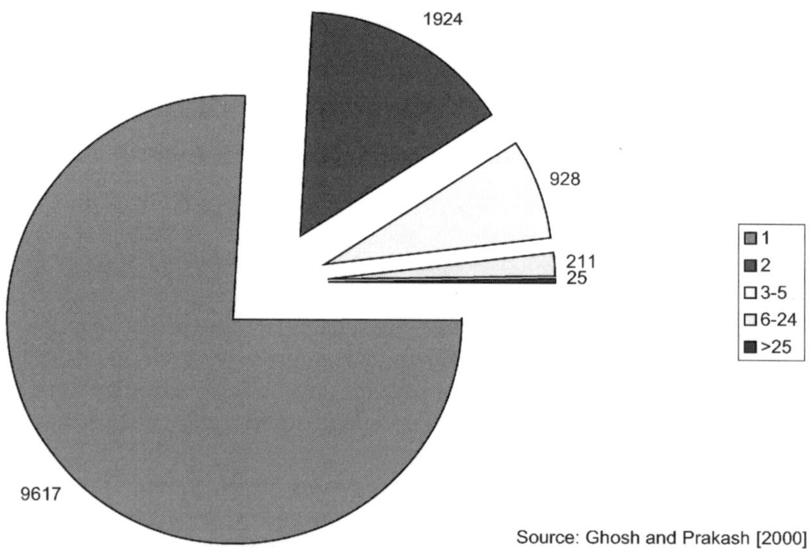

Fig. 1. Distribution of Contributions by Participant.

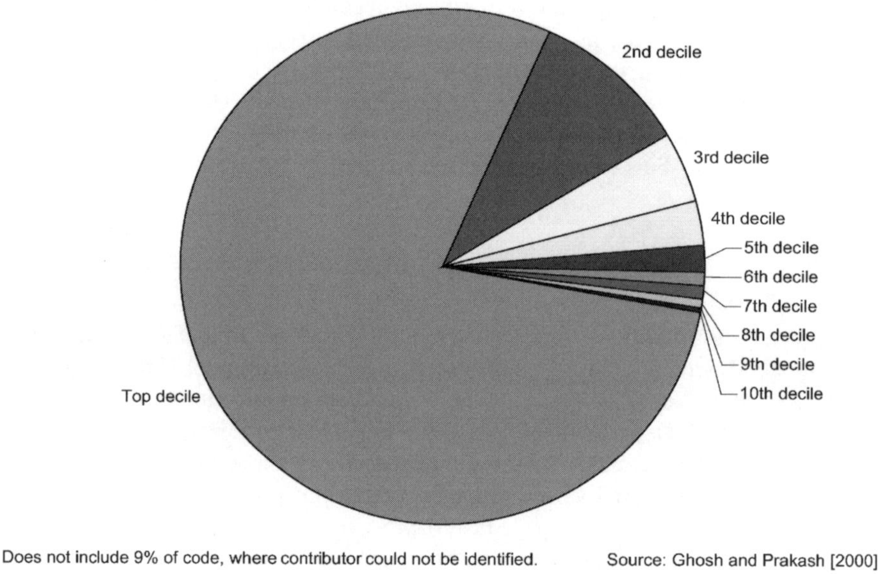

Fig. 2. Distribution of Code Contributed, by Decile.

constituting 3149 distinct projects (Ghosh & Prakash, 2000). The distribution of contributions is shown in Fig. 1. More than three-quarters of the nearly 13 thousand contributors made only one contribution; only one in twenty-five had more than five contributions. Yet the top decile of contributors accounted for fully 72% of the code contributed to the open source projects, and the top two deciles for 81% (see Fig. 2). This distribution would be even more skewed if those who simply reported errors, or "bugs," were considered: for every individual who contributes code, five will simply report errors (Valloppillil, 1998). To what extent this distribution is unique to open source software is unclear: the same skewness of output is also observed among programmers employed in commercial software development facilities (e.g. see Brooks, 1975; Cusumano, 1991), but it is unclear whether these distributions are similar in their properties.

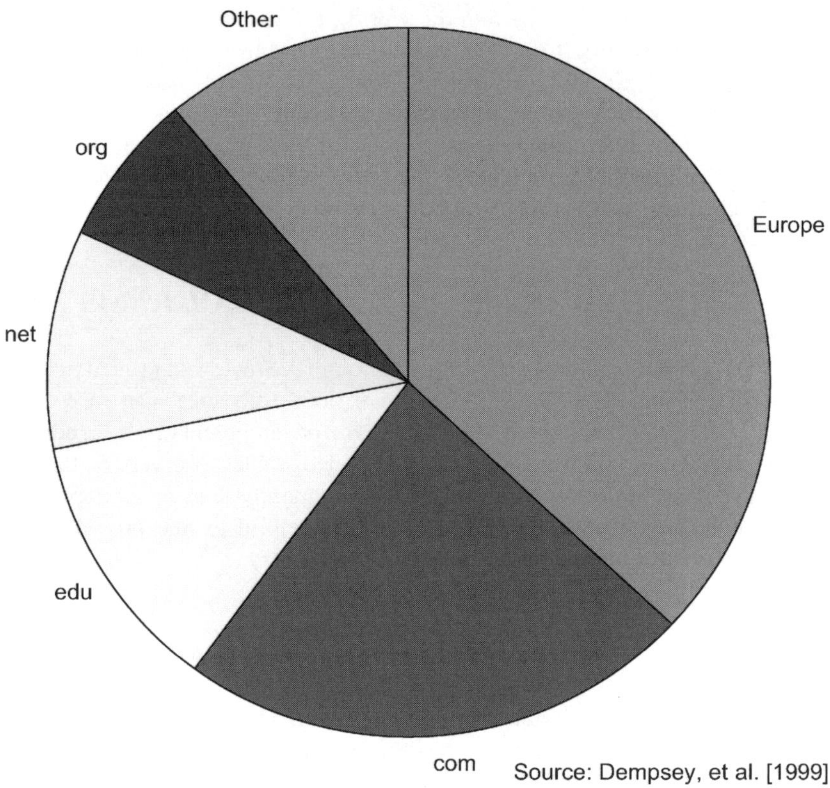

Fig. 3. Suffix of Linux Contributors.

The overall picture that we drew from our interviews and from the responses we received in reaction to the first draft of the paper is that the open source process is quite elitist. Important contributors are few and ascend to the "core group" status, the ultimate recognition by one's peers. The elitist view is also supported by Mockus et al.'s (1999) study of contributions to Apache. For Apache, the (core) "developers mailing list" is considered as the key list of problems to be solved, while other lists play a smaller role. The top 15 developers contribute 83 to 91% of changes (problem reports by way of contrast offer a much less elitist pattern).

Some evidence consistent with the suggestion that contributions to open source projects are being driven by signaling concerns can be found in the analysis of contributors to a long-standing archive of Linux postings maintained at the University of North Carolina by Dempsey et al. (1999). These authors examine the suffix of the contributors' e-mail addresses. While the location of many contributors cannot be precisely identified (for instance, contributors at ".com" entities may be located anywhere in the world), the results are nonetheless suggestive. As Fig. 3 depicts, 12% of the contributors are from entities with a suffix ".edu" (typically, U.S. educational institutions), 7% from ".orgs" (traditionally reserved from U.S. non-profits), fully 37% are from Europe (e.g. with suffixes such as ".de" and ".uk"), and 11% have other suffixes, many of which represent other foreign countries. This suggests that many of the contributions are coming from individuals outside the major software centers.

3. THE ORIGINS OF THE FOUR PROGRAMS

Each of the four case studies was developed through the review of printed materials and interviews (as well as those posted on various web sites) and face-to-face meetings with one or more key participants in the development effort. In addition, we held a number of conversations with knowledgeable observers of the open source movement. In Sections 4 and 5, we will frequently draw on examples from the four cases. Nonetheless, we felt it would be helpful to first provide a brief overview of the development projects.

3.1. Apache

The development of Apache began in 1994. Brian Behlendorf, then 21, had the responsibility for operating one of the first commercial Internet servers in the country, that powering *Wired* magazine's HotWired web site. This server, like most others in the country, was at the time running the Unix-based software written at the National Center for Supercomputer Applications (NCSA) at the University of Illinois. (The only competitive product at the time was the server developed at the joint European particle physics research facility CERN.) The NCSA had distributed its source code freely and had a development group actively involved in refining the code in consultation with the pioneering users. As Behlendorf and other users wrote emendations, or "patches," for the NCSA server, they would post them as well to mailing lists of individuals interested in Internet technology.

Behlendorf and a number of other users, however, encountered increasing frustrations in getting the NCSA staff to respond to their suggestions. (During this time, a number of the NCSA staff had departed to begin Netscape, and the University was in the process of negotiating a series of licenses of its software with commercial companies.) As a result, he and six other pioneering developers decided to establish a mailing list to collect and integrate the patches to the NCSA server software. They agreed that the process would be a collegial one. While a large number of individuals would be able to suggest changes, only a smaller set would be able to actually make changes to the physical code. In August 1995, the group released Apache 0.8, which represented a substantial departure from earlier approaches. A particular area of revision was the Application Program Interface (API), which allowed the development of Apache features to be very "modular." This step enabled programmers to make contributions to particular areas without affecting other aspects of the programs.

At the time that Apache was introduced, there was little in the way of competitive products: in fact, the absence of a good commercial alternative was a powerful motivation for the launching of the project. A variety of commercial software vendors, most notably Microsoft and Netscape, have subsequently targeted server software. Despite this competition, Apache has retained its dominant position. The November 2000 Netcraft survey (2000) of nearly 24 million Internet domains found that Apache had a dominant position: 59.7% of the site used this server software. The closest competitors, Microsoft's IIS and Netscape's Enterprise software, were at 20.2 and 6.7% respectively.[10]

In 1999, the Apache Software Foundation was established to oversee the development and diffusion of the program. The current status of Apache, as well as the other two open source projects that we focused on, is summarized in Table 1.

3.2. Linux

Linux, an amalgam of "Linus" and "Unix," was created by Linus Torvalds in 1991. Unlike the other case studies considered here, Torvalds was motivated to pursue this project by intellectual curiosity, rather than by a pressing practical need.

Table 1. The Open Source Programs Studied.

Program	Apache	Perl	Sendmail
Nature of program	World wide web (HTTP) server	System administration and programming language	Internet mail transfer agent
Year of introduction	1994	1987	1979 (predecessor program)
Governing body	Apache Software Foundation	Selected programmers (among the "perl-5-porters") (formerly, The Perl Institute)	Sendmail Consortium
Competitors	Internet Information Server (Microsoft)	Java (Sun)	Exchange (Microsoft)
	Various servers (Netscape)	Python (open source program)	IMail (Ipswitch)
		Visual Basic, ActiveX (Microsoft)	Post.Office (Software.com)
Market penetration	55% (September 1999) (of publicly observable sites only)	Estimated to have 1 million users	Handles ~80% of Internet e-mail traffic
Web site	www.apache.org	www.perl.org	www.sendmail.com

A 21-year-old graduate student, he sought to build the "kernel" – or core element – of a truly open source operating system.

Torvalds based his system on Minix, a public domain Unix system for personal computers. After approximately six months of development, a friend allowed him to post the operating system on a university server. He began encouraging contributions in a series of postings to on-line bulletin boards, such as one that posed the question "are you without a project and just dying to cut your teeth on an [operating system] you can try to modify to your needs?"

Torvalds initially distributed Linux under a licensing agreement that restricted any payment for the program, as well as required that all programs distributed or used with Linux be freely available. After half a year, however, he relaxed these restrictions. The number of users grew rapidly, from about one hundred after one year to half-a-million in 1994.

From the beginning, Torvalds retained clear leadership of the Linux project. He rapidly moved to writing less code and coordinating the software development project, assessing contributions and arbitrating disputes. Over time, a set of lieutenants have assumed responsibility for most of the decision-making, but Torvalds still retains authority for making the ultimate decisions. While employed at California-based semiconductor manufacturer Transmeta, Torvalds continues to devote about half his time to the Linux project.

While the origin of Linux was largely driven by intellectual curiosity on the part of Torvalds and his peers, the program has evolved into one that represents a significant competitor to Microsoft's Windows operating system. While the number of Linux users is difficult to determine because of the numerous channels through which the program is distributed, estimates range from 7 to 16 million users worldwide.

Reflecting its widespread diffusion, Linux has attracted a large share of the commercial investment in open source projects. A number of firms dedicated to supporting Linux have been established: pioneers included VA Linux, founded in 1993, and Red Hat, established in 1995. These commercial firms sell Linux software "packages," which are often far easier to install and operate than free versions available, provide technical support to end users and computer resellers, and sell complementary proprietary products. In addition, a number of established computer hardware and software firms have made extensive investments in Linux development.

3.3. Perl

Perl, or the Practical Extraction and Reporting Language, was created by Larry Wall in 1987. Wall, a programmer with Burroughs (a computer mainframe manufacturer now part of Unisys) had already written a number of widely adopted software programs. These included a program for reading postings on on-line newsgroups and a program that enabled users to readily update old source code with new patches.

The specific genesis of Perl was the large number of repetitive system administration tasks that Wall was asked to undertake while at Burroughs. In particular, Wall was required to synchronize and generate reports on two Unix-based computers as part of a project that Burroughs was undertaking for the U.S. National Security Agency. He realized that there was a need for a program language that was somewhere between the Unix shell language and the C language (suitable for developing complex programming applications). The Perl language sought to enable programmers to rapidly undertake a wide variety of tasks, particularly relating to system administration. The program was first introduced in 1987 via the Internet. It has become widely accepted as a language for developing scripts for Apache web servers, and is incorporated in a number of other programs.

Perl is administered on a rotating basis: the ten to twenty programmers (the number fluctuates over time) who have been most actively involved in the program take turns managing different aspects of the project. Wall himself has joined the staff of O'Reilly & Associates, a publisher specializing in manuals documenting open source programs. While he is no longer actively contributing to the programming, he remains active in managing the project.

As in the case of Apache, Perl's success has attracted competition from commercial developers. In particular, Sun's Java and Microsoft's ActiveX, both of which were introduced well after the diffusion of Perl, incorporate many of the same features. Rough estimates suggest that the number of Perl users is about one million. Some observers believe (see, for instance, the conversations archived at http://www.mail-archive.com/advocacy%40perl.org) that the growth usage of Perl has largely stabilized, and that many of the new users are turning to Java. As is often the case in this sector, confirming these claims is exceedingly difficult.

Two efforts to establish a Perl-related foundation have foundered. For instance, the Perl Institute had been intended to ensure that less glamorous tasks, such as documentation, were undertaken, in order to enhance the long-run growth of Perl. The failure of these efforts, however, may have reflected more about the specifics of the individual personalities involved than the prospects of the program itself.

3.4. Sendmail

Sendmail was originally developed in the late 1970s by Eric Allman, a graduate student in computer science at the University of California at Berkeley. As part of his responsibilities, Allman worked on a variety of software development and system administration tasks at Berkeley.

One of the major challenges that Allman faced was the incompatibility of the two major computer networks on campus. The approximately one dozen Unix-based computers had been originally connected through "BerkNet," a locally developed program that provided continuous interconnection. These computers, in turn connected to those on other campuses through telephone lines, using the UUCP protocol (Unix-to-Unix Copy Protocol). Finally, the Arpanet, the direct predecessor to the Internet, was introduced on the Berkeley campus around this time. Each of the networks used a different communications protocol: for instance, each person had multiple e-mail addresses, depending on the network from which the message was sent. To cope with this problem, Allman developed in 1979 a program called "Delivermail," which provided a way to greatly simplify the addressing problem. In an emendated form that allowed it to address a large number of domains, it was released two years later as "Sendmail."

Sendmail was soon adopted as the standard method of routing e-mail on the Arpanet. As the network grew, however, its limitations became increasingly apparent. A variety of enhanced versions of Sendmail were released in the 1980s and early 1990s which were incompatible with each other – in the argot of the open source community, the development of the program "forked." In 1993, Allman, who had returned to working at Berkeley after being employed at a number of software firms, undertook a wholesale rewrite of Sendmail. The development was sufficiently successful that the incompatible versions were largely abandoned in favor of the new version. While a variety of competitive products had appeared, such as Software.com's Post Office, Microsoft's Exchange, and Ipswitch's Imail, the open source program appeared to have a dominant competitive position. Observers have attributed this to the presence of an installed base of users and the ease of customizing the program. The program was estimated to handle about 75% of all Internet e-mail traffic in 2000.

In 1997, Allman established Sendmail, Inc. The company, which has been financed by a leading venture capital group Benchmark Capital, is seeking to sell Sendmail-related software enhancements (such as more user-friendly interfaces) and services. At the same time, the company seeks to encourage the continuing development of the software on an open source basis. For instance, Sendmail, Inc. employs two engineers who work almost full time on contributions to the open source program, which is run by the non-profit Sendmail Consortium.

4. WHAT DOES ECONOMIC THEORY TELL US ABOUT OPEN SOURCE?

This section and the next use economic theory to shed light on the three key questions: Why do people participate?[11] Why are there open source projects in the first place? And how do commercial vendors react to the open source movement?

4.1. What Motivates Programmers?

A programmer participates in a project, whether commercial or open source, only if she/he derives a net benefit (broadly defined) from engaging in the activity. The net benefit is equal to the immediate payoff (current benefit minus current cost) plus the delayed payoff (delayed benefit minus delayed cost).

A programmer working on an open source software development project incurs a variety of benefits and costs. The programmer incurs an opportunity cost of her/his time. While she/he is working on this project, she/he is unable to engage in another programming activity. This opportunity cost exists at the extensive and intensive margins. First, a programmer who would work as an independent on open source projects would forgo the monetary compensation she/he would receive if she/he were working for a commercial firm or a university. Second, and more to the point, for a programmer with an affiliation with a commercial company, a university or research lab, the opportunity cost is the cost of not focusing on her/his primary mission. For example, the academic's research output may sag, and the student's progress towards a degree slow down; these involve delayed costs. The size of this opportunity cost of not focusing on the primary mission of course depends on the extent of monitoring by the employer and more generally, the pressure on the job.

Two immediate benefits may counter this cost. First, the programmer, when fixing a bug or customizing an open source program, may actually improve rather than reduce her/his performance in the mission endowed upon her/his by her/his employer. This is particularly relevant for system administrators looking for specific solutions for their company. Second, the programmer compares how enjoyable the mission set by the employer and the open source alternative are. A "cool" open source project may be more fun than a routine task.

The delayed reward covers two distinct, although hard-to-distinguish, incentives. The *career concern incentive* refers to future job offers, shares in commercial open source-based companies,[12] or future access to the venture capital market.[13] The *ego gratification incentive* stems from a desire for peer recognition. Probably most programmers respond to both incentives. There are some differences between the two. The programmer mainly preoccupied by peer recognition may shun future monetary rewards, and may also want to signal her talent to a slightly different audience than those motivated by career concerns. From an economic perspective, however, the incentives are similar in most respects. We will group the career concern incentive and the ego gratification incentive under a single heading: *the signaling incentive*.

Economic theory (e.g. Holmström, 1999) suggests that this signaling incentive is stronger,

(i) the more visible the performance to the relevant audience (peers, labor market, venture capital community);
(ii) the higher the impact of effort on performance; and
(iii) the more informative the performance about talent.

The first condition gives rise to what economists call "strategic complementarities." To have an "audience," programmers will want to work on software projects that will attract a large number of other programmers. This suggests the possibility of multiple equilibria. The same project may attract few programmers because programmers expect that other programmers will not be interested; or it may flourish as programmers (rationally) have faith in the project.

The same point applies to forking in a given open source project. Open source processes are in this respect quite similar to academic research. The latter is well known to exhibit fads: see the many historical examples of simultaneous discoveries discussed by Merton (1973). Fields are completely neglected for years, while others with apparently no superior intrinsic interest attract large numbers of researchers. Fads in academia are frowned upon for their inefficient impact on the allocation of research. It should not be ignored, however, that fads also have benefits. A fad can create a strong signaling incentive: researchers working in a popular area may be highly motivated to produce a high-quality work, since they can be confident that a large audience will examine their work.[14]

Turning to the *leadership* more specifically, it may still be a puzzle that the leader initially turns valuable code to the community.[15] Despite the substantial status and career-concerns benefits of being a leader of an important open source project, it would seem that most should not resist the large monetary gains from taking a promising technology private. We can only conjecture as to why this is not the case. One possibility is that taking the technology private may meet layers of resistance within the leader's corporation. To the extent that the innovation was made while working in-house, the programmer must secure a license from the employer;[16] and her division, which does not want to lose a key programmer, may not be supportive of her demand. Another possibility is that the open source process may be a more credible way of harnessing energies when, say, fighting against a dominant player in the industry.

4.2. Comparison Between Open Source and Closed Source Programming Incentives

To compare programmers' incentives in the open source and proprietary settings, we need to examine how the fundamental features of the two environments shape the incentives just reviewed. We will first consider the relative short-term rewards, and then turn to the deferred compensation.

Commercial projects have an edge on the current-compensation dimension because the proprietary nature of the code generates income. This makes it privately worthwhile for private companies to offer salaries.[17] This contention is the old argument in economics that the prospect of profit encourages investment, which is used, for instance, to justify the awarding of patents to encourage invention.

By way of contrast, an open source project may well lower the cost for the programmer, for two reasons:

(i) "*Alumni effect*": Because the code is freely available to all, it can be used in schools and universities for learning purposes; so it is already familiar to programmers. This reduces their cost of programming for UNIX, for example.[18]
(ii) *Customization and bug-fixing benefits*: The cost of contributing to an open source project can be offset if the activity brings about a private benefit (bug fixing, customization) for the programmer and her firm. Note again that this factor of cost reduction is directly linked to the openness of the source code.[19]

Let us now turn to the delayed reward (signaling incentive) component. In this respect too, the open source process has some benefits over the closed source approach. As we noted, signaling incentives are stronger, the more visible the performance and the more attributable the performance to a given individual. Signaling incentives therefore may be stronger in the open source mode for three reasons:

(i) *Better performance measurement*: Outsiders can only observe inexactly the functionality and/or quality of individual elements of a typical commercially developed program, as they are unable to observe the proprietary source code. By way of contrast, in an open source project, the outsiders are able to see not only what the contribution of each individual was and whether that component "worked," but also whether the task was hard, if the problem was addressed in a clever way, whether the code can be useful for other programming tasks in the future, and so forth.
(ii) *Full initiative*: The open source programmer is her own boss and takes full responsibility for the success of a subproject. In a hierarchical commercial firm, however, the programmer's performance depends on her supervisor's interference, advice, etc. Economic theory would predict that the programmer's performance is more precisely measured in the former case.[20]
(iii) *Greater fluidity*: It may be argued that the labor market is more fluid in an open source environment. Programmers are likely to have less idiosyncratic, or firm-specific, human capital that limits shifting one's efforts to a new program or work environment. (Since many elements of the source code are shared across open source projects, more of the knowledge they have accumulated can be transferred to the new environment.)

These theoretical arguments also provide insights as to *who* is more likely to contribute and *what tasks* are best suited to open source projects. Sophisticated users derive direct benefits when they customize or fix a bug in open source software.[21] A second category of potential contributors consists of individuals with strong signaling incentives; these may use open source software as a port of entry. For instance, open source processes may give a talented system administrator at a small academic institution (who is also a user!) a unique opportunity to signal her talent to peers, prospective employers, and the venture capital community.[22]

As to the tasks that may appeal to the open source community, one would expect that tasks such as those related to the operating systems and programming languages, whose natural audience is the community of programmers, would give rise to strong signaling incentives. (For instance, the use of Perl is largely restricted to system administrators.) By way of contrast, tasks aiming at helping the much-less-sophisticated end user – e.g. design of easy-to-use interfaces, technical support, and ensuring backward compatibility – usually provide lower signaling incentives.[23]

4.3. Evidence on Individual Incentives

A considerable amount of evidence is consistent with an economic perspective.

First, user benefits are key to a number of open source projects. One of the origins of the free software movement was Stallman's inability to improve a printer program because Xerox refused to release the source code. In three of the four scenarios described in Section 3, the project founders were motivated by information technology problems that they had encountered in their day-to-day work. For instance, in the case of Apache, the initial set of contributors was almost entirely system administrators who were struggling with the same types of problems as Behlendorf. In each case, the initial release was "runnable and testable": it provided a potential, even if imperfect, solution to a problem that was vexing considerable numbers of data processing professionals.

Second, it is clear that giving credit to authors is essential in the open source movement. This principle is included as part of the nine key requirements in the "Open Source Definition" (Open Source Initiative, 1999). This point is also emphasized by Raymond (1999b), who points out "surreptitiously filing someone's name off a project is, in cultural context, one of the ultimate crimes."

More generally, the reputational benefits that accrue from successful contributions to open source projects appear to have real effects on the developers. This is acknowledged within the open source community itself. For instance, according to Raymond (1999b), the primary benefits that accrue to successful contributors of open source projects "good reputation among one's peers, attention and cooperation from others, ... [and] higher status [in the] ... exchange economy." Thus, while some of benefits conferred from participation in open source projects may be less concrete in nature, there also appear be quite tangible – if delayed – rewards.

The Apache project provides a good illustration of these observations. The project makes a point of recognizing all contributors on its web site, even those

Table 2. Commercial Roles Played by Selected Individuals Active in Open Source Movement.

Individual	Role and Company
Eric Allman	Chief Technical Officer, Sendmail, Inc. (support for open source software product)
Brian Behlendorf	Founder, President, and Chief Technical Officer, Collab.Net (management of open source projects)
Keith Bostic	Founder and President, Sleepycat Software
L. Peter Deutsch	Founder, Aladdin Enterprises (support for open source software product)
William Joy	Founder and Chief Scientist, Sun Microsystems (workstation and software manufacturer)
Michael Tiemann	Founder, Cygnus Solutions (open source support)
Linus Torvalds	Employee, Transmeta Corporation (chip design company)
Paul Vixie	President, Vixie Enterprises (engineering and consulting services)
Larry Wall	Employee, O'Reilly & Associates (software documentation publisher)

who simply identify a problem without proposing a solution. Similarly, the organization highlights its most committed contributors, who have the ultimate control over the project's evolution. Moreover, it appears that many of the skilled Apache programmers have benefited materially from their association with the organization. Numerous contributors have been hired into Apache development groups within companies such as IBM, become involved in process-oriented companies such as Collab.Net which seek to make open source projects more feasible (see below), or else moved into other Internet tools companies in ways that were facilitated by their expertise and relationships built up during their involvement in the open source movement. Meanwhile, many of the new contributors are already employed by corporations, and working on Apache development as part of their regular assignments.

There is also substantial evidence that open source work may be a good stepping stone for securing access to venture capital. For example, the founders of Sun, Netscape, and Red Hat had signaled their talent in the open source world. In Table 2, we summarize some of the subsequent commercial roles played by individuals active in the open source movement.

4.4. Organization and Governance

Favorable characteristics for open source production are: (a) its modularity (the overall project is divided into much smaller and well-defined tasks ("modules") that individuals can tackle independently from other tasks); and (b) the existence of fun challenges to pursue.[24] A successful open source project also requires a credible leader or leadership, and an organization consistent with the nature of the process. Although the leader is often at the origin a user who attempts to solve a particular program, the leader over time performs less and less programming. The leader must provide a "vision," attract other programmers, and, last but not least, "keep the project together" (prevent it from forking or being abandoned).

4.4.1. Initial Characteristics

The success of an open source project is dependent on the ability to break the project into distinct components. Without an ability to parcel out work in different areas to programming teams who need little contact with one another, the effort is likely to be unmanageable. Some observers argue that the underlying Unix architecture lent itself well to the ability to break development tasks into distinct components. It may be that as new open source projects move beyond their Unix origins and encounter new programming challenges, the ability to break projects into distinct units will be less possible. But recent developments in computer science and programming languages (e.g. the development of object-oriented programming) have encouraged further modularization, and may facilitate future open source projects.

The initial leader must also assemble a critical mass of code to which the programming community can react. Enough work must be done to show that the project is doable and has merit. At the same time, to attract additional programmers, it may be important that the leader does not perform too much of the job on his own and leaves challenging programming problems to others.[25] Indeed, programmers will initially be reluctant to join a project unless they identify an exciting challenge. Another reason why programmers are easier to attract at an early stage is that, if successful, the project will keep attracting a large number of programmers in the future, making early contributions very visible.

Consistent with this argument, it is interesting to note that each of the four cases described above appeared to pose challenging programming problems.[26] When the initial release of each of these open source programs was made, considerable programming problems were unresolved. The promise that the project was not near a "dead end," but rather would continue to attract ongoing participation from programmers in the years to come, appears to be an important aspect of its appeal.

In this respect, Linux is perhaps the quintessential example. The initial Linux operating system was quite minimal, on the order of a few tens of thousands of lines of code. In Torvalds' initial postings in which he sought to generate interest in Linux, he explicitly highlighted the extent to which the version would require creative programming in order to achieve full functionality. Similarly, Larry Wall attributes the much of the success of Perl to the fact that it "put the focus on the creativity of the programmer." Because it has a very limited number of rules, the program has evolved in a variety of directions that were largely unanticipated when Wall initiated the project.

4.4.2. Leadership

Another important determinant of project success appears to be the nature of its leadership. In some respects, the governance structures of open source projects are quite different. In a number of instances, such as Linux, there is an undisputed leader. While certain aspects are delegated to others, a strong centralization of authority characterizes these projects. In other cases, such as Apache, a committee will resolve the disputes by voting or a consensus process.

At the same time, leaders of open source projects share some common features. Most leaders are the programmers who developed the initial code for the project (or made another important contribution early in the project's development). While many make fewer programming contributions, having moved on to broader project management tasks, the individuals that we talked to believed that the initial experience was important in establishing credibility to manage the project. The splintering of the Berkeley-derived Unix development programs has been attributed in part to the absence of a single credible leader.

But what does the leadership of an open source project do? It might appear at first sight that the unconstrained, quasi-anarchistic nature of the open source process leaves little scope for a leadership. This, however, is incorrect. While the leader has no "formal authority" (she is unable to instruct anyone to do anything), she has substantial "real authority" in successful open source projects.[27] That is, her "recommendations," broadly viewed, tend to be followed by the vast majority of programmers working on the project. These recommendations include the initial "vision" (agenda for work, milestones), the subsequent updating of goals as the project evolves, the appointment of key leaders, the cajoling of programmers so as to avoid attrition or forking, and the overall assessment of what has been and should be achieved. (Even though participants are free to take the project where they want as long as they release the modified code, acceptance by the leadership of a modification or addition provides some certification as to its quality and its integration/compatibility with the overall project. The certification of quality is quite crucial to the open source project because the absence of liability raises concerns among users that are stronger than for commercial software, for which the vendor is liable.)

The key to a successful leadership is the programmers' trust in the leadership: that is, they must believe that the leader's objectives are sufficiently congruent with theirs and not polluted by ego-driven, commercial, or political biases. In the end, the leader's recommendations are only meant to convey her information to the community of participants. The recommendations receive support from the community only if they are likely to benefit the programmers, that is only if the leadership's goals are believed to be aligned with the programmers' interests.

For instance, the leadership must be willing to accept meritorious improvements, even though they do not fit within the leader's original blueprint. Trust in the leadership is also key to the prevention of forking. While there are natural forces against forking (the loss of economies of scale due to the creation of smaller communities, the hesitations of programmers in complementary segments to port to multiple versions, and the stigma attached to the existence of a conflict), other factors may encourage forking. User-developers may have conflicting interests as to the evolution of the technology. Ego (signaling) concerns may also prevent a faction from admitting that another approach is more promising, or simply from accepting that it may socially be preferable to have one group join the other's efforts even if no clear winner has emerged. The presence of a charismatic (i.e. trusted) leader is likely to substantially reduce the probability of forking in two ways. First, indecisive programmers are likely to rally behind the leadership's preferred alternative. Second, the dissenting faction may not have an obvious leader of its own.

A good leadership should also clearly communicate its goals and evaluation procedures. Indeed, the open source organizations go to considerable efforts to make the nature of their decision-making process transparent: the process by which the operating committee reviews new software proposals is frequently posted and all postings archived. For instance, on the Apache web site, it is explained how proposed changes to the program are reviewed by the program's governing body, whose membership is largely based on contributions to the project. (Any significant change requires at least three "yes" votes – and no vetoes – by these key decision-makers.)

5. COMMERCIAL SOFTWARE COMPANIES' REACTIONS TO THE OPEN SOURCE MOVEMENT

This section examines the interface between open and closed source software development. Challenged by the successes of the open source movement, the commercial software corporations may employ one of the following two strategies. The first is to emulate some incentive features of open source processes in a distinctively closed source environment. Another is to try to mix open and closed source processes to get the best of both worlds.

5.1. Why Don't Corporations Duplicate the Open Source Incentives?

As we already noted, owners of proprietary code are not able to enjoy the benefits of getting free programmer training in schools and universities (the alumni effect); nor can they easily allow users to modify their code and customize it without jeopardizing intellectual property rights.

Similarly, and for the reasons developed in Section 4, commercial companies will never be able to fully duplicate the visibility of performance reached in the open source world. At most can they duplicate to some extent some of the signaling incentives of the open source world. Indeed, a number of commercial software companies (e.g. video game companies, Qualcomm for the Eudora email program) list people who have developed the software. It is an interesting question why others do not. To be certain, commercial companies do not like their key employees to become highly visible, lest they be hired away by competitors.[28] But, to a large extent, firms also realize that this very visibility enables them to attract talented individuals and provides a powerful incentive to existing employees.[29]

To be certain, team leaders in commercial software build reputations and get identified with proprietary software just as they can on open source projects; but the ability of reputations to spread beyond the leaders is more limited, due to the non-verifiability of claims about who did what.[30]

Another area in which software companies might try to emulate open source development is the promotion of widespread code sharing within the company. This may enable them to reduce code duplication and to broaden a programmer's audience. Interestingly, existing organizational forms may preclude the adoption of open source systems within commercial software firms. An internal Microsoft document on open source (Valloppillil, 1998) describes a number of pressures that limit the implementation of features of open source development within Microsoft. Most importantly, each software development group appears to be largely autonomous. Software routines developed by one group are not shared with others. In some instances, the groups seek to prevent being broken up by not documenting a large number of program features. These organizational attributes, the document suggests, lead to very complex and interdependent programs that do not lend themselves to development in a "compartmentalized" manner nor to widespread sharing of source code.[31]

5.2. The Commercial Software Companies' Open Source Strategies

As should be expected, many commercial companies have undertaken strategies to capitalize on the open source movement. In a nutshell, they expect to benefit from their expertise in some segment whose demand is boosted by the success of a complementary open source program. While improvements in the open source software are not appropriable, commercial companies can benefit indirectly in a complementary proprietary segment.[32]

5.2.1. Living Symbiotically Off an Open Source Project

One such strategy is straightforward. It consists of commercially providing complementary services and products that are not supplied efficiently by the open source community. Red Hat and VA Linux for example, exemplify this "reactive" strategy.[33]

In principle a "reactive" commercial company may want to encourage and subsidize the open source movement, for example by allocating a few programmers to the open source project.[34] Red Hat will make more money on support if Linux is successful. Similarly, if logic semiconductors and operating systems for personal computers are complements, one can show by a revealed preference argument that Intel's profits will increase if Linux (which unlike Windows is free) takes over the PC operating system market. Sun may benefit if Microsoft's position is weakened; Oracle may wish to port its database products to a Linux environment in order to lessen its dependence on Sun's Solaris operating system, and so forth. Because firms do not capture all the benefits of the investments, however, the free-rider problem often discussed in the economics of innovation should apply here as well. Subsidies by commercial companies for open source projects should remain limited unless the potential beneficiaries succeed in organizing a consortium (which will limit the free-riding problem).

5.2.2. Code Release

A second strategy is to take a more proactive role in the development of open source software. Companies can release existing proprietary code and create some governance structure for the resulting open source process. For example, Hewlett-Packard recently released its Spectrum Object Model-Linker to the open source community in order to help the Linux community port Linux to Hewlett Packard's RISC architecture.[35] This is similar to the strategy of giving away the razor (the released code) to sell more razor blades (the related consulting services that HP will provide).

When can it be advantageous for a commercial company to release proprietary code under an open source license? The first condition is, as we have noted, that the company expects to thereby boost its profit on a complementary segment. A second is that the increase in profit in the proprietary complementary segment offsets any profit that would have been made in the primary segment, had it not been converted to open source. Thus, the temptation to go open source is particularly strong when the company is too small to compete commercially in the primary segment or when it is lagging behind the leader and about to become extinct in that segment.[36,37]

Various efforts by corporations selling proprietary software products to develop additional products through an open source approach have been undertaken. One of the most visible of these efforts was Netscape's 1998 decision to make "Mozilla," a portion of its browser source code, freely available. This effort encountered severe difficulties in its first year, only receiving approximately two dozen postings by outside developers. Much of the problems appeared to stem from the insufficiently "modular" nature of the software: reflecting its origins as a proprietary commercial product, the different portions of the program were highly interdependent and interwoven. Netscape eventually realized it needed to undertake a major restructuring of the program, in order to enhance the ability of open source programmers to contribute to individual sections. It is also likely that Netscape raised some suspicions by not initially adopting the right governance structure. Leadership by a commercial entity may not internalize enough of the objectives of the open source community. In particular, a corporation may not be able to credibly commit to keeping all source code in the public domain and to adequately highlighting important contributions.[38]

For instance, in the Mozilla project, Netscape's unwillingness to make large amounts of browser code public was seen as an indication of its questionable commitment to the open source process. In addition, Netscape's initial insistence on the licensing terms that allowed the corporation to relicense the software developed in the open source project on a proprietary basis was viewed as problematic (Hammerly, Paquin & Walton, 1999). (The argument is here the mirror image of the standard argument in industrial economics that a firm may want to license its technology to several licensees in order to commit not to expropriate producers of complementary goods and services in the future: see Shepard (1987) and Farrell and Gallini (1988).) Netscape initially proposed the "Netscape Public License," a cousin to the BSD license that allowed Netscape to take pieces of the open source code and turn them back into a proprietary project again. The licensing terms, however, may not have been the hindering factor, since the terms of the final license are even stricter than those of the GPL. Under this new license (the "Mozilla Public License"), Netscape cannot relicense the modifications to the code.

5.2.3. Intermediaries

In this light, it is tempting to interpret the creation of organizations such as Collab.Net as efforts to certify corporate open source development programs, just as investment banks and venture capitalists play a certification role for new firms. Collab.Net, a new venture funded by the venture capital group Benchmark Partners, will organize open source projects for corporations who wish to develop part of their software in this manner. Collab.Net will receive fees for its online marketplace (SourceXchange, through which corporations will contact open source developers), for preparing contracts, for helping select and monitor developers, and for settling disputes. Hewlett Packard released the core of its E-speak technology (which enable brokering capabilities) to open source[39] and posted six projects related to this technology.

Hewlett Packard's management of the open source process seems consistent with Dessein (1999). Dessein shows that a principal with formal control rights over an agent's activity in general gains by delegating his control rights to an intermediary with preferences or incentives that are intermediate between his and the agent's. The partial alignment of the intermediary's preferences with the agent's fosters trust and boosts the agent's initiative, ultimately offsetting the partial loss of control for the principal. In the case of Collab.Net, the congruence with the open source developers is obtained through the employment of visible open source developers (for example, the president and chief technical officer is Brian Behlendorf, one of the cofounders of the Apache project) and the involvement of O'Reilly, a technical book publisher with strong ties to the open source community.

6. FOUR OPEN ECONOMIC QUESTIONS ABOUT OPEN SOURCE

There are many other issues posed by open source development that require further thought. This section will highlight a number of these as suggestions for future work.

6.1. Which Technological Characteristics are Conducive to a Smooth Open Source Development?

This paper has identified a number of attributes that make a project a good or poor candidate for open source development. But it has stopped short of providing a comprehensive picture of determinants of a smooth open source development. Let us mention a few topics that are worth further investigation.

6.1.1. Role of Applications and Related Programs

Open source projects differ in the functionalities they offer and in the number of add-ons that are required to make them attractive. As the open source movement comes to maturity, it will confront some of the same problems as commercial software does, namely the synchronization of upgrades and the efficient level of backward compatibility. A user who upgrades a program (which is very cheap in the open source case) will want either the new program to be backward compatible or applications to have themselves been upgraded to the new version.[40] We know from commercial software that both approaches to compatibility are costly; for example, Windows programmers devote a lot of time to backward compatibility issues, and encouraging application development requires fixing applications programming interfaces about three years before the commercial release of the operating system. A reasonable conjecture could be that open source programming would be appropriate when there are fewer applications or when IT professionals can easily adjust the code so as to ensure compatibility themselves.

6.1.2. Influence of Competitive Environment

Based on very casual observation, it seems that open source projects sometimes gain momentum when facing a battle against a dominant firm, although our examples show open source projects can do well even in the absence of competition.[41] To understand why this might be the case (assuming this is an empirical fact, which remains to be established!), it would be useful to go back to the economics of cooperative joint ventures. The latter are known to work better when the members have similar objectives.[42] The existence of a dominant competitor in this respect tends to align the goals of the members, and the best way to fight an uphill battle against the dominant player is to remain united. To be certain, open source software development works differently from joint venture production, but it also relies on cooperation within a heterogeneous group; the analogy is well worth pursuing.

6.1.3. Project Lifespan

One of the arguments offered by open source advocates is that because their source code is publicly available, and at least some contributions will continue to be made, its software will have a longer duration. (Many software products by commercial vendors are abandoned or no longer upgraded after the developer is acquired or liquidated, or even when the company develops a new product to replace the old program.) But another argument is that the nature of incentives being offered open source developers – which as discussed above, lead them to work on highly visible projects – might lead to a "too early" abandonment of projects that experience a relative loss in popularity. An example is the XEmacs project, an open source project to create a graphical environment with multiple "windows" that originated at Stanford. Once this development effort encountered an initial decline in popularity, many of the open source developers appeared to move onto alternative projects.

6.2. Optimal Licensing

Our discussion of open source licensing has been unsatisfactory. Some licenses (e.g. BSD and its close cousin the Apache license) are relatively permissive, while others (e.g. GPL) force the user to distribute any changes or improvements (share them) if they distribute the software at all.

Little is known about the trade-off between encouraging add-ons that would not be properly supplied by the open source movement and preventing commercial vendors (including open source participants) from free riding on the movement or even "hijacking it." An open source project may be "hijacked" by a participant who builds a valuable module and then offers proprietary APIs to which application developers start writing. The innovator has then built a platform that appropriates some of the benefits of the project. To be certain, open source participants might then be outraged, but it is unclear whether this would suffice to prevent the hijacking. The open source community would then be as powerless as the commercial owner of a platform above which a "middleware" producer superimposes a new platform.[43]

The exact meaning of the "viral" provisions in the GPL license, say, or more generally the implications of open source licenses have not yet been tested in court. Several issues may arise in such litigation: for instance, who has standing for representing the project if the community is fragmented, and how a remedy would be implemented (e.g. the awarding of damages for breach of copyright agreement may require incorporating the beneficiaries).

6.3. Coexistence of Commercial and Open Source Software

On a related note, the existence of commercial entities living symbiotically off the efforts of open source programmers as well as participating in open source projects raises new questions.

The flexible open source licenses allow for the coexistence of open and closed source code. While it represents in our view (and in that of many open source participants) a reasonable compromise, it is not without hazards.

The coexistence of commercial activities may alter the programmers' incentives. Programmers working on an open source project may be tempted to stop interacting and contributing freely if they think they have an idea for a module that might yield a huge commercial payoff. Too many programmers may start focusing on the commercial side, making the open source process less exciting. Although they refer to a different environment, the concerns that arise about academics' involvement in start-up firms, consulting projects, and patenting may be relevant here as well. While it is too early to tell, some of these same issues may appear in the open source world.[44]

6.4. Can the Open Source Process be Transposed to Other Industries?

An interesting final question is whether the open source model can be transposed to other industries. Could automobile components be developed in an open source mode, with GM and Toyota performing an assembler function similar to that of Red Hat for Linux? Many industries involve forms of cooperation between commercial entities in the form of for-profit or not-for-profit joint ventures. Others exhibit user-driven innovation or open science cultures. Thus a number of ingredients of open source software are not specific to the software industry. Yet, no other industry has yet produced anything quite like open source development. An important research question is whether other industries ever will.

Although some aspects of open source software collaboration (such as electronic information exchange across the world) could easily be duplicated, other aspects would be harder to emulate. Consider, for example, the case of biotechnology. It may be impossible to break up large projects into small manageable and independent modules and there may not sufficient sophisticated users who can customize the molecules to their own needs. The tasks that are involved in making the product available to the end user involve much more than consumer support and even friendlier user interfaces. Finally, the costs of designing, testing, and seeking regulatory approval for a new drug are enormous.

More generally, in many industries the development of individual components require large- team work and substantial capital costs, as opposed to (for some software programs) individual contributions and no capital investment (besides the computer the programmer already has). Another obstacle is that in mass-market industries users are numerous and rather unsophisticated, and so deliver few services of peer recognition and ego gratification. This suggests that the open source model may not easily be transposed to other industries, but further investigation is warranted.

Our ability to answer confidently these and related questions is likely to increase as the open source movement itself grows and evolves. At the same time, it is heartening to us how much of open source activities can be understood within existing economic frameworks, despite the presence of claims to the contrary. The literatures on "career concerns" and on competitive strategies provide lenses through which the structure of open source projects, the role of contributors, and the movement's ongoing evolution can be viewed.

NOTES

1. The media like to portray the open source community as wanting to help mankind, as it makes a good story. Many open source advocates put limited emphasis on this explanation.

2. These are summarized in Darwall and Lerner (2000).

3. It should be noted that these changes are far from universal. In particular, many information technology and manufacturing firms appear to be moving to less of an emphasis on basic science in their research facilities (for a discussion, see Rosenbloom & Spencer, 1996).

4. This history is of necessity highly abbreviated and we do not offer a complete explanation of the origins of open source software. For more detailed treatments, see Browne (1999), DiBona, Ockman and Stone (1999), Gomulkiewicz (1999), Levy (1984), Raymond (1999a), and Wayner (2000).

5. Programmers write source code using languages such as Basic, C, and Java. By way of contrast, most commercial software vendors only provide users with object, or binary, code. This is the sequence of 0s and 1s that directly communicates with the computer, but which is difficult for programmers to interpret or modify. When the source code is made available to other firms by commercial developers, it is typically licensed under very restrictive conditions.

6. It should be noted, however, that some projects, such as the Berkeley Software Distribution (BSD) effort, did take alternative approaches during the 1980s. The BSD license also allows anyone to freely copy and modify the source code (as long as credit was given to the University of California at Berkeley for the software developed there, a requirement no longer in place). It is much less constraining than the GPL: anyone can modify the program and redistribute it for a fee without making the source code freely available. In this way, it is a continuation of the university-based tradition of the 1960s and 1970s.

7. Two main open source projects (GNOME and KDE) are meant to remedy Linux's limitations on desktop computers (by developing mouse and windows interfaces).

8. For example, Torvalds (1999) argues that the Linux model works best with developer-type software. Ghosh (1999) views the open source process as a large repeated game process of give-and-take among developer-users (the "cooking pot" model).

9. To be certain, backward compatibility efforts may sometimes be exerted by status-seeking open source programmers. For example, Linux has been made to run on Atari machines, a pure bravado effort since no one uses Ataris anymore.

10. A complication is introduced by the fact that firewall-protected servers may be quite different in nature. For instance, a survey of both protected and unprotected servers in the summer of 1996 by Zoma Research concluded that open source server programs, including Apache, accounted for only 7% of all installations, far less than the contemporaneous Netcraft estimate.

11. We focus primarily on programmers' contributions to code. A related field of study concerns field support, which is usually also provided free of charge in the open source community. Lakhani and von Hippel (2000) provide empirical evidence for field support in the Apache project. They show that providers of help often gain learning for themselves, and that the cost of delivering help is therefore usually low.

12. Linus Torvalds and others have been awarded shares in Linux-based companies that went public. Most certainly, these rewards were unexpected and did not affect the motivation of open source programmers. If this practice becomes "institutionalized," such rewards will

in the future be expected and therefore impact the motivation of open source leaders. More generally, leaders of open source movements may initially not have been motivated by ego gratification and career concerns. Like Behlendorf, Wall and Allman, the "bug fixing" motivation may have originally been paramount. The private benefits of leadership may have grown in importance as the sector matured.

13. Success at a commercial software firm is likely to be a function of many attributes. Some of these (e.g. programming talent) can be signaled through participation in open source projects. Other important attributes, however, are not readily signaled through these projects. For instance, commercial projects employing a top-down architecture require that programmers work effectively in teams, while many open source projects are initiated by relatively modest pieces of code, small enough to be written by a single individual.

14. Dasgupta and David (1994) suggest an alternative explanation for these patterns: the need to impress less-informed patrons who are likely to be impressed by the academic's undertaking research in a "hot" area. These patterns probably are driven by academic career concerns. New fields tend to be relatively more attractive to younger researchers, since older researchers have already invested in established fields and therefore have lower marginal costs of continuing in these fields. At the same time, younger researchers need to impress senior colleagues who will evaluate them for promotion. Thus, they need the presence of some of their seniors in the new fields.

15. Section 5 will discuss *companies*' incentives to release code.

16. Open source projects may be seen as imposing less of a competitive threat to the firm. As a result, the firm may be less inclined to enforce its property rights on innovations turned open source. Alternatively, the firm may be unaware that the open source project is progressing.

17. To be certain, commercial firms (e.g. Netscape, Sun, O'Reilly, Transmeta) supporting open source projects are also able to compensate programmers, because they indirectly benefit financially from these projects. Similarly, the government and not-for-profit corporations have done some subsidizing of open source projects. Still, there should be an edge for commercial companies.

18. While we are here interested in private incentives to participate, note that this complementarity between apprenticeship and projects is socially beneficial. The social benefits might not increase linearly with open source market share, however, since the competing open source projects may end up competing for attention in the same common pool of students.

19. To be certain, commercial companies leave Application Programming Interfaces for other people to provide add-ons, but this is still quite different from opening the source code.

20. On the relationship between empowerment and career concerns, see Ortega (2000). In Cassiman's (1998) analysis of research corporations (for-profit centers bringing together firms with similar research goals), free riding by parent companies boosts the researchers' autonomy and helps attracting better talents. Cassiman argues that it is difficult to sustain a reputation for respecting the autonomy of researchers within firms. Cassiman's analysis looks at real control, while our argument here results from the absence of formal control over the OS programmer's activity.

21. A standard argument in favor of open source processes is their massive parallel debugging. Typically, commercial software firms can only ask users to point at problems:

beta testers do not fix the bugs, they just report them. It is also interesting to note that many commercial companies do not discourage their employees from working on open source projects. In many cases where companies encourage such involvement, programmers use open source tools to fix problems. Johnson (1999) builds a model of open source production by a community of user-developers. There is one software program or module to be developed, which is a public good for the potential developers. Each of the potential developers has a private cost of working on the project and a private value of using it; both of which are private information. Johnson shows that the probability that the innovation is made need not increase with the number of developers, as free-riding is stronger when the number of potential developers increases.

22. An argument often heard in the open source community is that people participate in open source projects because programming is fun and because they want to be "part of a team." While this argument may contain a grain of truth, it is somewhat puzzling as it stands; for, it is not clear why programmers who are part of a commercial team could not enjoy the same intellectual challenges and the same team interaction as those engaged in open source development. (To be sure, it may be challenging for programmers to readily switch employers if their peers in the commercial entity are not congenial.) The argument may reflect the ability of programmers to use participation in open source projects to overcome the barriers that make signaling in other ways problematic.

23. Valloppillil (1998) further argues that reaching commercial grade quality often involves unglamorous work on power management, management infrastructure, wizards, etc., that makes it unlikely to attract open source developers. Valloppillil's argument seems a fair description of past developments in open source software. Some open source proponents do not confer much predictive power on his argument, though; they predict, for example, that open source user interfaces such as GNOME and KDE will achieve commercial grade quality.

24. Open source projects have trouble attracting people initially unless they leave fun challenges "up for grabs." On the other hand, the more programmers an open source project attracts, the more quickly the fun activities are completed. The reason why the projects need not burn out once they grow in ranks is that the "fixed cost" that individual programmers incur when they first contribute to the project is sunk and so the marginal cost of continuing to contribute is smaller than the initial cost of contributing.

25. For example, Valloppillil's (1998) discussion of the Mozilla release.

26. It should be cautioned that these observations are based on a small sample of successful projects. Observing which projects succeed or fail and the reasons for these divergent outcomes in an informal setting such as this one is quite challenging.

27. The terminology and the conceptual framework are here borrowed from Aghion-Tirole (1997).

28. instance, concerns about the "poaching" of key employees was one of the reasons cited for Steve Jobs' recent decision to cease giving credit to key programmers in Apple products (Claymon, 1999).

29. For the economic analysis of employee visibility, see Gibbons (1997) and Gibbons and Waldman's (1999) review essays. Ronde (1999) models the firms' incentives to "hide" their workers from the competition in order to preserve their trade secrets.

30. Commercial vendors try to address this problem in various ways. For example, Microsoft developers now have the right to present their work to their users. Promotions to "distinguished engineer" or to a higher rank more generally as well as the granting of

stock options as a recognition of contributions also make the individual performance more visible to the outside world.

31. Cusumano and Selby (1995), however, document a number of management institutions at Microsoft that attempt to limit these pressures.

32. Another motivation for commercial companies to interface with the open source world may be public relations. Furthermore, firms may temporarily encourage programmers to participate in open source projects to learn about the strengths and weaknesses of this development approach.

33. Red Hat provides support for Linux-based products, while VA Linux provides hardware products optimized for the Linux environment. In December 1999, their market capitalizations were $17 and $10 billion respectively, though they have subsequently declined significantly.

34. Of course, these programmers also increase the company's ability to learn from scientific and technical discoveries elsewhere and help the company with the development of the proprietary segment.

35. Companies could even (though probably less likely) encourage "ex nihilo" development of new pieces of open source software.

36. See, for example, the discussion of SGI's open source strategy in Taschek (1999).

37. It should also be noted that many small developers are uncomfortable doing business with leading software firms, feeling them to be exploitative, and that these barriers may be overcome by the adoption of open source practices by the large firms. A rationalization of this story is that, along the lines of Farrell and Katz (2000), the commercial platform owner has an incentive to introduce substitutes in a complementary segment, in order to force prices down in that segment and to raise the demand for licenses to the software platform. When, however, the platform is available through (say) a BSD-style license, the platform owner has no such incentives, as he cannot raise the platform's price. Vertical relationships between small and large firms in the software industry are not fully understood, and would reward further study.

38. An interesting question is why corporations do not replicate the modular structure of open source software in commercial products more generally. One possibility may be that modular code, whatever its virtues for a team of programmers working independently, is not necessarily better for a team of programmers and managers working together.

39. Some of the E-speak code remains proprietary to Hewlett Packard; so will some applications and utilities developed in the future. It should also be noted that HP can profit by providing services to E-speak users, which, while not proprietary, should be an arena in which HP has a natural advantage.

40. The former solution may be particularly desirable if the user has customized last generation's applications.

41. Wayner (2000) argues that the open source movement is not about battling Microsoft or other leviathans and notes that in the early days of computing (say, until the late seventies) code sharing was the only way to go as "the computers were new, complicated, and temperamental. Cooperation was the only way that anyone could accomplish anything." This argument is consistent with the hypothesis stated below, according to which the key factor behind cooperation is the alignment of objectives and this alignment may come from the need to get a new technology of the ground, from the presence of a dominant firm, or from other causes.

42. See, e.g. Hansmann (1996).

43. The increasing number of software patents being granted by the U.S. Patent and Trademark Office provide another avenue through which such a "hijacking" might occur. In a number of cases, industry observers have alleged that patent examiners – not being very familiar with the unpatented "prior art" of earlier software code – have granted unreasonably broad patents, in some cases giving the applicant rights to software that was originally developed through open source processes.

44. A related phenomenon that would reward academic scrutiny is "shareware." Many of packages employed by researchers (including several used by economists, such as MATLAB, SAS, and SPSS) have grown by accepting modules contributed by users. The commercial vendors co-exist with the academic user community in a positive symbiotic relationship. These patterns provide a useful parallel to open source.

ACKNOWLEDGMENTS

The assistance of the Harvard Business School's California Research Center, and Chris Darwall in particular, was instrumental in the development of the case studies and is gratefully acknowledged. We also thank a number of practitioners – especially Eric Allman, Mike Balma, Brian Behlendorf, Keith Bostic, Tim O'Reilly, and Ben Passarelli – for their willingness to generously spend time discussing the open source movement. George Baker, Jacques Crémer, Rob Merges, Bernie Reddy, Pierre Régibeau, Bernard Salanié, many open source participants, seminar participants at the American Economics Association annual meetings, European Economic Association Bolzano meetings, and Harvard, and three anonymous referees provided helpful comments. Harvard Business School's Division of Research provided financial support. The Institut D'Economie Industrielle receives research grants from a number of corporate sponsors, including French Telecom and the Microsoft Corporation. This is an expanded version of Lerner and Tirole (2002), for which Blackwell Publishing holds the copyright and has granted permission for this use. All opinions and errors remain our own.

REFERENCES

Aghion, P., & Tirole, J. (1997). Formal and real authority in organizations. *Journal of Political Economy, 105*, 1–29.

Brooks, F. (1975). *The mythical man month: Essays on software engineering*. Reading, MA: Addison-Wesley.

Browne, C. B. (1999). Linux and decentralized development. http://www.firstmonday.dk/issues/issue3_3/browne/index.html (accessed December 17, 1999).

Caminer, D., Aris, J., Hermon, P., & Land, F. (1996). *User driven innovation: The world's first business computer*. New York: McGraw-Hill.

Cassiman, B. (1998). *The organization of research corporations and researcher ability*. Unpublished Working Paper, University Pompeu Fabra.

Claymon, D. (1999). Apple in Tiff with programmers over signature work. *San Jose Mercury News* (December 2).

Cockburn, I., Henderson, R., & Stern, S. (1999). Balancing incentives: The tension between basic and applied research. Working Paper 6882, National Bureau of Economic Research.

Cusumano, M. A. (1991). *Japan's software factories: A challenge to U.S. management*. New York, NY: Oxford University Press.

Cusumano, M. A., & Selby, R. W. (1995). *Microsoft secrets: How the world's most powerful software company creates technology, shapes markets, and manages people*. New York: Free Press.

Darwall, C., & Lerner, J. (2000). A note on open source software. *Harvard Business School Note* 9-201-078.

Dasgupta, P., & David, P. (1994). Towards a new economics of science. *Research Policy*, *23*, 487–521.

Dempsey, B. J., Weiss, D., Jones, P., & Grenberg, J. (1999). *A quantitative profile of a community of open source Linux developers*. Unpublished Working Paper, School of Information and Library Science, University of North Carolina at Chapel Hill, http://www.metalab.unc.edu/osrt/develpro.html (accessed November 1).

Dessein, W. (1999). *Authority and communication in organizations*. Unpublished Working Paper, Université Libre de Bruxelles.

DiBona, C., Ockman, S., & Stone, M. (Eds) (1999). *Open sources: Voices from the open source revolution*. Sebastopol, CA: O'Reilly.

Farrell, J., & Gallini, N. (1988). Second sourcing as a commitment: Monopoly incentives to attract competition. *Quarterly Journal of Economics*, *103*, 673–694.

Farrell, J., & Katz, M. L. (2000). Innovation, rent extraction, and integration in systems markets. *Journal of Industrial Economics*, *48*, 413–432.

Gambardella, A. (1995). *Science and innovation: The U.S. pharmaceutical industry during the 1980s*. Cambridge, UK: Cambridge University Press.

Ghosh, R. A. (1999). Cooking pot markets: An economic model for the trade in free goods and services on the Internet. http://www.firstmonday.dk/issues/issue3_3/ghosh/index.html (accessed December 17).

Ghosh, R., & Prakash, V. V. (2000). The Oribiten free software survey: May 2000. http://www.orniten.org/ofss/01.html (accessed May 10).

Gibbons, R. (1997). Incentives and careers in organizations. In: D. Kreps & K. Wallis (Eds), *Advances in Economic Theory and Econometrics* (Vol. 2). Cambridge University Press.

Gibbons, R., & Waldman, M. (1999). Careers in organizations: Theory and evidence. In: O. Ashenfelter & D. Card (Eds), *Handbook of Labor Economics* (Vol. 3B, Chap. 36). New York: North Holland.

Gomulkiewicz, R. W. (1999). How Copyleft uses license rights to succeed in the open source software revolution and the implications for article 2B. *Houston Law Review*, *36*, 179–194.

Hammerly, J., Paquin, T., & Walton, S. (1999). Freeing the source: The story of Mozilla. In: C. DiBona, S. Ockman & M. Stone (Eds), *Open Sources: Voices from the Open Source Revolution* (pp. 197–205). Sebastopol, CA: O'Reilly.

Hansmann, H. (1996). *The ownership of enterprise*. New York: Belknap Harvard.

Henderson, R., & Cockburn, I. (1994). Measuring competence? Exploring firm effects in pharmaceutical research. *Strategic Management Journal*, *15*(Winter Special Issue), 63–84.

Holmström, B. (1999). Managerial incentive problems: A dynamic perspective. *Review of Economic Studies*, *66*, 169–182.

Johnson, J. P. (1999). *Economics of open-source software*. Unpublished Working Paper, Massachusetts Institute of Technology.

Lakhani, K., & von Hippel, E. (2000). How open source software works: "Free" user-to-user assistance. Working Paper 4117, Sloan School of Management, Massachusetts Institute of Technology.

Lerner, J., & Tirole, J. (2002). Some simple economics of open source. *Journal of Industrial Economics*, *52*, 197–234.

Levy, S. (1984). *Hackers: Heroes of the computer revolution*. Garden City, NY: Anchor Press/Doubleday.

Merton, R. K. (1973). *The sociology of science: Theoretical and empirical investigations*. Chicago, IL: University of Chicago Press.

Mockus, A., Fielding, R., & Herbsleb, J. (2000). A case study of open source software program movement: The Apache server. herbsleb@research.bell-labs.com (accessed December 30).

Nadeau, T. (1999). Learning from Linux. http://www.os2hq.com/archives/linmemo1.htm (accessed November 12).

Netcraft (2000). The Netcraft web server survey http://www.netcraftcom/survey (accessed December 24).

Open Source Initiative (1999). Open source definition. http://www.opensource.org/osd.html (accessed November 14).

Ortega, J. (2000). *Power in the firm and managerial career concerns*. Unpublished Working Paper, Universidad Carlos III de Madrid.

Raymond, E. S. (1999a). The cathedral and the bazaar. http://www.tuxedo.org/-esr/writings/catherdral-bazaar (accessed November 9).

Raymond, E. S. (1999b). Homesteading the Noosphere: An introductory contradiction. http://www.tuxedo.org/-esr/writings/homesteading (accessed November 11).

Ronde, T. (1999). *Trade secrets and information sharing*. Unpublished Working Paper, University of Mannheim.

Rosenberg, N. (1976). *Perspectives on technology*. Cambridge: Cambridge University Press.

Rosenbloom, R. S., & Spencer, W. J. (Eds) (1996). *Engines of innovation: U.S. industrial research at the end of an era*. Boston: Harvard Business School Press.

Shepard, A. (1987). Licensing to enhance demand for new technologies. *Rand Journal of Economics*, *18*, 360–368.

Stallman, R. (1999). The GNU operating system and the free software movement. In: C. DiBona, S. Ockman & M. Stone (Eds), *Open Sources: Voices from the Open Source Revolution* (pp. 53–70). Sebastopol, CA: O'Reilly.

Taschek, J. (1999). Vendor seeks salvation by giving away technology. http://www.zdnet.com/pcweek/stories/news/0,4153,404867,00.html (accessed December 17).

Torvalds, L. (1999). Interview with Linus Torvalds: What motivates free software developers? http://www.firstmonday.dk/issues/issue3_3/torvalds/index.html (accessed December 17).

U.S. Department of Labor, Bureau of Labor Statistics (2000). *Occupational Outlook Handbook: 2000–2001 Edition*. Washington: Government Printing Office.

Valloppillil, V. (1998). *Open source software: A (new?) development methodology [also referred to as The Halloween Document]*. Unpublished Working Paper, Microsoft Corporation, http://www.opensource.org/hallowen/halloween1.html (accessed November 9, 1999).

von Hippel, E. (1988). *The sources of invention*. New York: Oxford University Press.

Wayner, P. (2000). *Free for all: How Linux and the free software movement undercut the high-tech titans*. New York: Harper Collins.

SUBMARINES AND TECHNOLOGICAL INNOVATION: U.S. CONTINUATION PATENTING IN SOFTWARE AND BIOTECHNOLOGY TECHNOLOGIES IN THE 1980s AND 1990s

Stuart J. H. Graham and David C. Mowery

ABSTRACT

This chapter examines the role of "continuations" (procedural revisions of patent applications) within software patents and overall patenting in the United States during 1987–1999. Our research represents the first effort of which we are aware to analyse data on continuations in software or any other patent class, and as such provides information on the effects of 1995 changes in the U.S. patent law intended to curb "submarine patenting." Our analysis of all U.S. patents issued 1987–1999 shows that the use of continuations grew steadily in overall U.S. patenting through 1995, with particularly rapid growth in continuations in software patenting. Sharp reversals in these growth rates after 1995 suggest that changes in the U.S. patent law were effective. Continuations were used more intensively by packaged-software firms prior to the effective date of the 1995 changes in patent law than by other patentees, and both software and non-software patents subject to continuation tend to be more valuable.

Intellectual Property and Entrepreneurship
Advances in the Study of Entrepreneurship, Innovation and Economic Growth, Volume 15, 71–101

ISSN: 1048-4736/doi:10.1016/S1048-4736(04)01503-6

1. INTRODUCTION

Research in "knowledge management" has grown significantly during the past few years in the study of innovation. An important part of "knowledge management" involves firms' protection of knowledge assets, and firm strategies for the use of patents accordingly have received increased attention. Nonetheless, this research has devoted little attention to the ways in which the characteristics of national patent systems and the characteristics of specific fields of technological innovation interact in the development and implementation of firms' patent strategies. Another important issue in knowledge management for managers and policymakers alike concerns the operation of national patent systems in novel areas of technology that have little well-defined "prior art" in patents or publications and are highly cumulative, meaning that innovation in a given area is closely linked with other areas and prior generations that may themselves be subject to increased patenting. Since patent-based prior art is scarce in such areas, patent offices face serious difficulties in evaluating the "novelty" and "non-obviousness" of patent applications. Inventors and entrepreneurs for their part are operating in an environment of high uncertainty in which the artifacts on which they rely for their innovations may prove to be patented, thus exposing them to litigation and the invalidation of their patents.

Computer software and biotechnology are fields in which patenting in the United States has grown rapidly since 1980 (Audretsch, 2001; Graham & Mowery, 2003). As we note below, much of this increased patenting reflects changes in U.S. policy toward software and biotechnology patents, increasing their economic value. A portion of the growth in patenting in these new technology areas during the 1980s and 1990s also may reflect increased "defensive patenting," as firms patent in order to retain "freedom to innovate" and to acquire a form of "currency" for cross-licensing negotiations (see Hall & Ziedonis, 2001, for a discussion of this phenomenon in the semiconductor industry).

This chapter builds on earlier work on software patenting (Graham & Mowery, 2003; Graham et al., 2002) that highlighted the dimensions of growth in such patenting. By focusing on the role of "continuations" (revisions of patent applications) in biotechnology and software patenting during the 1980s and 1990s, we hope to shed additional light on the strategic motives behind the growth of U.S. patenting in these sectors. This research also represents the first effort of which we are aware to analyze data on continuations in these or any other patent classes, and as such provides some information on the effects of the 1995 changes in U.S. patent law that sought to reduce the use by inventors of "submarine patents.

An inventor seeking a "submarine" patent submits an application, thereby establishing a priority date for her application, and files numerous continuations on the initial application. Prior to 1999, the chain of applications remained secret throughout the period of USPTO review for all patent applications – and remains secret after 1999 for U.S.-only filings. The long-hidden patent, which prior to June of 1995 was valid for 17 years from its date of issue, would issue covering technologies in widespread industrial use.[1] These patents are often termed "submarines," because like a submarine, they can surface unexpectedly and create havoc.

Immediately below, we summarize earlier work on the dimensions and characteristics of growth in software and biotechnology patenting in the United States during the 1980s and 1990s. We follow this discussion with an examination of our data on continuations, and summarize our findings in the conclusion.

2. SOFTWARE PATENTING: OVERVIEW

2.1. Growth in Software Patenting

Change in the structure of markets for computer software in the U.S. and global economies since the early 1980s has increased the economic importance of formal instruments for protection of software-related intellectual property (Graham & Mowery, 2003; Mowery, 1999). For much of the short life of the computer industry, computer software was custom-designed for a given machine and application by the manufacturer of the hardware or by the user, thus limiting the importance of intellectual property protection. Only during the 1970s did independent software firms begin to play a significant role in developing and marketing "standard" software for mainframes and minicomputers. With the development and widespread adoption of the desktop computer in the early 1980s, a large and rapidly growing market for "packaged" computer software emerged that exhibited few if any of the "custom-engineering" characteristics of the products in other segments of the software industry. Instead, packaged software markets came to resemble those for entertainment or books, in which "hits" were extraordinarily profitable. The growth in markets for packaged software meant that formal protection of the intellectual property contained in the packages was of much greater importance.

2.2. The Evolution of Intellectual Property Protection in the U.S. Software Industry

In 1980, the United States Congress selected copyright as the appropriate protection for computer software (Samuelson, 1984). A series of federal court decisions during the late 1980s and early 1990s initially strengthened, and then weakened, the power of copyright to protect intellectual property in packaged software. These shifts in the judicial treatment of copyright, along with a supportive decision from the U.S. Supreme Court upholding the patentability of software (*Diamond v. Diehr*),[2] led software developers to expand their use of patents in the 1980s. The economic value of these patents was highlighted in several high-profile cases during the 1990s, including the 1994 *Stac Electronics* decision that found Microsoft guilty of patent infringement.[3] The United States Patent and Trademark Office (USPTO) also adopted a more favorable posture toward applications for software patents during this period, and some observers (e.g. Aharonian, 2000) have questioned the ability of patent examiners to identify "novelty" in this area of newly patentable subject matter.

2.3. Trends in Software Patenting, 1987–1997

In earlier work (Graham & Mowery, 2003), we showed that software patenting by both packaged-software "specialists" and electronics system firms in the United States grew significantly after the mid-1980s in the face of a decline in the use of copyright protection for software.[4] There are at least three broad explanations for the rapid growth in software patenting during this period. One explanation hypothesizes that increased patenting is the consequence of greater returns from innovation or the strengthening of the patent right itself (Kortum & Lerner, 1995). Another interpretation argues that increased patenting, especially by large firms such as Microsoft and IBM, reflects the growing importance of "defensive patenting." Firms engaged in defensive patenting seek patents less to support the commercial development of specific invention than as a means of avoiding costly litigation (see Hall & Ziedonis, 2001 for a discussion of "defensive patenting" in semiconductors). "Defensive patenters" apply for a large number of patents for exchange in cross-licensing agreements, thus preserving their "freedom to innovate."

Alternatively, growth in software patents may reflect a decline in the rigor of USPTO review of the increased number of software-related patent applications that followed the changes in the legal treatment of software patents. Lacking patent-based prior art to guide their evaluation of a much larger flow of applications, this explanation argues that USPTO examiners now issue larger quantities of lower-quality patents. Both of these explanations suggest that the "quality" of software patents should have declined during the 1980s and 1990s. But the "defensive patenting" explanation predicts that the patents assigned to large software firms should exhibit particularly significant declines in quality, whereas the "weakened review" explanation predicts an across-the-board decline in the quality of all issued software patents.

In earlier work, we found no evidence of a decline in "quality" in major software firms' software patents.[5] This finding may mean that defensive motives are not driving the patenting behavior of these firms, if one assumes that such defensive patents on average exhibit a lower measure of "quality" than those issuing to other patentees. To probe the implications of this finding, we extended our analysis of the causes of increased software patenting by analyzing the use by software-patent applicants of "continuations."

2.4. Continuations in Software Patenting

The use of continuations by applicants for U.S. patents compounds the effects of uncertainty surrounding intellectual property issues in software. The continuation is a procedure available under U.S. patent law that permits an applicant to refile a pending patent application,[6] permitting an applicant to avoid the implementation of a patent examiner's issue decision by renewing at least some portion of an original application while retaining the filing date of the original application. There are three types of continuing patent applications: The continuation, the continuation-in-part, and the division.[7] The "continuation" application discloses the identical invention claimed in the prior "parent" application, with a requirement that no extraneous matter be disclosed in the follow-up application.[8] The "continuation-in-part" application allows the addition of some new matter to the original application – although the benefit of early priority is awarded only for the original disclosures contained in the new application. A "division" occurs when the original application contains more than one independent invention.

The USPTO allows the applicant to "elect" one of these inventions while filing second and subsequent disclosed inventions separately, generally later in time. Like the continuation and continuation-in-part applications, these later applications disclosing the non-elected inventions that satisfy legal requirements also benefit from the earliest filing date.[9] All of the three different "types" of continuations thus allow a delay in the prosecution and final issue decision for a U.S. patent application.

An essential feature of the continuation process permits the applicant to adopt the date of an earlier application still pending within the Patent Office.[10] So long as the original "parent" application and the follow-up "continuation" application disclose the same invention, the applicant can preserve the parent's filing date.[11] Because the costs to filing a continuation are low, and because there is no limit to the number of times that a parent application may be continued and abandoned, the continuation process may be used to prevent the patent examiner from officially reaching a decision to issue or deny issuance of a patent. Strings of continuations may thus develop from a single application, with the issue decision being postponed in some extreme situations for decades,[12] enabling the invention to remain hidden for extended periods of time. Prior to 1999, all U.S. patent applications remained secret until the issue of the patent, and an applicant thus could maintain the secrecy of their invention and application throughout this extended period. The continuation is unavailable in other industrial nations' patent systems, including those of Europe and Japan.[13]

Revisions in U.S. patent law that became effective in 1995 changed the term of patent protection from 17 years from the date of patent issue to 20 years from the date on which the application was filed. These legislative changes sought to reduce the incentives for inventors to pursue so-called "submarine" patents. An inventor seeking a "submarine" patent submitted an application, thereby establishing a priority date for his application, and filed numerous continuations on the initial application, which remained secret throughout the period of USPTO review of repeated revisions ("continuations") in the original application. The submarine patent, long hidden from view and (prior to June of 1995) valid for 17 years from its date of issue, would issue covering technologies widely in use. In cases in which these technologies were covered by patents applied for after the priority date established by the original "submarine" application, the owner of the original patent could pursue infringement actions against other patentholders.

After 1995, continuations still offered some strategic benefits to an applicant by allowing an extended period of secrecy prior to issue of a patent, but even this benefit was reduced in significance by 1999 legislation mandating the publication of most patent applications 18 months after their submission.[14] The "publication" requirement has removed some but not all benefits from continuations. Applicants may still make changes to individual claims reported in the published disclosure, and thus use the continuation procedure to keep individual claims secret, consistent with the 20-year time limit enacted in 1995 (Glazier, 1997).[15]

In software technologies, the impact of the 1995 and 1999 regime changes may be less significant than in other technologies. Because the product life cycle for software is short, the "tradeoff" embodied in the 1995 changes in patent policy, which forced patentees to choose between *ex ante* prosecution and *ex post* protection, is of little significance. The costs to a software inventor of pursuing continuations and thereby opting for a shorter patent term should be smaller than they are for innovators in sectors such as pharmaceuticals that have far longer product life cycles. Furthermore, because returns to investment in software are often predicated upon being first-to-market and creating a critical mass of adopters, the ability of continuations and "submarine patents" to create uncertainty about the legitimacy of a new product may have an especially strong chilling effect on commercial innovation in this field. As a result, the effects of the 1995 and 1999 changes in patent policy on the incentives for software inventors to pursue "submarine patents" remain unclear.

2.5. *The Threat from Submarine Patents: Rambus Inc.*

One example of the strategic use by a patent applicant of continuations is the case of Rambus. When a U.S. patent for a "Synchronous Memory Device" issued to Rambus, Inc.[16] in September 1999, semiconductor chip manufacturers recognized the competitive threat: within sixteen months, Rambus was demanding royalties of up to 3.5% of sales of products using this patent from 7 of the top 10 global DRAM manufacturers. The patent wrote directly on the communication standard that the industry had adopted in 2000, forcing manufacturers to choose between acceptance of Rambus's licensing terms and litigation that could result in an injunction that would close multibillion fabrication facilities (see Hall & Ziedonis, 2001, for a discussion of the risks of patent litigation for semiconductor firms). These manufacturers knew that the patent covered the communication technology because Rambus had helped them write the industry standard beginning in 1991. Rambus relied on continuations in its licensing hold-up strategy, keeping its patent application hidden from its manufacturer-partners for nine years while the industry standard was being formalized.

When Rambus finally allowed the patent to issue from the continuation application process, it claimed that competitors were infringing, prompting firms who also participated in the standard-setting process – Hyundai Electronics, Infineon Technologies, and Micron Technology – to take Rambus to the court. In January 2003, however, the Court of Appeals for the Federal Circuit (CAFC) in January 2003 upheld Rambus's claims. Repeating the logic of its 1988 *Kingsdown* decision,[17] the Court supported Rambus's use of the continuation to alter claims in a pending application that cover new uses for a technology being developed by competitors or customers.

The *Rambus* case highlights one way in which continuations may be employed strategically within industry standard-setting processes. Moreover, *Rambus* suggests that this use of continuations is supported by the federal courts as a strategic tool. According to the logic of the *Rambus* decision, a patent applicant may make changes to the application's claims over a number of years by employing the continuation, and then allow the patent to issue and demand royalties, from software firms or non-software firms adopting the standard.

The *Rambus* case is not a software case per se, but continuations have been used in software patenting as well. In the legal documents put forth by Microsoft in

Reiffin vs. Microsoft Corporation,[18] the inventor Reiffin was shown to have filed an application on a multi-threading computer process in 1982, using continuations and abandonments in 1985, 1990, and again in 1994 before patents issued in 1997, more than fifteen years after the original filing.[19] According to the trial court, Reiffin attempted to sell and/or license the multi-threading technology following the issue of these patents to several large computer companies, including IBM, Sun Microsystems, Apple, Intel, and Microsoft. Although the motives for these actions are disputed, the *Reiffin* case demonstrates the uncertainty that late-surfacing patents can introduce into the marketplace for software.

2.6. Other Strategic Uses of Continuations

Another strategic motive for use by applicants of the continuation procedure has been proposed by Quillen and Webster (2001), who argue that continuations can be used to "wear down" an examiner and allow a patent of dubious merit to issue. If Quillen and Webster's characterization is accurate, a high share of continuations in a given technology field's patents may indicate an unusually high incidence of "junk" patents. But continuation applications also may be common in new technologies simply because neither applicant nor examiner understand the prior art and/or their interpretations of novelty and non-obviousness requirements differ significantly. Significant information asymmetries between examiner and applicant may lead to a larger share of continuations in patenting in new fields of inventive activity such as software or biotechnology.

3. PATENTING AND CONTINUATIONS IN BIOTECHNOLOGY

Like software, change in technology strategies and market structure within the biotechnology industry have increased the strategic and economic value of intellectual property. Lerner (1995) suggests that inventor mobility and the role of university research in biotechnology innovation tend to weaken the effectiveness of trade secrecy, which leads innovators in biotechnology to rely more heavily on patents than their counterparts in other industries. To the extent that intellectual-property strategies in biotechnology resemble those in the much larger and more diverse pharmaceuticals industry, Cohen, Nelson and Walsh's (2001) survey results support Lerner's observation: Product innovators in "drugs" report that patenting is more effective at capturing value than in any of 34 other sectors, while the reported effectiveness of secrecy in "drugs" does not exceed the mean of all 35 sectors included in the survey.

Patenting in biotechnology has also played an important role in an unusual feature of this sector, the large number of alliances between small "research boutiques," new firms focused largely on drug discovery, and established, large pharmaceutical firms. Sharp (1999) and Audretsch (2001) identify three phases in the development of alliance strategies involving U.S. biotechnology firms. The modern biotechnology industry traces its origins to the discovery in 1973 of the Cohen/Boyer technique at Stanford University and the University of California, San Francisco for recombinant DNA.[20] This first phase of industry evolution was associated with rapid entry by firms, which numbered 200 by 1982 (Argyres & Liebeskind, 2002). During this first phase of the industry's development (1973 to the mid-1980s), large incumbent pharmaceutical firms avoided alliances with the entrant biotechnology firms, but instead invested in R&D to monitor the burgeoning technologies. During phase two (mid-1980s to early1990s), the large pharmaceutical firms pursued acquisitions, mergers, and alliances with smaller biotechnology specialists. In the most recent phase (early1990s to the present), a form of vertical specialization has become common, with the large incumbents producing mass-marketed drugs from the substances synthesized by the smaller "research boutiques."

Biotechnology patents may reduce the contracting hazards faced by small patentholding firms when entering alliances with large incumbent firms in pharmaceuticals. Patents offer institutional support and protection mechanisms for patentholders (which in many cases are smaller firms) in these alliance contracts, the terms of which are necessarily incomplete. The continuation procedure provides the option of "serial prosecution" of a patent for a new molecule derived from biotechnology research, which may offer additional support for such agreements. Serial prosecution, the practice of patenting a cumulative technology "in series" through a sequence of patent applications, offers alliance partners the opportunity of gradually resolving uncertainty over the ownership of the fruits of the innovative activities of one or both of them while retaining secrecy over the novel substance or process. The use of continuations to support serial prosecution may also serve to build more effective thickets around a technology by creating longer periods of secrecy, thus allowing an inventor the opportunity to discover and patent technologies that lie near the focal invention in the technology space. Moreover, to the extent that "stacks" of patents signal the strength of a patentholder's ownership of intellectual property to competitors or to juries – as in the semiconductor industry – the serial continuation process, which permits the redrafting of claims over time, gives continuation applicants additional opportunities to "manage" their application portfolios and maximize the number of patents issuing from the patent office.

3.1. The Evolution of Patent Protection in the U.S. Biotechnology Industry

At the genesis of the technology, great uncertainty surrounded the legal status of patents in biotechnology. In 1973, the year of invention of the Cohen/Boyer process, legal doctrines were unsettled regarding the patentability of natural substances,[21] although the U.S. Supreme Court's *Funk Brothers Seed Co. v. Kalo Inoculant Co.* (1948)[22] had apparently held that natural phenomena were unpatentable. In 1980, however, the Supreme Court clarified the legal landscape in the *Diamond v. Chakrabarty*[23] decision, making the results of the application of "human ingenuity" to natural products patentable. Throughout the 1980s, the USPTO used the *Chakrabarty* logic to extend patenting to other products of "nature," allowing the patenting of subject-matter such as man-made multi-cellular plants[24] and genetically altered oysters.[25] In 1987, the U.S. Patent Office promulgated a rule declaring that non-naturally occurring, man-made living microorganisms, plants, and animals were patentable subject matter,[26] and biotechnology patents have since been upheld consistently by the courts.

As was the case in software, the *Chakrabarty* decision and its subsequent application by the USPTO created a new area of patentable subject matter, thus increasing patent examiners' uncertainty and presenting the Patent Office with application disclosures for which it lacked both in-house expertise and clear patent prior art. The resulting increase in USPTO's backlog of biotechnology patent applications (Barton, 1995) may have lengthened the examination lags for biotechnology patents during the 1980s and 1990s (Wright, 1997). As was the case with software, the "new technology" of biotechnology created both technological and legal uncertainty, adding delay and other costs to the administration of the patent system.

3.2. A Biotechnology "Submarine": Monsanto Co.'s "Plant Cell" Patent

When United States Patent 6,174,724 entitled "Chimeric genes suitable for expression in plant cells" issued and was assigned to the Monsanto Company on January 16, 2001, researchers working in genetically modified food technologies were uneasy. Monsanto's broad patent was recognized by scientists to cover certain marker-gene systems commonly used in biotechnology research. The patent, which was first filed in the USPTO on January 17, 1983, was kept pending in the Patent Office for eighteen years through a series of four continuations and abandonments. As a result, when the patent finally issued in 2001, Monsanto's patented method for employing antibiotic resistance markers in transforming plants was a process that had been used in developing almost all of the commercially available genetically modified crops that were then available. The Rural Advancement Foundation International pointed to the unduly long period of examination, the secrecy surrounding the pending application, and the fact that many researchers had been long using the patented technology when it declared that Monsanto was holding a "submarine patent."[27]

4. RESEARCH QUESTIONS

Surprisingly, the large literature on patenting strategies in high-technology industries has largely overlooked the role of continuations in these strategies. In particular, the overall importance of continuations before 1995 in software and biotechnology patenting and the effects of the statutory changes in 1995 on the incidence and characteristics of continuations in these fields of patenting have not been examined. Because biotechnology, like computer software, is a "novel" technology for which uncertainties over patentability were resolved only in the mid-1980s, comparison of patenting and continuation practice in software and biotechnology may shed additional light on the motives for and effects of these practices.

We address the following questions in our analysis of continuations in software and biotechnology patenting:

(1) How does the share of continuations in issued software and biotechnology patents compare with this share for all patents before and after 1995?
(2) How does this share differ among leading packaged-software firms before and after 1995?
(3) How do continuations contribute to the duration of "pendency" (the length of the USPTO review process) for patent applications? This issue is of particular relevance in light of the findings of Johnson and Popp (2001) that longer-pendency patents tend to be cited more heavily by other patents and therefore may be more technologically and economically important.
(4) Do issued software and biotechnology patents subject to continuations differ in terms of their forward citations (a measure of the "importance" of these patents) or their probability of litigation?

The examination of these questions should shed additional light on the effects of the 1995 changes in U.S. patent law and on the relative importance of strategic motives and field-specific uncertainty on the growth in U.S. software and biotechnology patenting since 1985.

4.1. Data on Continuations in Software and Non-Software Patents, 1987–1999

We computed the incidence of continuations as the shares of all U.S. software patents[28] and non-software patents issuing each year during 1987–1999 that had at least one continuation in their application lineage and plotted these with continued patents in all other technology classes (Fig. 1).[29] Continuations account for roughly 18% of all issued patents in 1987, climbing to a peak of 28% in 1997 and declining to 21% in 1999. The downward trend in the share of non-software continuations after 1997 may reflect the fact that patents issuing during and after this year are the first to reveal the effects of the 1995 changes in U.S. patent law.

Figure 1 also displays trends in continuations within issued software patents, highlighting some interesting contrasts. Continuations were less common in software patenting than in overall patenting during 1987–1992; 11–20% of patents issuing each year included at least one continuation. After 1992, however, the incidence of continuations in software patents grew to levels exceeding those in overall patenting. By 1996, continuations accounted for nearly 40% of all issued software patents, substantially above the share (27%) within all patents in that year with continuations. The incidence of continuations in software patenting declined after 1996 more rapidly than in all patents, and the shares of continuations in the two categories (20%) are essentially equal by 1999.[30] The post-1997 decline in the share of continuations in U.S. software and non-software patents suggests that the 1995 legislation reversed growth in continuations within overall U.S.

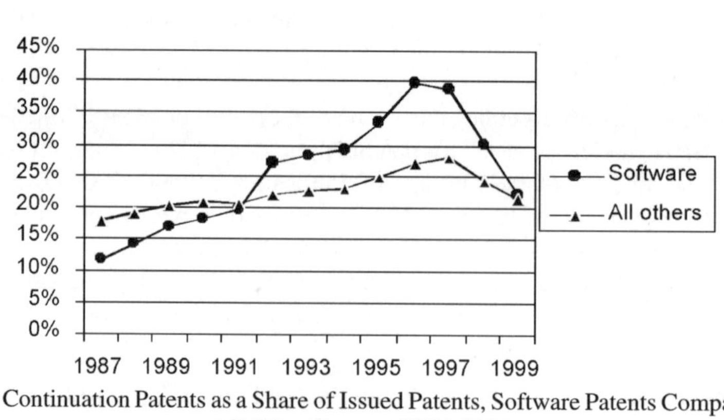

Fig. 1. Continuation Patents as a Share of Issued Patents, Software Patents Compared with All Other Patents, 1987–1999.

patenting, although the share of continuations has been reduced only to levels comparable to those observed in the mid-1980s.

The combination of rapid growth in patenting and high levels of continuation practice in software patenting during the 1990s may reflect: (1) the operation of the "attrition strategies" proposed by Quillen and Webster; (2) information asymmetries in applicant/examiner negotiations over the validity of software patent applications; or (3) frequent revisions in their patent applications by corporations (primarily the packaged-software firms) with little experience in patenting. Alternatively, firms may use continuation strategies for patents that protect "foundational" technologies underpinning several product generations. For these technologies, the delays in patent issue, coupled with the longer period of secrecy associated with continuations, may provide important competitive benefits that trade secret protection does not provide. The decline in continuations after 1995 also suggests that the "strategic" motives for continuations that we discussed earlier were no more significant in software than in other fields in the wake of the 1995 policy changes, although this hypothesis requires additional examination.

We also analyzed the patenting and continuation practices of the four largest patentholders among the leading specialized U.S. producers of packaged software, Microsoft, Adobe Systems, Novell, and Borland International (now Inprise).[31] Figure 2 displays the share of continuations in the patents assigned to each of these firms during 1991–1999 (as three-year moving averages). Microsoft, which has the largest number of software patents among these firms, displays a relatively flat continuation rate that ranges between 24 and 31%. Continuation trends for Adobe Systems and Novell present interesting contrasts with one another and with

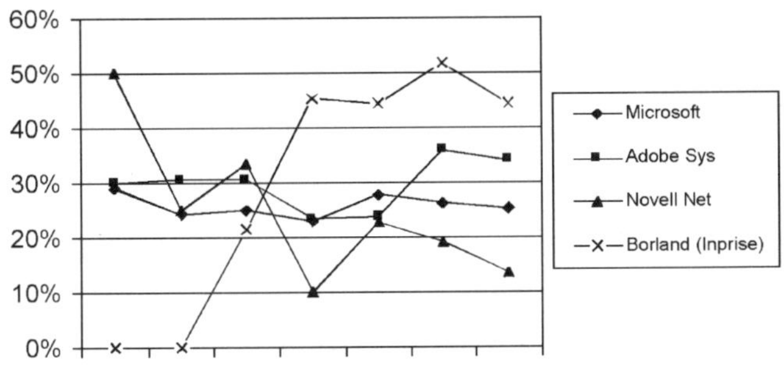

Fig. 2. Continuation Firm-Patents as a Share of All Issued Firm-Patents, Top Patenting Packaged-Software Firms (Microsoft, Adobe, Novell, Borland), 3-Year Moving Average, 1991–1999.

Microsoft. The annual continuation shares for these firms show that Adobe has increased its use of continuations in patenting practice, while Novell has markedly reduced the use of continuations. Indeed, Novell's use of continuations declined during the period of significant growth in its overall patenting and patenting propensities (Graham & Mowery, 2003). The clear outlier among these four large packaged-software firms is Borland International (Inprise), a firm that exhibits a significantly higher propensity to patent (Graham & Mowery, 2003) and a significantly more intensive use of continuations in its patent applications than do these other three firms after 1994. This trend may reflect the firm's unhappy experience in litigation.[32]

4.2. *Data on Continuations in Biotechnology Patents, 1987–1999*

We computed the share of continuations in biotechnology patents[33] in two samples of patents issuing each year during 1987–1999 that show at least one continuation in their application lineage. Since there is no "biotechnology-specific" patent class, we created these samples from corporate patents held by the "Top 100" and "Top 10" firms specializing in biotechnology, based on 1997 revenues reported in Med Ad News (1998). In order to limit these firm patents to biotechnology (and eliminate from our sample patents filed in other technologies, such as electronics or mechanicals), we restricted the sample into the primary international patent classifications for Human Necessities (classes A-), processes (classes B-), and Chemistry (classes C-).[34] Our data show that biotechnology continuation rates for the "biotechnology" patents within this sample of 100 firms are much higher than those for all other patents, accounting for at least 27% and as much as 67% of all patents issued during the period (Fig. 3). Like the pattern in software, these biotechnology continuations peak at 67% in 1995 and decline to 56% by

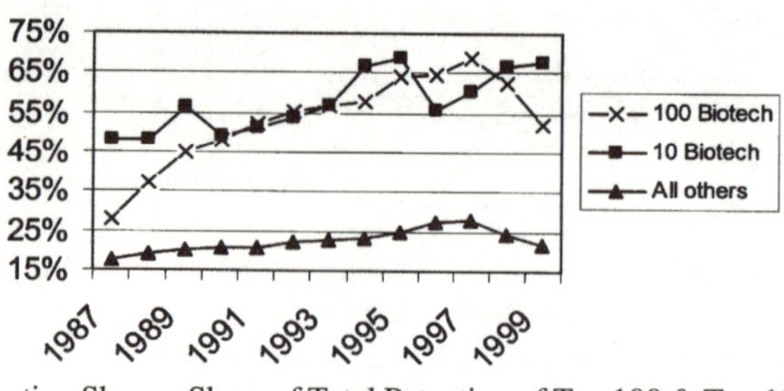

Fig. 3. Continuation Shares, Share of Total Patenting of Top 100 & Top 10 Biotechnology Firms.

1996. In contrast to software patents, however, the "continuation rate" among biotechnology patents remains well above the rate for overall patents by the end of the 1987–1999 period.

Patenting by the "Top 10" biotechnology firms differs from the pattern within both the larger sample of biotechnology-firm patents and our software patent data. Patenting by the top 10 biotechnology firms, which accounted for 30% of overall patenting and 31% of continuation patenting by the "Top 100" during 1987–1999, are initially higher than those for the fuller sample, accounting for at least 47% of issued patents in 1987. In contrast to the trends in software and the broader biotechnology sample, patent continuations by these leading biotechnology firms increase through the 1987–1999 period, accounting for 68% of issued patents in 1999. As is true of our broader sample of "biotechnology patents," the patents issuing to the top 10 biotechnology firms display a higher share of continuations than is true of all patents; this share is increasing by the end of the 1987–1999 period, and by 1999 is even higher than that for the top 100 biotechnology firms' patents. In both of our samples of biotechnology patents, and especially that for the largest biotechnology firms, continuations remain much more important after the 1995 regime change than is true of overall patenting (Fig. 4).

In order to investigate the continuation patenting behavior of these large firms, we plot the share of continuation patents within all biotechnology patents during 1991–1999 (as a 3-year moving average) for each of the largest four biotechnology firms (by 1997 revenues). These four firms – Amgen, Biogen, Genentech, and Genzyme – differ substantially in their use of continuations. The two largest patenters among the sample, Biogen and Genentech (which respectively account for 41 and 35% of the overall biotechnology patents for these four firms issuing during 1995–1999), both increase their use of continuations throughout the 1991–1999 period to levels above 70% and exhibit no marked decline after the 1995 regime change. These two firms demonstrate a continuation rate

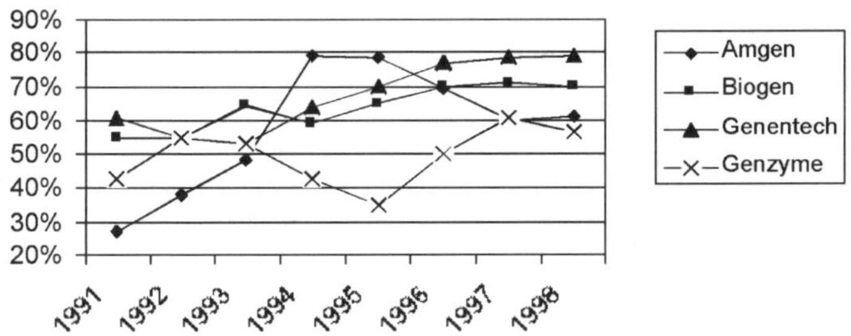

Fig. 4. Continuations as a Share of All Firm-Issued Patents, Top Four Biotechnology Firms (1997 Revenues), 1991–1999.

at the end of the period that is higher than the continuation rates associated within the two large firms with substantially fewer patent applications (Amgen and Genzyme). Amgen's continuation rate peaks at 79% in the 1993–1996 period, but this firm's share of continuations subsequently stabilizes at roughly 60% during 1997–1999, a share that is similar to that of Genzyme during the post-1995 period.

These firm-specific contrasts in the use of continuations by large biotechnology firms, as well the differences in continuation rates between biotechnology and those in other patent classes, may reflect the high levels of uncertainty regarding the technological and legal validity of biotechnology patents. This greater uncertainty may result in more frequent revisions by applicants, as we suggested earlier. Alternatively, the higher levels of uncertainty and greater complexity of the review process within USPTO may tend to favor the use by applicants of the "attrition strategy" proposed by Quillen and Webster. Alternatively, these trends may reflect strategic behavior by biotechnology patent applicants, although we lack a complete explanation for such behavior.

4.3. Continuations and Software Patent "Grant Lags"

How do continuations affect the duration of the review period for a software patent application (referred to as the "pendency" period or the "grant lag")? This issue is an important one, since one of the primary mechanisms that make submarine patents possible and potentially litigious is the duration of the grant lag associated with continuations. In addition, as we note below, recent empirical analysis (Johnson & Popp, 2001) suggests that longer-pendency patents receive more citations in subsequent patents, indicating higher economic or technological value.

We find a positive relationship between pendency and the number of continuations in a patent's application history. For software patents issuing between 1984 and 1999, a continuation is associated with an average additional lag of 24 months from application to issue. Figure 5 plots both the mean and the marginal number of weeks of additional pendency associated with each additional filing in the continuation-application lineage. Although each additional continuation extends the review process for patent applications, the magnitude of this extension in the review process declines with each additional continuation. The first continuation filed by an applicant adds more than 18 months of review lag and non-disclosure, but the "incremental pendency" drops by almost a third, to approximately 13 weeks, for four continuations. Nonetheless, the average pendency lag for an issued software patent with one continuation is 51 months, substantially less than

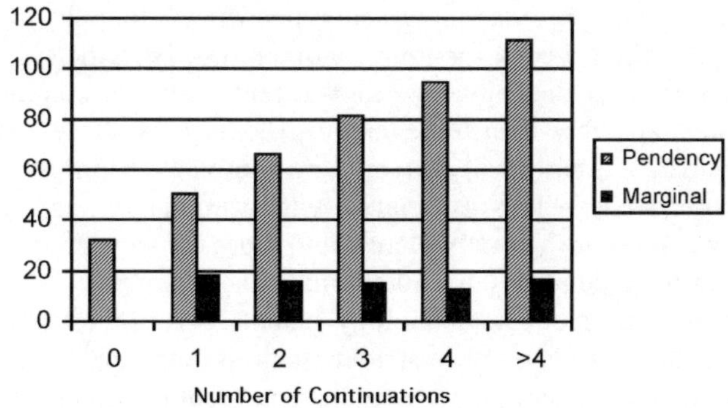

Fig. 5. Mean Application Pendency and Marginal Effects of Continuation, Software Patents Issued 1984–1999.

the average pendency lag of 111 months for a software patent issuing with five or more continuations in its review.

Biotechnology patents reveal a similar relationship between pendency and the number of continuations in a patent's application history. For patents in the international patent classes "human necessities, processes, and chemicals" assigned to the top 100 biotechnology firms between 1984 and 1999, a continuation application is associated with an average additional lag of 41 months from application to issue. Figure 6 plots both the mean and the marginal number of weeks of additional pendency associated with each additional filing in the continuation-application lineage. As is true for software patents, the marginal effects of additional continuations in this sample of biotechnology patents decline. The first continuation filed by an applicant adds more than 19 months of review

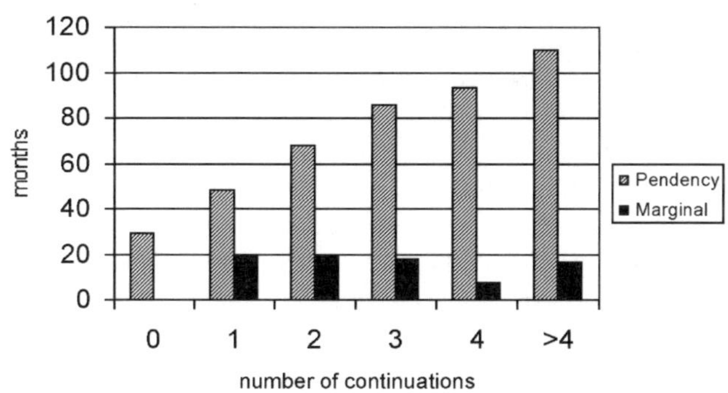

Fig. 6. Mean Application Pendency and Marginal Effects of Continuation, Biotechnology Patents Issued 1984–1999 (Months Between Filing and Issue).

lag and non-disclosure, but the "incremental pendency" drops by more than one half, to less than eight weeks for four continuations. Nonetheless, the average pendency lag for an issued biotechnology patent with one continuation is 49 months, substantially less than the average pendency lag of 110 months for a biotechnology patent issuing with five or more continuations in its review.

The mean pendency delays associated with continuations in our sample of biotechnology patents are virtually identical to those for software patents, but the difference in average pendency for all continued software patents (24 additional months) and all continued biotechnology patents (41 additional months) is a consequence of the average number of continuations in the two groups of patents: Only 7.5% of software patents show more than one continuation in the application chain, while for biotechnology patents a full 32% of applications show more than one continuation.

These apparent differences in the behavior of biotechnology and software patentees may result from differences between the product lifecycles in the two sectors. Product lifecycles in biotechnology tend to be much longer than those in computer software, and profits from product innovations typically are realized much later than is true for software. Because more of the returns from product innovation therefore are likely to be realized in the later portion of the term of a biotechnology patent, in contrast to software, biotechnology patent applicants faced particularly strong incentives to use continuations to extend patent life prior to the 1995 regime change. Moreover, the fact that the share of continuations in our biotechnology patent sample has scarcely declined since 1995, in contrast to the software patent sample, suggests that continuations may still play an important role in extending patent life and/or application secrecy for biotechnology patentees.

4.4. "Long Chain" Continuations Before and After 1995 in Software and Biotechnology

These differences between the product lifecycles for computer software (short) and biotechnology (long) suggest that rapid issue of a patent on an invention is more valuable for software products than for biotechnology inventions. As a result, we expect long-chain continuations to be less common in software patents than in biotechnology. In addition, software products, unlike biotechnology products, are not required to navigate the time-consuming Food and Drug Administration regulatory trials, nor is the payoff to R&D investment "backloaded." Indeed, the economic returns to software inventions are if anything "front loaded," making rapid issue of a patent more valuable.

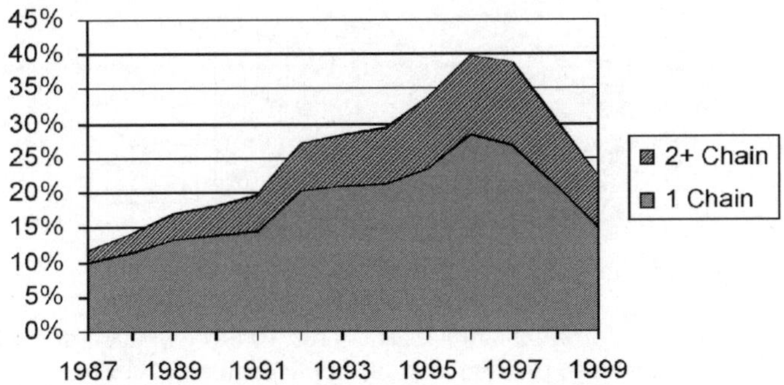

Fig. 7. Continuations in Software Patents, Shares of Single (1) and Multiple (2+) Continuations in Application Chain, 1987–1999.

We compared the share of "long-chain" continuations (more than one continuation in an issued patent's application lineage) in computer software (Fig. 7) with that for biotechnology (Fig. 8). "Long chain" continuation shares in software account for 2% of all software patents in 1987, peak at 15% in 1997, and decline to 10% of all software patenting in 1999. We contrast these trends with "long chains" in the full sample of biotechnology continuations (filed by the top 100 biotechnology firms, Fig. 8), the share of which is 7% in 1987, rising to 44% in 1997 before declining to 30% by 1999. We also compare the frequency of "long-chain" continuations in our overall sample of biotechnology patents with long-chain continuations in the 30% of this sample of patents held by the leading 10 biotechnology firms (Fig. 9): For this smaller sample, the share of "long chain" continuations is never less than 21% (1992) and rises to 52% in 1998, with no discernable decline after 1995. Indeed, the share of "long chain" continuations increases slightly (2–3%) throughout the 1995–1999 period. Consistent with the

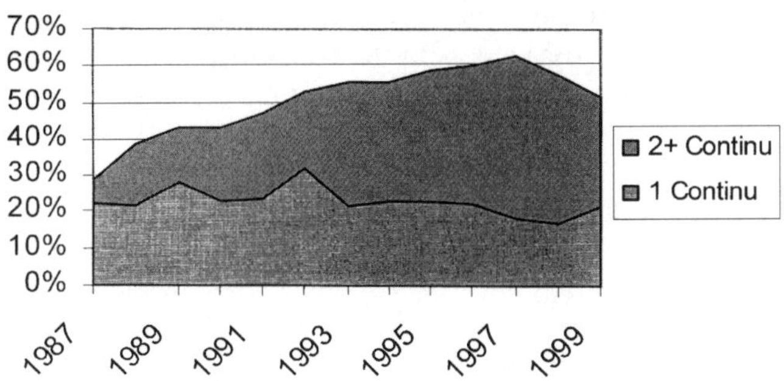

Fig. 8. Continuations in Biotechnology Patents (Top 100 Firms), Share of Single (1) and Multiple (2+) Continuations in Application Chain, 1987–1999.

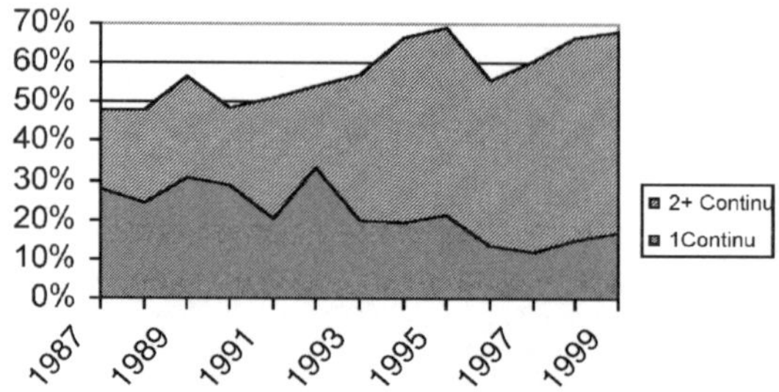

Fig. 9. Continuations in Biotechnology Patents (Top 10 Firms), Share of Single (1) and Multiple (2+) Continuations in Application, 1987–1999.

expectations, "long chain" continuations are less prevalent in software patenting than in biotechnology patents. These results generally support the product-life cycle hypothesis that predicts greater use of continuations in biotechnology than in software. Nevertheless, the increase during 1987–1999 in "long chain" continuations for biotechnology patents is intriguing and puzzling.

We probed the use of multiple continuations further in an analysis of two groups in each of our technology sectors, creating samples of patents issuing within three years of their final administrative application date for 1994 (the year prior to the regime change) and 1996 (the year after the regime change). The share of software patents issuing with a continuation-application lineage declined from 41% within our sample of 1994 applications to 28% for the 1996 applications. In biotechnology patenting, however, the pattern was reversed: The share of biotechnology patents (defined as patents issued to the top 100 biotechnology firms) issuing with a continuation-application lineage increased from 57% for 1994 applications to 64% for 1996 applications.

The effects of continuations on average pendency within the pre- and post-1995 samples of software and biotechnology patents, however, change in complex ways. Average pendency for patents in the two software-patent samples differs significantly,[35] with mean pendency for a continued software patent declining from 64 months for our 1994 patents to 59 weeks for our 1996 software patents. But mean pendency for a non-continued software patent increases from 28 weeks in our 1994 sample to 30 weeks for our 1996 sample – the average additional pendency delay associated with a continuation therefore declines. Mean pendency in biotechnology patents increases from 25 months (1994 issues) to 30 months (1996 issues) for non-continued biotechnology patents, and from 62 months (1994 issues) to 66 months (for 1996 issues) for continued patents in this sector. The average contribution to pendency of continuations in software patents thus declined by more (from 36 to 29 months) than was the case in our biotechnology patent sample (from 37 to 36 months). This change is consistent with the speculative hypothesis that the 1995 changes in patent policy discouraged continuations in software patenting more substantially than was true in biotechnology. But additional analysis is necessary to examine this hypothesis more thoroughly.

4.5. The "Importance" and "Originality" of Continuation Software Patents: Citations and Litigation Data

Johnson and Popp (2001) analyzed the relationship between patent "importance" (measured as the number of citations to a patent by subsequent patents) and the length of time elapsed between the application and issue of the patent, concluding that longer-pendency patents are more heavily cited. Grant lags of up to five years are associated with significantly higher levels of citation, while the greatest marginal effects in citations occur for an increase in pendency from two to three years. While Johnson and Popp find a relationship between the existence of a "related" (continued) application and forward citation count, their aggregated technology-class controls do not allow us to examine the same relationship in software patenting. Accordingly, we compared the "importance" of software patents with continuations against that of software patents with no continuations, utilizing two measures of importance: forward citations, and subsequent litigation.

The number of forward citations has been widely used as a measure of patent importance, supported by findings that the number of forward citations is a significant predictor of the market value of an innovation (Hall, Jaffe & Trajtenberg, 2000). Lemley suggests that costly patent litigation is more likely for economically valuable and/or "important" patents (2001), and studies have demonstrated that the likelihood of litigation is significantly correlated with increasing incidence of forward citations (Lanjouw & Schankerman, 1997).

For patents issuing during the years prior to the regime change (1987–1994), we find a significant difference[36] between the number of forward citations to patents that had no continuation application in their lineage (0.92) and citations to patents using the continuation application (0.97).[37] This difference is driven by the forward citations to patents with a string of less than five continuations in their application lineage: patents with five or more continuation applications attracted significantly fewer forward citations (0.59) than those with no continuations.

We cannot compute comparable citation statistics for the post-1995 patents because of truncation in our patent sample. But we examined the effects of the 1995 changes in patent law on the continuations-citations relationship by analyzing forward citation counts for software patents with final application dates of 1994 and 1996 issuing within three years, computing a ratio of forward citations received by these patents to those received by all patents issued in the same year.[38] Non-continued software patents applied for in 1994 were cited on average 2.5 times more intensively than all patents issuing in the same year, a ratio that exceeds that for continued software patents applied for in the same year (2.2). When we compare continued and non-continued software patents applied for in 1996, however, the "relative citation ratio" for continued software patents (2.4) slightly exceeds that for those without continuations (2.1). The results for the 1996 application-year patents thus are broadly consistent with those for the larger sample of pre-1995 patents for which a more detailed citations analysis is feasible – before and after the 1995 changes in U.S. patent law, continuations are associated with higher levels of forward citations to patents.

Our analysis of citations also examined the "originality" of continued and non-continued software patents. The originality measure was first proposed by Trajtenberg et al. (1997) as a proxy for the technological scope of a patent. The measure is based on a Herfindahl concentration index of the technology classes comprising a patent's backward citations, providing an index of the "breadth" of the fields of knowledge on which a patent draws (Hall et al., 2001).[39] Patents with higher "originality scores" draw on a broader swath of existing technological fields, and therefore may be more novel technological advances. Analysis of software patenting during the 1987–1994 periods reveals significant differences in the originality score for patents with and without continuations. Although non-continued software patents show a mean originality measure of 0.44, continued patents exhibit a significantly higher mean score (0.50).[40] These statistically significant differences hold throughout all years in our sample, and thus are not driven by change over time in citation practices. Moreover, patents with longer continuation histories tend to have higher "originality" scores. The mean originality scores for a patent with one, two, and three or greater continuations in its application lineage are 0.50, 0.53, and 0.55, respectively.[41]

These results suggest that patents with longer continuation histories draw on a wider array of technologies, a characterization that is broadly consistent with higher levels of both "newness" and complexity, this assertion must be qualified. It is possible, for example that the more prolonged and repeated reviews of continued patent applications by the applicant, the applicant's agent, and the patent examiner may lead to the revelation of additional "prior art." If this additional patent-based prior art comes from broader technology fields, a larger originality score would result. Our analysis does not unable us to reject either explanation, and we leave this question for future research.

4.6. The Probability of Litigation for Software and Biotechnology Patents with Continuations

Finally, we investigated litigation rates for continued software and biotechnology patents. As we noted earlier, continuations lengthen the lags associated with patent issue considerably, and contribute to the finding of Johnson and Popp (2001) that longer pendency periods are positively correlated with patent "importance." Inasmuch as litigation rates are another measure of patent importance, how does continuation affect the probability of litigation for biotechnology patents?

Figure 10 plots the litigation rates (as the share of suits being filed through December 2000) for patents by issue year for 1987–1999 for software patents with and without continuation applications in their lineage, and all other non-software patents. Because of the limitations of our data and the lag in suit filings, our plots show considerable right censoring. The available data demonstrate that litigation is much more common for software patents than in non-software patent classes, regardless of continuation status. Litigation rates approach 40 per 1000 patents among non-continued software patents for patents issued in 1991, well above a peak of less than 10 per 1000 patents in all other technologies in the issue years 1992–1993. The litigation rate for continued software patents is even higher, peaking at more than 70 suits per 1000 patents in the 1991 issue year, and dominating the litigation rate for non-continued software patents for patents

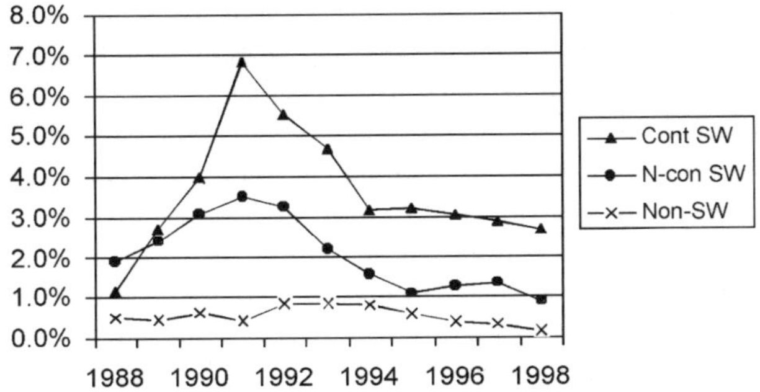

Fig. 10. Litigation Rates for Assigned Patents, by Issue Year, Comparing Continued Software Patents, Non-Continued Software Patents, and All Non-Software Patents, 1987–1999 (3-Year Moving Average).

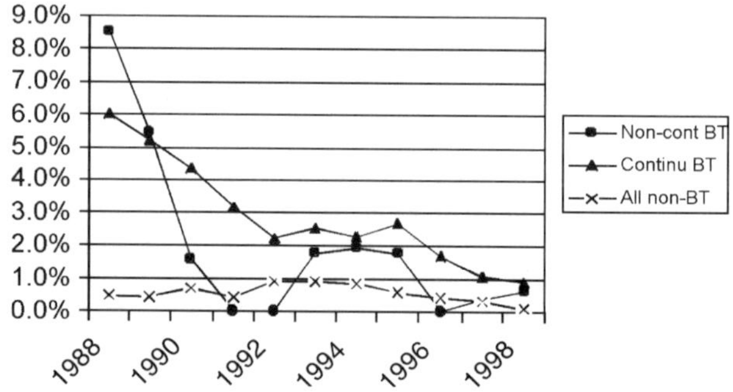

Fig. 11. Litigation Rates for Assigned Patents, by Issue Year, Comparing Continued Biotechnology Patents, Non-Continued Biotechnology Patents, and All Non-Biotechnology Patents, 1987–1999 (3-Year Moving Average).

issued 1990–1999. These findings suggest that software patents with a continuation lineage are particularly prone to litigation.

We plotted the litigation rates for patents by issue year, 1987–1999, for biotechnology patents with and without continuation applications in their lineage, and compared these rates with those for all non-biotechnology patents (Fig. 11). Litigation rates are much higher for biotechnology patents than non-biotechnology patents regardless of continuation status. Litigation rates approach 90 per 1000 patents among non-continued biotechnology patents in 1991, well above a peak of less than 10 per 1000 patents in all other technologies in 1992–1993. The litigation rate for continued biotechnology patents is higher than that for non-continued biotechnology patents for all but the earliest years of our sample, peaking at 60 suits per 1000 patents in 1987. Moreover, the litigation rate for continued biotechnology patents exceeds that for continued software patents throughout the 1990–1999 period. These findings suggest that biotechnology patents with a continuation lineage are particularly prone to litigation, exhibiting a higher litigation rate than continued software patents.

Overall, the results of the analysis of litigation data are consistent with those for patent citations. But we cannot exclude the possibility that continued patents are more likely to be the subject of litigation simply because of the widespread use by "submarine" patentholders of litigation, rather than because of some intrinsically greater technological or economic importance of these patents. Furthermore, insofar as litigation is often is triggered by the divergent expectations of the parties regarding the value of the suit (Priest & Klein, 1984), increased continuation practice by the patentee may reflect an uncertainty in the patentability or value of the underlying technology that is not resolved in the patent prosecution process, but instead carries over into the post-issue validity determination between competitors in the courts.

5. CONCLUSION

In spite of the internationally unique status of continuations in the U.S. patent system and the occasional use of continuations for strategic purposes, continuations have received virtually no attention in scholarly research on patent policy and strategy. This paper has presented the results of an exploratory analysis of the use of continuations in U.S. software and biotechnology patenting during the 1980s and 1990s, a period of expanded patenting in both fields of technology. We find that continuations grew rapidly as a fraction of total software and biotechnology patenting during the early 1990s, a period characterized by rapid expansion in patenting in both of these fields. These parallel trends are consistent with the use by patent applicants of continuations as a response to uncertainty over the standards employed by examiners, perhaps as a result of the rapid growth of patenting in a field with limited patent-based prior art. Just as with overall trends in patenting and in patent propensities, we observe considerable firm-specific differences in the use of continuations within software and biotechnology patenting during the 1990s. Large biotechnology firms use continuations more intensively in their patenting than do smaller firms, which may reflect more extensive defensive patenting by these firms in the biotechnology sector.

This analysis also suggests that the changes in U.S. patent law that became effective in 1995 have reversed the rapid growth in continuations in software and patenting overall, but have had much less effect on the incidence of continuations in biotechnology. This reversal in the growth trend for continuations suggests that a substantial portion of the expanding use of continuations in the early 1990s in software and overall U.S. patenting was driven by various forms of "submarine" patent strategies. Inasmuch as these legislative changes were justified in part as a way to discourage submarine patenting, they appear to have been successful. Nonetheless, the reversal in the growth of continuations has hardly eliminated continuations, instead reducing their share in both software and non-software patents to levels observed in the late 1980s. In addition, in biotechnology, where patents potentially retain great value through their full 20-year terms after the 1995 policy changes, continuations remain important.

Since we lack data on patents issuing after the 1999 "publication requirements" became effective, we are not able to evaluate the effects on software or biotechnology of these policies. It seems plausible, however, that the 1999 publication requirement is not likely to reduce continuation rates in biotechnology to the levels observed by 1999 in software or in overall patenting, for reasons noted above.

Continuations contribute significantly to the length of patent "grant lags," although the marginal contribution of additional continuations to these lags declines. Trends in the issue to packaged-software and biotechnology firms of multiple-continuation patents before and after 1995 suggest that these multiple-continuation patents may be the primary locus of submarine patenting strategies. Although the 1995 changes in patent law appear to have discouraged such strategies in software, there is little evidence of such effects in biotechnology patenting. We find that software patents applied for before 1995 that issue with continuations are cited modestly but significantly more intensively than software patents with no continuations. Data on post-1995 applications with continuations reveals a similar difference. Software and biotechnology patents with continuations also are much more likely to be the focus of litigation, although this effect is stronger for biotechnology patents.

These results are intriguing but do not support strong conclusions about the motives for and effects of the rapid growth in patenting within the U.S. computer software and biotechnology industries during the 1980s and 1990s. There is some evidence that continuations were used by packaged-software firms more intensively than by patent applicants in other fields during the pre-1995 period, but continuations in software patenting decline after 1995. In biotechnology, by contrast, continuations remain significant after 1995. We are not able to determine whether these longitudinal and sector-specific differences in the use of continuations reflect asymmetries in information or expectations between applicants and patent examiners, rather than the conscious use of secrecy-lengthening "submarine" patent strategies. Our evidence on litigation and citations suggests that patents subject to continuations also tend to be more "important," but this finding is not dispositive with respect to an interpretation of the causes of continuations in patenting.

The use of continuations in patenting has received surprisingly little scholarly attention within the rapidly expanding literature on patenting strategy and policy. The results of our exploratory analysis of patenting in software and biotechnology suggest significant sector-specific differences and suggest further that the effects of recent changes in U.S. policy on patent terms differ considerably among these sectors. In future research, we plan to extend our analysis to cover more sectors and to probe more deeply into the reasons for these longitudinal and sector-specific differences in continuation behavior.

NOTES

1. When these inventions were covered by patents applied for after the priority date established by the original "submarine" application, the owner of the original patent might pursue infringement actions against other late-filing patentholders.

2. 450 U.S. 175 (1981). Prior to the *Diamond v. Diehr* decision, software was considered unpatentable in U.S. law (see *Gottschalk v. Benson*, 409 U.S. 63 (1972)).

3. 93–0413 (S. D. Cal. decided Feb. 23, 1994).

4. Lacking a clear a priori definition of "software-related" patent classes, we focused on the following eleven main groups in the International Patent Class classification scheme:

G06F Electric Digital Data Processing:

3/ Input arrangements for transferring data to be processed...
5/ Methods or arrangements for data conversion...
7/ Methods or arrangements for processing data...
9/ Arrangements for programme control...
11/ Error detection; Error correction; Monitoring...
12/ Accessing, addressing or allocating within memory systems or architectures...
13/ Interconnection of, or transfer of information or other signals...
15/ Digital computers in general...

G06K Recognition of Data

9/ Methods or arrangements for reading or recognising printed or written characters...
15/ Arrangements for producing a permanent visual presentation of the output data

H04L Electric Communication Technique

9/ Arrangements for secret or secure communication

Since they exist throughout the 1984–1997 period, these patent classes provide a useful basis for examining time trends in U.S. software patenting. They do not cover all software patents, but they do provide imperfect but reliable longitudinal coverage of a segment of the overall software industry. The share of U.S. patents accounted for by patents in these IPC groups more than doubled during 1987–1997, from 1.7% of all patents in 1987 to 3.8% of all patents in 1997. See Graham and Mowery (2003).

5. Graham and Mowery (2003) analyzed trends in citations to the software patents obtained by our sample of packaged-software specialists and electronic-system firms, focusing on the possibility that the expansion in patenting might have been associated with the issue of patents that obtained fewer "forward citations," relative to other software patents. We found no evidence of a decline in the relative intensity with which these firms' software patents were cited by other software patents in the 1987–1997 period. This evidence must be treated with caution – it is possible, for example, that "forward citations" to all software patents are declining during this period, perhaps in reaction to the lessened rigor of USPTO of all applications in this area, and the patents assigned to these firms simply declined by a smaller amount.

6. 35 USC § 120 (1991). Benefit of earlier filing date in the United States.

7. The continuation and continuation-in-part are authorized under 35 USC § 120 "Benefit of earlier filing date in the United States." Divisions are authorized under 35 USC § 121 "Divisional applications."

8. The fee required to request a continuing application is no higher than that for filing a first application – $710 in 2001, and $355 for small entities. 37 C. F. R. 17(e).

9. Continuation applications have been available to patentees in the United States at least since 1863. In *Godfrey v. Eames*, 68 U.S. 317 (1863), the U.S. Supreme Court interpreted the Patent Act of 1836 to allow continuation applications, in that case when the original application was abandoned on the same day that the new continued application was filed.

The use of continuing applications was also upheld by the Supreme Court in *Crown Cork & Seal Co. v. Gutmann Co.*, 304 U.S. 159 (1938) and *General Talking Pictures Corp. v. Western Electric Co.*, 304 U.S. 175 (1938).

10. The application must meet certain conditions. Under 35 USC § 120, a patent application is entitled to adopt the filing date of a "parent" application when: (1) both applications disclose the same invention; (2) both applications are filed by the same inventor; (3) both applications are simultaneously co-pending; (4) the earlier application meets the disclosure requirements of 35 USC § 112; and (5) the later application contains a specific reference to the earlier application. *Sampson v. Ampex Corp.* (1971, DC NY), 333 F. Supp. 59, aff'd. (2nd Cir. NY) 463 F2d 1042. The language of 35 USC § 120 was initially adopted in the Patent Act of 1952, and legislative history suggests that the section was intended to write existing legal practice on the issue into the statute. *In re Henriksen*, 55 C. C. P. A. 1384 (1968).

11. The use here of "continuation" includes also the "continuation-in-part" process by which particular claims or groups of claims can be continued instead of the entire application and all disclosures therein. Because the use of the continuation-in-part process to shroud certain critical claims under a veil of secrecy is, in effect, virtually identical to secreting the entire application, both are considered "continuations" for this analysis. Divisions also allow applicants to manage the issue date of patents, but will not be considered "continuations" for this analysis.

12. See, for instance, Jerome Lemelson's U.S. patent 5,283,641, Apparatus and methods for automated analysis, issued February 1, 1994. The priority date of this patent was December 24, 1954, and there were no fewer than eleven continuations and divisions in this patent's chain while it languished in the patent office for 39 years.

13. There is a form of "division" available through the European Patent Office, although with the 18-month publication requirement having been a part of the European Patent Convention since its inception, the opportunity for inventors to "hide" technology advances has been strictly limited.

14. The U.S. patent law now requires the publication of all applications after 18 months, but applicants are granted exemptions from this requirement by filing a statement that they have no intention of filing the application in a foreign jurisdiction that requires an 18-month publication of its patent applications.

15. For example, an applicant may use the continuation procedure to make subtle changes in one or more claims consistent with the original disclosure, but not identical to the original published claims. The continuation may thus delay issue of the patent while the applicant observes the uses to which the technology is put in the marketplace, and alters the claim accordingly. Thus, when the patent ultimately is allowed to issue, the technology disclosed in the patent may be of no surprise to competitors, but the precise boundaries of the claims may be a surprise, thus allowing the patentee to demand royalties from competitors infringing on the altered claim. The 18-month disclosure rule may therefore reduce uncertainty over the existence of a pending patent, but the applicant may, through the use of continuation practice and ongoing negotiations between applicant and examiner, keep the precise boundaries of the patent's claims hidden from the public for an extended time, thus leaving considerable uncertainty unresolved.

16. Rambus Incorporated (Nasdaq: RMBS) styles itself an "intellectual property company": the company's primary activities include the design, development, and licensing of semiconductor-chip connection technologies that allow memory devices to perform at the higher speeds demanded by new generations of processors and controllers. Rambus holds

over one-hundred U.S. and foreign patents, and has licensed its technologies widely within the semiconductor industry.

17. *Kingsdown Medical Consultants and E. R. Squibb & Sons v. Hollister Inc.*, 863 F.2d 867 (CAFC, 1988).

18. *Reiffin v. Microsoft Corporation*, 158 F.Supp.2d 1016 (N. D.Cal., 2001).

19. U.S. Patents 5,694,603 "Computer memory product with preemptive multithreading software" and 5,694,604 "Preemptive multithreading computer system with clock activated interrupt."

20. The Cohen-Boyer invention was followed closely by the Milstein/Kohler technique for producing monoclonal antibodies at the University of Cambridge in 1975, which was not patented.

21. Compare, for instance, *Park-Davis & Co. v. H. K. Mulford & Co.*, 196 F. 496 (2d Cir. 1912)(declaring that a process to produce purified adrenaline was patentable) with *In re King*, 107 F.2d 618 (CCPA, 1939)(declaring that a process to produce purified lemon juice was not patentable).

22. 333 U.S. 127.

23. 447 U.S. 303.

24. *Ex Parte Hibberd*, 227 USPQ 443 (USPTO Bd. of Appeals and Interferences, 1985).

25. *Ex Parte Allen*, 2 USPQ2d 1425 (USPTO Bd. of Appeals and Interferences, 1987).

26. 1077 Official Gazette 24 (April 21, 1987).

27. Rural Advancement Foundation International (2001). "Monsanto's 'Submarine Patent' Torpedoes Ag Biotech," Press Release, April 27.

28. Defined in Note 3, *supra*.

29. These data thus represent only the terminal patent, the patent that emerges at the end of the chain of continuations. Because some applications may have been abandoned, this terminal patent may be the only evidence of a continuation chain. The data in Fig. 1 are based upon issue date rather than upon application date because under continuation practice, the recorded application date for what is essentially the final application in a chain of continuations is a somewhat artificial measure. Issue dates are a more meaningful metric of the use by patentees of the procedure because the issue date is one over which the applicant exercises somewhat more control than in the "normal" single-application patent case, by virtue of their use of the continuation procedure. We note that there will always be a certain amount of uncertainty about the date of issue, although the patentee can minimize this uncertainty through the negotiation process with the USPTO examiner and the deft use of the continuation process itself.

30. The low rates of continuation in software patenting during the 1987–1992 period may reflect the fact that our measures of continuations necessarily lag trends in patent applications by at least 2–3 years. Continuation trends for the 1987–1992 period therefore reflect patenting practice during the early to mid-1980s, a period of relatively modest rates of patenting in software. But continuations grow as a share of software patenting during the period of rapid growth in such patenting by software and electronic systems firms (Graham & Mowery, 2003).

31. As reported in the Softletter100 for 1997, a trade publication.

32. Borland International was party in the 1980s and 1990s to a series of high-profile court actions against Lotus Development, these two software firms contesting the originality of their spreadsheet programs Quattro Pro and Lotus 1–2–3, respectively.

33. Because patent offices in neither the U.S. nor Europe make any distinction in their classifications between pharmaceutical and biotechnology patents, patent classes are not a meaningful mechanism for identifying biotechnology patents (Harhoff & Reitzig, 2001). Because we lack a clear definition of biotechnology patent classes, we focus on the patenting of large biotechnology firms. Using the top 100 biotechnology-only companies (based on 1997 revenues, as reported by Med Ad News (1998)), we compiled a list of all patents assigned to these 100 firms, and to the following "Top 10" biotechnology firms throughout 1987–1999: Amgen, Chiron, Genentech, Genzyme, Biogen, Nabi, Immunex, Centocor, NexStar, and Scios. Our two samples include, respectively, 5,237 and 1,557 patents.

34. Lanjouw and Schankerman (1997) use international patent classes (A-) and (C-) in their definition of "drugs" patents. We add international patent class (B-) to capture process patents as well.

35. At the 99% level of confidence, using a standard paired difference statistical test.

36. At the 95% level of confidence, using a standard paired difference statistical test.

37. These forward citation counts are calculated using a 2-year forward window, excluding self-citations.

38. The construction of our sample of "biotechnology patents," which is based on corporate patent portfolios rather than on patent classes, makes it impossible for us to replicate this citation analysis for the biotechnology patents.

39. The measure is computed as follows: Originality $= 1 = \sum_{j}^{n_i} s_{ij}^2$ where s_i is the percentage of citations made by the patent i that belong to patent class j from n_i patent classes (Hall et al., 2001).

40. At the 99% level of confidence, using a standard paired difference statistical test.

41. Again, significant at the 99% level of confidence, using a standard paired difference statistical test.

ACKNOWLEDGMENTS

An earlier version of this chapter was presented at the 2002 Summer DRUID conference, Copenhagen, Denmark, June 6–8, 2002. We appreciate the comments of conference participants on that draft. Research for this paper was supported by the National Research Council, the Andrew W. Mellon Foundation, and the Alfred P. Sloan Foundation.

REFERENCES

Aharonian, G. (2000). Worst software engineering patent of 1999. *Patent News* (January).

Argyres, N., & Porter Liebeskind, J. (2002). Governance inseparability and the governance of U.S. biotechnology industry. *Journal of Economic Behavior and Organization*, *47*, 197–219.

Audretsch, D. (2001). The role of small firms in U.S. biotechnology clusters. *Small Business Economics*, *17*, 3–15.

Barton, J. (1995). Adapting the intellectual property system to new technologies. *International Journal of Technology Management*, *2/3*, 151–172.

Glazier, S. (1997). *Patent strategies for business*. 3d. LBI Institute, Washington, DC.

Graham, S., Hall, B. H., Harhoff, D., & Mowery, D. C. (2002). Patent quality control: A comparison of U.S. patent re-examinations and European patent oppositions. In: W. Cohen & S. Merrill (Eds), *Patents in the Knowledge-Based Economy*. Washington, DC: Board on Science, Technology and Economic Policy (STEP), The National Academies.

Graham, S., & Mowery, D. C. (2003). Intellectual property protection in the U.S. software industry. In: W. Cohen & S. Merrill (Eds), *Patents in the Knowledge-Based Economy*. Washington, DC: Board on Science, Technology and Economic Policy (STEP), The National Academies.

Hall, B., Jaffe, A., & Trajtenberg, M. (2000). Market value and patent citations: A first look. NBER Working Paper No. 7741.

Hall, B., Jaffe, A., & Trajtenberg, M. (2001). The NBER patent citations data file: Lessons, insights, and methodological tools. NBER Working Paper No. 8498.

Hall, B., & Ziedonis, R. H. (2001). The determinants of patenting in the U.S. semiconductor industry, 1980–1994. *Rand Journal of Economics*, *32*, 101–128.

Harhoff, D., & Reitzig, M. (2001). Determinants of opposition against EPO grants: The case of biotechnology and pharmaceuticals. Working Paper, Ludwig-Maximilians-Universitat, Munich.

Johnson, D. K. N., & Popp, D. (2001). Forced out of the closet: The impact of the American Inventors Protection Act of 1999 on the timing of patent disclosure. NBER Working Paper No. 8374.

Lanjouw, J., & Schankerman, M. (1997). Stylized facts of patent litigation: Value, scope and ownership. NBER Working Paper No. 6267 (December).

Lemley, M. (2001). Rational ignorance at the patent office. *Northwestern University Law Review*, *95*(4).

Lerner, J. (1995). Patenting in the shadow of competitors. *Journal of Law and Economics*, *38*, 463–495.

Med Ad News (1998). Top 100 biotechnology companies. Engel Publishing Partners, 17 (July), 31–35.

Mowery, D. C. (1999). The computer software industry. In: D. C. Mowery & R. R. Nelson (Eds), *The Sources of Industrial Leadership*. New York: Cambridge University Press.

Priest, G. L., & Klein, B. (1984). The selection of disputes for litigation. *Journal of Legal Studies*, *13*, 1–55.

Quillen, C., & Webster, O. (2001). Continuing patent applications and performance of the United States patent office. *Federal Circuit Bar Journal* (August).

Sharp, M. (1999). The science of nations: European multinationals and American biotechnology. *Biotechnology*, *1*, 132–162.

Trajtenberg, M., Jaffe, A., & Henderson, R. (1997). University vs. corporate patents: A window on the basicness of invention. *Economics of Innovation and New Technology*, *5*(1).

Wright, J. (1997). Implications of recent patent law changes on biotechnology research and the biotechnology industry. *Virginia Journal of Law and Technology*, *1*, 2–11.

FIRM STRATEGIES AND TRENDS IN PATENT LITIGATION IN THE UNITED STATES☆

Deepak Somaya

ABSTRACT

Patent litigation has been rising rapidly in the United States since the mid-1980s, and particularly so in high-technology industries. The strategies pursued by firms with their patents have a significant influence on their decisions to file suit, and on the outcomes within litigation. The influence of strategic motivations on settlement outcomes is studied in two illustratively different industries – computers and research medicines. Evidence is found for two types of influences – the use of patents (as isolating mechanisms) to protect valuable strategic stakes, and their "defensive" role in obtaining access to external technologies (through mutual hold-up).

So here's my fantasy: We take all our patents and trademarks out in back, and we have a weenie roast. I watch the smoke twirl up in the sky, and I say, "Adios!" Then I go back in and write checks to pay all my legal bills. And when someone wants to play legal patty-ake [sic], we just say "peace" and go figure out how to get along. In my dreams.[1]

☆ This is an extended version of a paper published earlier in the *Strategic Management Journal*, 2003, 24(1): 17–38, under the title "Strategic Determinants of Decisions not to Settle Patent Litigation."

Intellectual Property and Entrepreneurship
Advances in the Study of Entrepreneurship, Innovation and Economic Growth,
Volume 15, 103–147

ISSN: 1048-4736/doi:10.1016/S1048-4736(04)01504-8

1. INTRODUCTION

Patent strategy is increasingly becoming an important part of corporate strategy in the knowledge economy, especially for technology firms that rely on intellectual property rights for their competitive advantage (Rivette & Kline, 2000). While firms may outsource many legal services for their patents, the strategic importance of patents means that their managers (like the CEO quoted above) must understand and exercise their legal options diligently. Understanding how managers make such tradeoffs about their patents is of central interest in this paper, specifically in the context of patent litigation. Recent years have seen a growing scholarly interest in patent litigation to unpack the role patents play in firm strategy and the economy. Prior research has studied firm valuation (Bhagat, Brickley & Coles, 1994; Lerner, 1995), predatory behavior (Lanjouw & Lerner, 2001; Lerner, 1995), and suit filing decisions (Lanjouw & Lerner, 1998; Lanjouw & Schankerman, 1997, 1998) in the context of patent litigation.

Patent enforcement entails substantial direct and indirect costs. Direct legal costs alone can run in the range of $1.0–3.0 Million (in 1997 dollars) for each side through trial (AIPLA, 1997). In addition, patent litigation also involves considerable organizational dislocation, absorbing the time and energy of key managers, lawyers, engineers, and scientists in the company. Not surprisingly, prior research has found that the filing of patent suits led to a 2–3.1% average decrease in the market value of the firms involved (Bhagat, Brickley & Coles, 1994; Lerner, 1995). This paper investigates systematic reasons for non-settlement, which can explain why firms are willing to continue fighting patent suits despite these costs.

In unstructured exploratory interviews, patent lawyers sometimes suggested that (arguably) irrational behavior plays a large part in litigation outcomes – egos get involved, poor legal advice is followed, or the parties lose perspective in the "brutality" of the lawsuit. However, in many instances, more systematic explanations were cited as important factors – namely, a patentee's desire to "take out a competitor" in a key market, to build a tough reputation, to protect its "crown jewels," or to extract royalties at a higher rate than was otherwise "affordable" by the non-patentee. Moreover, other factors were suggested as drivers of suit settlement – patentees may be forced to compromise in order to obtain access to complementary patents, or settlement may be induced by on-going business relations between the parties.

This paper investigates the strategic dimension of patent enforcement suggested by these systematic explanations, and suggests two main drivers of non-settlement in patent suits. First, patent suits are less likely to be settled if patentees, or their opponents, have strong strategic stakes in the litigated patent. This explanation rests on an understanding of patents as isolating mechanisms, i.e. mechanisms that prevent rivals from easily replicating valuable firm-specific assets deployed in the pursuit of competitive advantage (Lippman & Rumelt, 1982; Rumelt, 1984). Second, in systems-products industries, which combine multiple inventions in end-products, mutually blocking patent rights can encourage settlement in patent litigation due to the threat of counter-suits by rivals using their own patents. A key contribution of this paper is the linking of these strategies with formal law and economics theories of suit settlement.

Firm strategy in the patent area is still not well understood, and survey research suggests that patent strategies vary across different industries (Cohen, Nelson & Walsh, 2000; Levin, Klevorick, Nelson & Winter, 1987). This paper presents some broad distinctions between the features of patent litigation in different industries, and investigates the differences in depth in Computers and Research Medicines (Pharmaceuticals plus Medical Biotechnology). Prior research suggests that strategic stakes are likely to be more important in research medicines, and mutual hold-up and counter-suing strategies in computers.

The rest of this paper is organized as follows. Section 2 describes broad trends in patent litigation, and examines differences between industries. Section 3 develops a set of hypotheses relating strategic stakes and mutual hold-up to settlement decisions in patent litigation. Section 4 describes the two-industry sub-sample, and discusses observed litigation differences in computers and research medicines. Section 5 discusses empirical methods, variables, and sample statistics. Section 6 presents and discusses the main empirical findings of the paper. Section 7 concludes.

2. TRENDS AND INTER-INDUSTRY DIFFERENCES IN PATENT LITIGATION

When patent disputes occur, they are taken in the first instance to a federal district court that has jurisdiction over the dispute. Similarly, copyrights and trademarks, which are also protected by federal law, fall under the purview of the federal court system. Appeals against decisions made by the district courts in trademark and copyright disputes are heard by the Appeals Courts of the respective circuits into which federal districts are grouped. However, since October 1982, appeals in patent cases have been consolidated into the newly created Court of Appeals of the Federal Circuit. In addition, this court was given jurisdiction over some trademark appeals that arise out of decisions made by the Patent and Trademark Office, but not over other trademark disputes. Throughout this paper, the focus is on suits at the district court level.

2.1. Patent Litigation: Historical Trends and Technology Profile

Merz and Pace (1994) have previously reported the historical trend in patent litigation from 1970 to 1989. They uncovered a rapid growth in patent suits filed per year since the creation of the Federal Circuit, which some scholars have argued was responsible for a more "patent friendly" environment in the U.S. (Cohen, Nelson & Walsh, 2000; Kortum & Lerner, 1999; Merges & Nelson, 1990; Scotchmer, 1991). Figure 1 confirms their findings, and further extends the timeline to 2000. In addition, it also describes the corresponding trends in copyright and trademark suits, and patenting, over the same period (1970–2000). The litigation trends are based on data collected by the Federal Judicial Center (FJC) from the federal district courts since 1970 through summary data sheets that are filed by the court clerk.[2] Patent, copyright and trademark cases are identified by the clerk through unique "nature of suit" codes.

Consistent with the trend in patenting (Kortum & Lerner, 1999), patent litigation was virtually stagnant in the 1970s, but nearly tripled since the mid-1980s, recording a marginally faster rate of growth than even patenting itself. Taken in the broad historical perspective, these increases in patent litigation are by no means unique, as both trademark and copyright litigation grew rapidly till the mid-1980s, and there has been another boom in trademark litigation in the 1990s. Copyright suit filings have been in decline since 1994 however, and after 1998, they were lower than patent suit filings for the first time since 1975.

To investigate additional facets of patent litigation, another data source from the U.S. Patent and Trademark Office (USPTO) containing suits filed between 1975 and 1995 was also used.[3] These "USPTO data" have the advantage that they identify the patent(s) involved in each suit, which enables the use of data on patent characteristics from other sources. The USPTO suits were matched to FJC suits using various identifying information available in the two data sets. A small number of USPTO suits (less than 5%) could not be matched to any suit in the FJC data.[4] Among the remaining suits, the share of FJC suits found in the USPTO data varies both over time and across districts, even though the courts are required by law (35 USC § 290) to report all patent cases to the patent office. Figure 2 shows that the reporting was poor in the 1970s, but has stayed above 50% since 1982. Table 1 describes the circuit-wise differences in reporting patent suits, but these differences appear to be a result of clerical laxity rather than systematic bias.

For each suit in the FJC-USPTO matched dataset, the first listed patent (the "focal" patent) was classified into one of six broad U.S. patent classification-based technology groups used in prior work (see Jaffe, Fogarty & Banks, 1998), but with one difference – computer patents were moved into the "electronics, optical

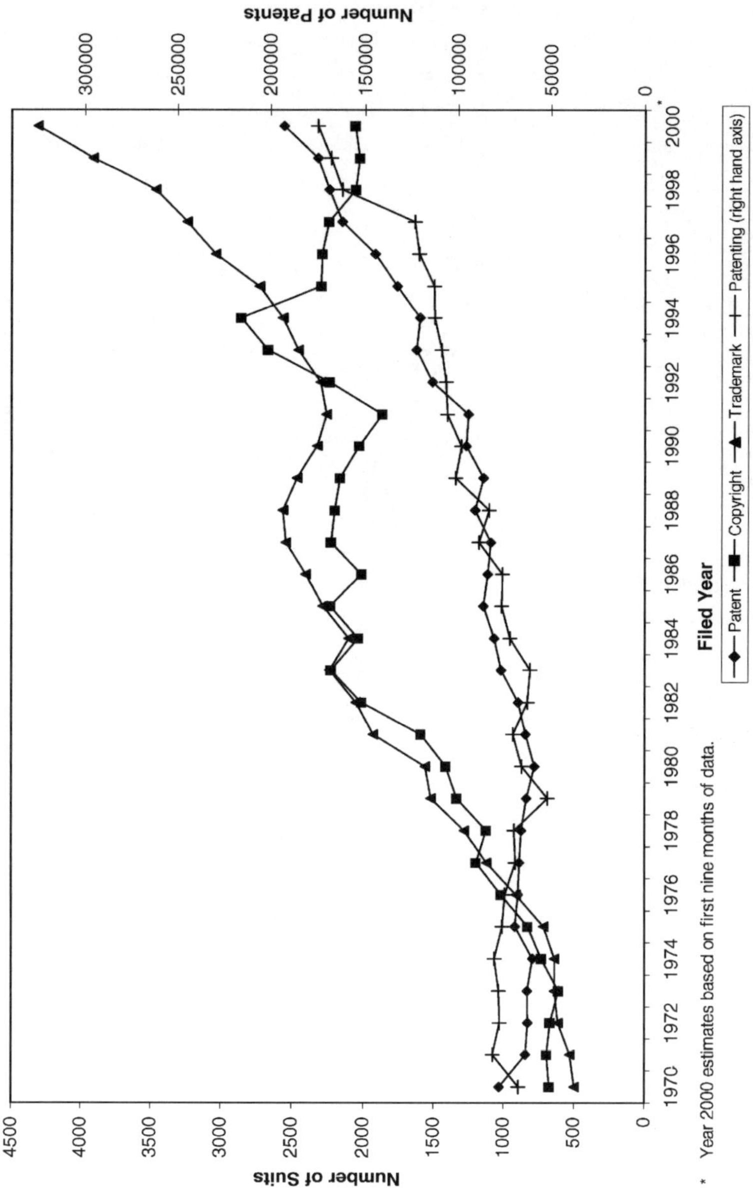

* Year 2000 estimates based on first nine months of data.

Fig. 1. Rates of Intellectual Property Litigation in the United States (1970–2000).

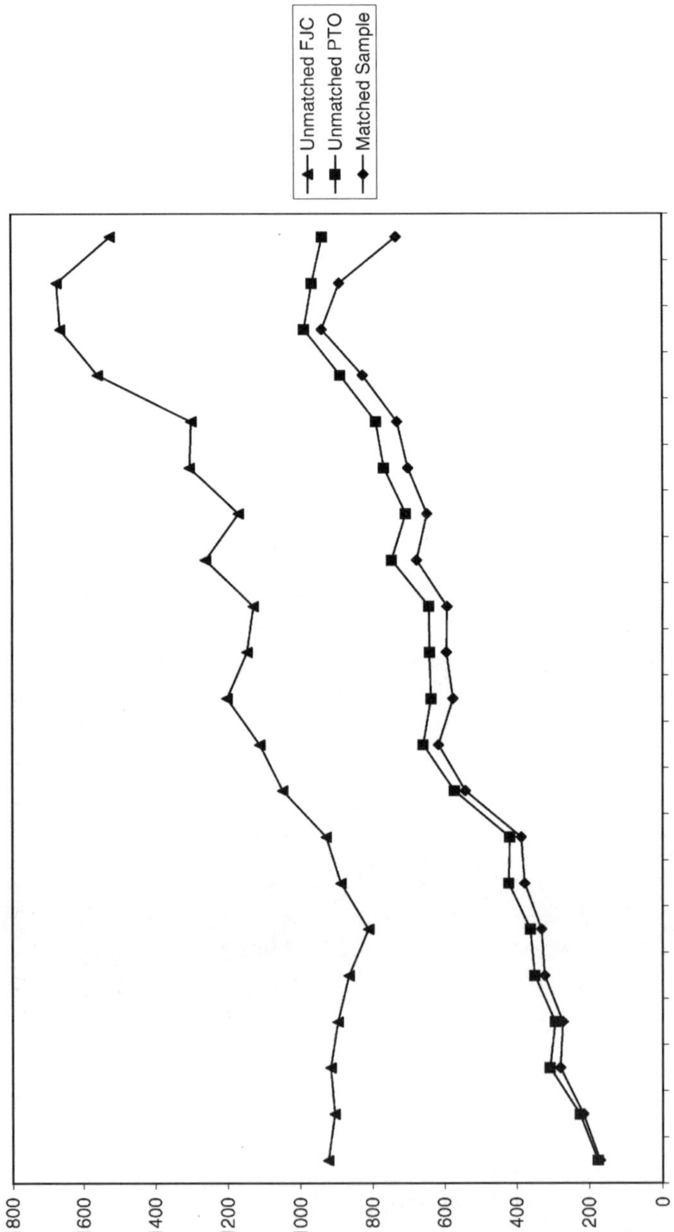

Fig. 2. Comparison Between USPTO and FJC Samples of Patent Litigation (1975–1995).

Table 1. Comparison of USPTO and Federal Judicial Center Data by Major Federal Jurisdictions.

Circuit/District Name	1975–1980		1981–1985		1986–1990		1991–1995	
	USPTO Suits	FJC Suits	USPTO Suits	FJC Suits	USPTO Suits	FJC Suits	USPTO Suits	FJC Suits
DC circuit	9 (7.1%)	127	19 (22.6%)	84	38 (32.5%)	117	40 (41.2%)	97
First circuit	50 (21.5%)	233	34 (16.0%)	212	34 (14.0%)	242	44 (11.7%)	376
Second circuit (rest of)	82 (22.8%)	359	85 (25.9%)	328	155 (40.9%)	379	76 (19.7%)	385
New York southern (dist. 8)	101 (28.3%)	357	197 (60.4%)	326	184 (55.8%)	330	135 (38.9%)	347
Third circuit	206 (38.4%)	537	411 (73.3%)	561	485 (72.7%)	667	525 (69.5%)	755
Fourth circuit	69 (22.5%)	306	112 (46.7%)	240	148 (55.8%)	265	263 (54.3%)	484
Fifth circuit	120 (33.2%)	361	209 (46.5%)	449	256 (53.1%)	482	349 (53.9%)	648
Sixth circuit	153 (29.1%)	526	264 (53.8%)	491	363 (67.6%)	537	432 (63.3%)	683
Seventh circuit (rest of)	65 (30.0%)	217	55 (31.1%)	177	134 (62.0%)	216	134 (42.8%)	313
Illinois northern (dist. 52)	206 (41.6%)	495	272 (68.9%)	395	320 (77.1%)	415	336 (71.8%)	468
Eighth circuit	87 (38.2%)	228	138 (55.6%)	248	164 (51.3%)	320	307 (69.9%)	439
Ninth circuit (rest of)	75 (31.4%)	239	109 (44.0%)	248	197 (51.0%)	386	333 (53.0%)	628
California central (dist. 71)	71 (37.8%)	188	157 (73.4%)	214	214 (67.5%)	317	297 (61.5%)	483
California northern (dist. 73)	170 (34.1%)	499	241 (47.8%)	504	321(51.7%)	621	498 (64.8%)	768
Tenth circuit	52 (26.3%)	198	85 (49.1%)	173	124 (69.3%)	179	250 (74.0%)	338
Eleventh circuit	105 (32.2%)	326	172 (52.6%)	327	221 (61.0%)	362	347 (65.5%)	530
Total	1621 (31.2%)	5196	2560 (51.4%)	4977	3358 (57.5%)	5835	4366 (56.4%)	7742

and nuclear" group from the "mechanical" group. The share of U.S. patent litigation in each technology group was estimated from the data by assuming that the USPTO sampling error in each "time period-jurisdiction group" is unbiased. For this, the matched data was divided into four roughly equal time periods, and fifteen jurisdictions comprising the four largest districts for patent litigation and the eleven federal circuits (excluding those four districts).

The technology-group trends for patent litigation and patenting (Fig. 3) show that litigation in design and chemical patents has been roughly constant over time, even though the numbers of patents issued have increased. By contrast, both patenting and patent litigation in two technology groups – drugs and medical technologies (DMT), and electronics, computers, optical and nuclear (ECON) – has grown very rapidly. Prior research has reported the increases in patenting in these high-technology areas (Hicks, Breitzman, Olivastro & Hamilton, 2001), but not the corresponding increases in patent litigation.

The overall growth trend in the ECON and DMT technology groups has meant that their shares of patenting and litigation increased substantially in later periods (Fig. 4). While suit filing and patenting rates also increased in the mechanical and "other" technology groups, both experienced decreases in their shares. The "other" group is somewhat unique in that its litigation share is much larger than (nearly twice) its patenting share. This may reflect the larger share of patent classes relating to a specific industry or application in this group (an artifact of the U.S. patent classification system).

2.2. Characteristics of Patent Litigation by Technology Class

Outcomes in patent suits are reported (where available) for all patent suits in the FJC data by five-year intervals by suit closing date in Table 2. Using the closing date rather than the filing date is appealing because the latter would imply making difficult inferences about suits – mostly filed in later years – that are still pending. Over time, there appears to be a secular trend towards resolution of patent suits before the later procedural stages of suit (motions, conference, and trial), and by dismissal or settlement rather than judgment in court. Correspondingly, judgments (which do not always dispose of the suit) have been entered for a smaller percentage of suits over time, and the mean duration (or length) of suits in court has also declined.[5] These trends are consistent with each other – earlier procedural resolution would imply a lower probability of disposition by (or occurrence of) a judgment, and also shorter suit durations. However, it is not clear what explains them. While similar but somewhat muted changes are also visible in trademark suits, that is not the case in copyright (not reported). Thus, it is likely that factors

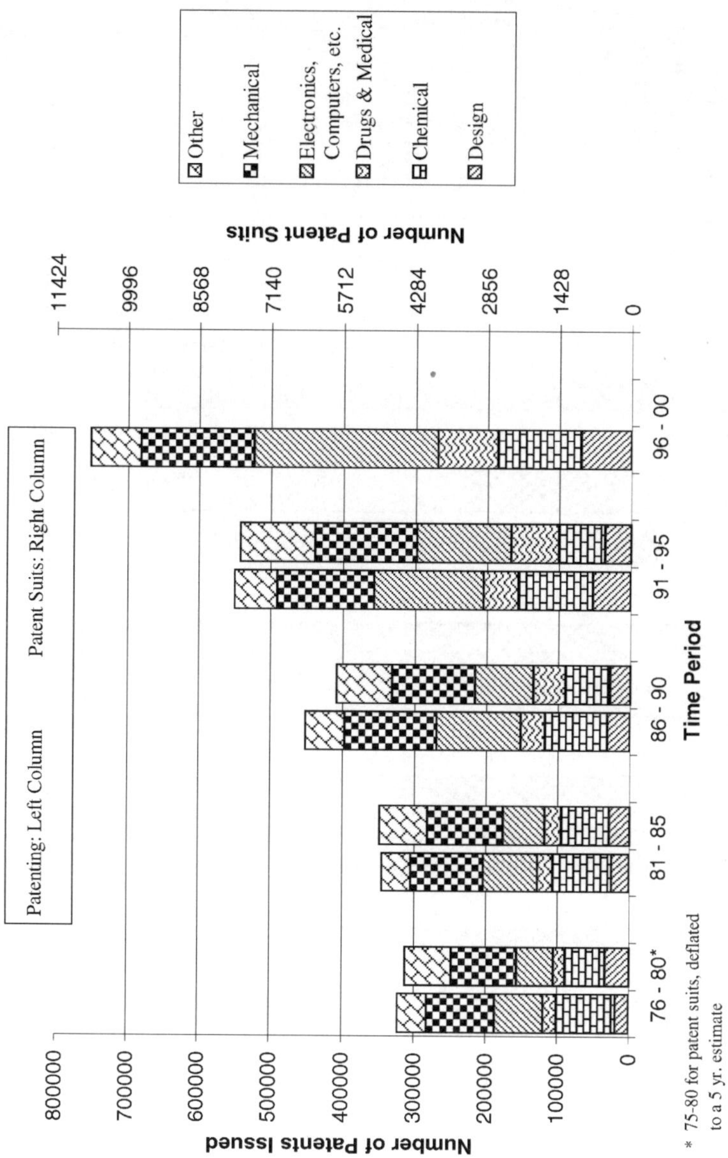

Fig. 3. Levels of U.S. Patenting and Patent Litigation by Technology Group (1976–2000).

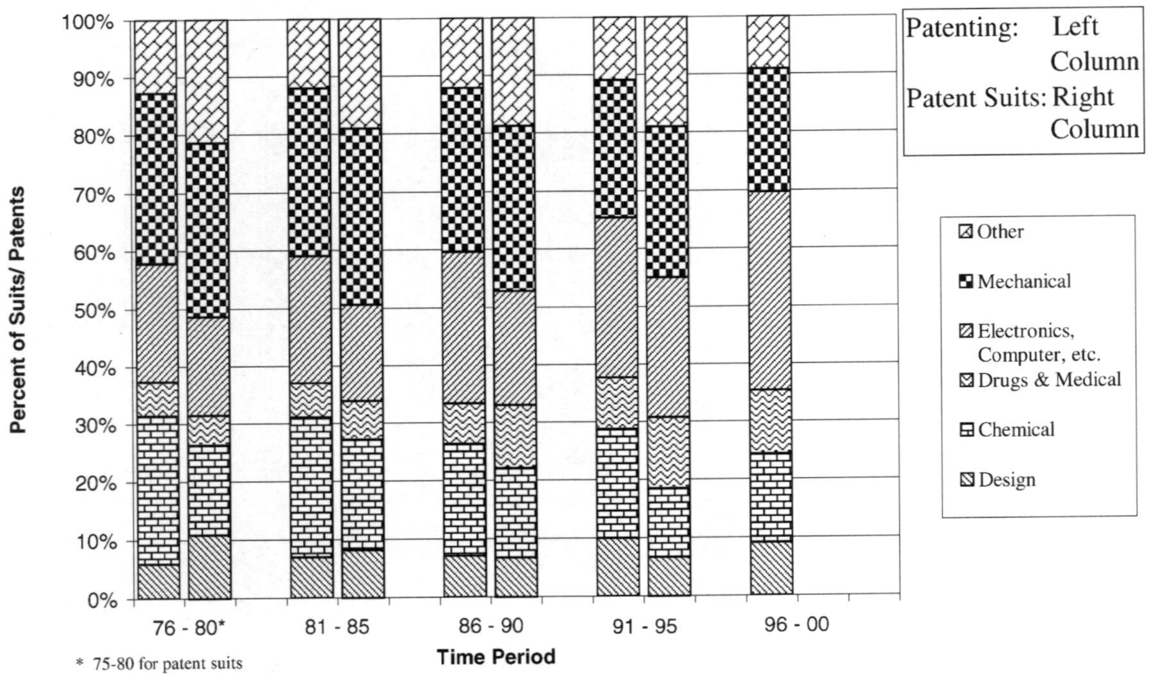

Fig. 4. Changing Composition of U.S. Patenting and Patent Litigation by Technology Group (1976–2000).

Table 2. Litigation Outcomes for Intellectual Property Suits in the United States (by Closing Date).

	Patent Litigation Outcomes by Years					
	1970–1975	1976–1980	1981–1985	1986–1990	1991–1995	1996–1900[a]
Total no. of suits	5423	4244	4842	5763	7181	9658
Procedural progress						
Before issue joined	26.5% (0.006)	28.3% (0.007)	27.6% (0.007)	29.2% (0.006)	33.8% (0.006)	34.8% (0.005)
After issue joined	29.3% (0.007)	26.8% (0.007)	24.7% (0.006)	23.8% (0.006)	30.9% (0.006)	32.5% (0.005)
Pre-trial motions	16.6% (0.005)	13.3% (0.005)	17.6% (0.006)	18.1% (0.005)	14.8% (0.004)	15.1% (0.004)
Pre-trial conference	15.6% (0.005)	20.5% (0.006)	19.3% (0.006)	20.0% (0.005)	13.5% (0.004)	12.2% (0.003)
Trial	12.0% (0.005)	11.1% (0.005)	10.9% (0.005)	8.9% (0.004)	7.0% (0.003)	5.3% (0.002)
Data available for	85.9%	97.5%	93.3%	93.5%	91.6%	91.0%
Disposition mechanism						
Dismissed or settled	65.8% (0.019)	63.9% (0.011)	64.8% (0.007)	67.6% (0.006)	72.0% (0.006)	76.0% (0.005)
Consent judgment	15.0% (0.014)	17.1% (0.008)	15.7% (0.005)	16.6% (0.005)	12.4% (0.004)	8.9% (0.003)
Court or jury judgment	19.1% (0.016)	18.9% (0.009)	19.5% (0.006)	15.8% (0.005)	15.6% (0.004)	15.1% (0.004)
Data available for	11.8%	46.8%	94.5%	93.6%	91.6%	91.0%
Adjudication						
Judgment entered	(n/a)	31.3% (0.010)	29.3% (0.007)	28.3% (0.006)	26.3% (0.005)	22.0% (0.004)
Judgment for plaintiff	(n/a)	58.7% (0.019)	59.3% (0.013)	64.1% (0.012)	58.1% (0.011)	53.9% (0.011)
Data available for	0.0%	48.8%	100.0%	100.0%	100.0%	100.0%
Duration of suits						
Mean (in days)**	392 (6.2)	372 (6.5)	334 (5.6)	298 (4.7)	256 (3.5)	243 (2.8)
Median (in days)	444	416	347	331	275	258

Note: Standard Errors in Parentheses.
[a]Includes only nine months for 2000.

affecting all intellectual property suits, or even all suits (including data quality issues), are not responsible for these trends.

One possible explanation lies in the changing technological composition of patent suits. The outcomes and broad characteristics for patent suits in each of the six technology groups discussed earlier are reported in Table 3. The tables divide the suits in the FJC-USPTO matched data into two time periods, again using their closing dates. In Table 3, design patent suits appear to be a bit of an anomaly – they are more likely to end in the early stages of suit, to have shorter durations, but also to experience some adjudication. This pattern is consistent with the perception that design patents are on average less valuable than utility patents, and also likely to be less complex, thus lending themselves to early resolution (and adjudication) in litigation.

Among utility patents, litigation outcomes in drugs and medical technologies (DMT) and electronics, computers, optical and nuclear (ECON) represent polar extremes. ECON patent suits are more likely to end in earlier stages of litigation, less likely to be disposed of by a judgment, and more likely to experience adjudication than any other utility technology group. DMT patents are typically at the other extreme, with large differences (generally, more than twice the standard error) between the percentage of DMT and ECON suits in each case. The two technology groups are also at opposite extremes in multi-patent litigation, with ECON having the highest percentage of multi-patent suits, and the highest average number of patents per suit.

Unfortunately, besides being available for fewer years, the USPTO-FJC matched sample also appears to be unrepresentative of the population of patent suits for suit outcomes, especially in the 1976–1985 time period. Therefore, a direct comparison of suit outcomes by technology group between the two time periods would be inappropriate. This limits our ability to examine how the changing mix of technologies over time, and especially the growth of ECON technologies, may have influenced the overall patent litigation trends discussed earlier.

A comparison of the *relative* changes among technology groups between the two time-periods suggests that the differences between ECON and other technology groups has not widened. At the same time, the growth of ECON litigation alone cannot account for the magnitude of changes at the aggregate level. Thus, we must tentatively conclude that the overall trends cut across technology groups, at least to some extent. Further, these changes cannot be explained by the fraction of foreign or individual owned patents in litigation, which have remained virtually unchanged.

The differences in suit outcomes and other characteristics between ECON and other technology groups (in particular DMT) are consistent with our knowledge about inter-industry differences in patent use (Cohen, Nelson & Walsh, 2000; Levin, Klevorick, Nelson & Winter, 1987). Variation across industries in the

Table 3. Outcomes and Characteristics of Patent Suits by Technology Sector in the United States.

	1976–1985							1986–1995						
	Design	Chemical	DMT	ECON	Mech.	Other	Total	Design	Chemical	DMT	ECON	Mech.	Other	Total
Total no. of suits (est.)	1026	1521	522	1425	2728	1864	9086	895	1889	1366	2757	3576	2461	12944
No. of observations	329	543	188	528	979	672	3239	502	1070	811	1574	2030	1420	7407
% of suits observed	32.1%	35.7%	36.0%	37.1%	35.9%	36.1%	35.6%	56.1%	56.7%	59.4%	57.1%	56.8%	57.7%	57.2%
Suit outcomes														
Resolved before motions	68.0%	55.4%	50.6%	60.2%	54.0%	56.8%	57.1%	64.7%	56.0%	56.0%	60.5%	58.2%	57.3%	58.4%
	(0.026)	(0.022)	(0.038)	(0.022)	(0.016)	(0.020)	(0.009)	(0.022)	(0.016)	(0.018)	(0.013)	(0.011)	(0.014)	(0.006)
Resolved during pre-trial motions and conference	28.2%	34.5%	36.9%	33.5%	35.7%	33.8%	34.0%	30.7%	34.6%	34.8%	33.9%	33.4%	36.0%	34.1%
	(0.025)	(0.021)	(0.036)	(0.021)	(0.016)	(0.019)	(0.009)	(0.021)	(0.015)	(0.018)	(0.012)	(0.011)	(0.013)	(0.006)
Resolved during or after trial	3.8%	10.1%	12.5%	6.3%	10.3%	9.4%	8.9%	4.6%	9.4%	9.3%	5.6%	8.4%	6.7%	7.4%
	(0.011)	(0.013)	(0.025)	(0.011)	(0.010)	(0.012)	(0.005)	(0.010)	(0.009)	(0.011)	(0.006)	(0.006)	(0.007)	(0.003)
Disposition of suit by court or jury judgment	16.3%	17.1%	22.0%	12.2%	16.4%	19.3%	16.8%	11.6%	15.7%	18.7%	12.6%	15.2%	12.7%	14.4%
	(0.024)	(0.018)	(0.032)	(0.016)	(0.013)	(0.017)	(0.007)	(0.015)	(0.012)	(0.015)	(0.009)	(0.008)	(0.009)	(0.004)
Judgment entered for	34.9%	28.1%	28.9%	24.1%	28.0%	34.2%	29.3%	29.3%	27.7%	33.3%	23.1%	28.6%	29.1%	27.9%
	(0.030)	(0.020)	(0.034)	(0.020)	(0.015)	(0.020)	(0.009)	(0.020)	(0.014)	(0.017)	(0.011)	(0.010)	(0.012)	(0.005)
Suit characteristics														
Mean duration (days)[a]	228	312	379	333	307	302	304	218	350	270	280	292	277	286
	(11.6)	(13.8)	(26.7)	(14.6)	(10.3)	(12.1)	(5.4)	(9.3)	(11.8)	(10.9)	(7.7)	(7.2)	(8.0)	(3.6)
Median duration (days)	238	330	409	351	317	322	320	223	373	282	306	313	290	304
Patents owned by foreign entities %	n/a	11.4%	16.9%	9.6%	13.8%	8.8%	11.7%	n/a	15.8%	12.9%	8.5%	12.7%	11.8%	12.1%
	n/a	(0.014)	(0.028)	(0.013)	(0.011)	(0.011)	(0.006)	n/a	(0.011)	(0.012)	(0.007)	(0.007)	(0.009)	(0.004)
Patents owned by individuals %	n/a	25.1%	32.8%	24.8%	32.7%	44.4%	32.6%	n/a	26.0%	24.1%	26.9%	31.5%	49.1%	32.3%
	n/a	(0.019)	(0.035)	(0.019)	(0.015)	(0.019)	(0.009)	n/a	(0.013)	(0.015)	(0.011)	(0.010)	(0.013)	(0.006)
Fraction of multi-patent suits	n/a	6.3%	4.7%	11.5%	5.3%	5.0%	6.7%	n/a	3.4%	3.8%	5.2%	4.0%	2.8%	4.0%
	n/a	(0.012)	(0.019)	(0.016)	(0.009)	(0.011)	(0.006)	n/a	(0.006)	(0.008)	(0.007)	(0.005)	(0.006)	(0.003)
Average number of patents per suit	n/a	1.297	1.164	1.576	1.242	1.245	1.316	n/a	1.142	1.186	1.262	1.182	1.080	1.180
	n/a	(0.066)	(0.067)	(0.096)	(0.045)	(0.062)	(0.031)	n/a	(0.030)	(0.046)	(0.041)	(0.031)	(0.021)	(0.016)

Note: Numbers in parantheses ( ) are standard errors. Years refer to closing dates. DMT = Drugs and Medical Technologies; ECON = Electronics, Computers, Opticals, and Nuclear; Mech. = Mechanical.

[a] Geometric mean (std. err. by delta method).

features of innovation and in the compulsions for commercializing innovation drive firms to use different strategies with their patents, which are in turn reflected in patent litigation. These differences are discussed in greater depth below, using computers and research medicines as two illustrative industries.

3. STRATEGIC DETERMINANTS OF PATENT SUIT SETTLEMENT DECISIONS

The analysis of patent litigation in this paper is rooted, by necessity, in theories of suit settlement from law and economics (Lanjouw & Lerner, 1998). In these theories, the primary incentive for suit settlement arises from the "bargaining surplus" created by the large costs entailed in patent litigation, which can be shared by the parties if the dispute is terminated.[6] Non-settlement, then, arises when other considerations overwhelm these basic incentives to settle.

Three primary causal mechanisms for non-settlement are suggested in the literature. First, sheer uncertainty may lead to different expectations between the parties about the likely result of litigation (Priest & Klein, 1984). Although these differences may arise from random chance, psychological studies of bargaining suggest that they may be exacerbated by self-serving biases in estimating likely outcomes (Babcock, Loewenstein, Issacharoff & Camerer, 1995; Loewenstein, Issacharoff, Camerer & Babcock, 1993). Second, private information may result in non-settlement due to strategic signaling games being played in pursuit of information rents (Bebchuk, 1984; P'ng, 1983; Schweizer, 1989; Spier & Spulber, 1993). These two explanations of non-settlement are called divergent expectations (DE) and asymmetric information (AI) theories respectively.

The third explanation rests on asymmetric stakes (AS) – i.e. one party has more to gain from winning the suit than the other party has to lose, thus wiping out the bargaining surplus.[7] In a simple AS model with lump-sum transfers, Lanjouw and Lerner (1998) suggest that non-settlement can result from the "infringing" firm's inability to adequately compensate the patentee because monopoly prices can not be sustained in the final goods market with two firms. Put differently, expected total market profits, which is the subject of patent disputes, depends on whether the parties settle (Meurer, 1989), thus creating asymmetric stakes. However, asymmetric stakes that arise when only simple lump-sum transfers are considered can be overcome by more complex contractual arrangements – like running royalties – that conserve the bargaining surplus. Therefore, this paper treats asymmetric stakes more generally – they are the result of difficulties in crafting and enforcing negotiated settlement contracts, which stem from the unique advantages conferred by the litigated patent.

Patent suits may be filed due to any of the three explanations for non-settlement discussed above. However, once the suit begins, AI and DE problems are likely to be mitigated in the course of early court procedures. Case preparation, repeated interaction between the parties, assignment of a judge to the case, and deposition and discovery proceedings provide very useful, and (generally) credible information to the litigants. As a result, not only are information asymmetries substantially reduced, but, in the face of looming litigation costs, beliefs about the outcome of the case are also likely to be carefully reevaluated.

Therefore in the empirical analysis, which focuses on non-settlement of patent suits after they are filed, asymmetric stakes, which are unlikely to be affected by early court procedures, are treated as the primary causal mechanism. Alternative explanations for the results are discussed in section five. This approach is corroborated by empirical findings that asymmetric stakes are disproportionately large in intellectual property litigation (Siegelman & Waldfogel, 1999; Waldfogel, 1995), and that they explain settlement behavior better in the later stages of suits (Waldfogel, 1998).

3.1. Strategic Stakes, and Non-Settlement in Patent Suits

Firms build on their patents, both technologically and commercially, to different extents. Moreover, patents function as isolating mechanisms for these associated assets – i.e. by protecting the patented technology, patents also raise barriers to the imitation of the related assets. Patents may thus isolate rents from co-specialized organizational assets (Teece, 1986), follow-on innovations, and investments in product development (Kitch, 1977). In addition, they can help build strategic positions (Teece, Pisano & Shuen, 1997) that can confer advantages in future rounds of competition (Stefanadis, 1997). This is illustrated, for example, by Swiss pharmaceutical giant Hoffman La-Roche's leveraging of its U.S. market position in Valium into an exclusive license for Glaxo's eventual bestseller Zantac, or by Eli Lilly's use of its dominant position in the U.S. insulin market to obtain a similar franchise on Genentech's recombinant version (called Humulin).

The greater the assets associated with a litigated patent, the more difficult it becomes to craft suit settlement contracts that capture the benefits conferred by the patent. In this paper, asymmetric stakes manifested in this form are called strategic stakes, based on the common understanding that competitive advantage is generally built on firm-specific assets that cannot be easily contracted for in the market (e.g. see Teece, 1998, p. 57). When asymmetric stakes in this form are high, the patentee is generally unwilling to accept a competitor within the commercial space of the patent. The alternative, a possible injunction giving the patentee exclusive rights in the patented technology then becomes an attractive option. Here, a patentee's strategic stake is measured by self-citations to the litigated patent, which reflect the extent to which the patentee has built on its own technology. The citing patents are an indication not only of the related technological assets that may be isolated by the litigated patent, but also the patentee's level of interest in the associated commercial space.

H1. The higher the patentee's strategic stake in the litigated patent (as measured by its self-citations) the more likely is non-settlement in a patent suit.

Strategic stakes are also likely to be high for patents sued early in their lifetime. If the patent is due to expire shortly after the suit, patentees are less likely to gain strategically from exclusivity in the patent's domain. Since the uncertainties over the remaining life of the patent are relatively small, it would be much easier to contract for its potential benefits and thus settle the suit. Moreover, suits filed late in the life of a patent are more likely to center on damages for past infringement, than on building the patentee's competitive advantage for the future.

H2. The earlier a patent is sued in its lifetime the more likely is non-settlement in the suit.

In many instances, settlement in patent suits may be driven, not by an accommodation of the non-patentee through a license, but by the non-patentee's abdication of the patent domain in the face of poorer expected returns from fighting the suit. In other words, a settlement can result if the expected cost of litigation plus the expected transfer (away) for the non-patentee is larger than the loss of rents from on-going "infringement." If the non-patentee has high strategic stakes in the domain of the litigated patent – as measured by its citations to the patent – it is likely to have greater expected returns from fighting the suit by virtue of its continued rent stream from that domain. The non-patentee would therefore be unwilling to merely stop infringement and forego its own commercial opportunities, thus increasing the chances of non-settlement.

H3. The higher the strategic stakes of the non-patentee in the technological domain of a litigated patent (as measured by its citations to the patent) the more likely is non-settlement in the suit.

3.2. Mutual Hold-Up Threats in Litigation

Systems product industries, such as the computer industry, make products that include numerous patented inventions, which may in turn be owned by many different firms. Firms in these industries must therefore devise mechanisms to obtain access to each other's patents in order to retain their ability to commercialize products. Such mechanisms include patent pools (Bittlingmayer, 1988; Merges, 1996; Taylor, 1992), the cross-licensing of patent portfolios (Grindley & Teece, 1997), and even tacit truces on patent litigation between firms (Von Hippel, 1988).

In general, these mechanisms operate on the principle of reciprocal access to patents. In patent litigation, the need for such reciprocal access translates into the possibility of mutual hold up for the parties. When faced with a patent suit, a systems product firm may be able to threaten its rival with a patent "counter-suit" using one or more of its own patents. Such a counter-suit would, in effect, prevent either firm from commercializing the end product. This mutual hold-up imposes additional costs of pursuing suit on both parties, increasing the bargaining surplus available, and encouraging settlement.

Thus, in systems product industries, a firm's patents become tools in negotiating a settlement to patent suits filed by one's rivals. By the same token, rival patent portfolios, and their mutual blocking ability, are central to cross-licensing deals, and a key driver behind patent pools. In recent years, firms in a number of systems product industries, notably semiconductors, have responded to stronger patent rights in the United States by building "defensive" patent portfolios to ensure their own commercial freedom (Hall & Ziedonis, 2001).

Firms in systems product industries are not always engaged in building entire end products in an integrated fashion. Some firms may specialize in narrow areas of technology, and commercialize their innovations through components or licensed technologies that are integrated into the end product by other firms (Linden & Somaya, 2003; Somaya & Teece, 2003). This obviates the need for broad reciprocal access to all patents in a systems product domain. Moreover, firms may have inherent strengths and weaknesses in their patent portfolio. Therefore, the ability of a firm to force suit settlement through mutual hold up is constrained by the extent to which its patent portfolio "reads" on the suing firm's commercial interests. The (in-) citations by the suing firm's (other) patents to patents in its rival's portfolio is a measure of the extent to which its need for reciprocal access can encourage settlement of the suit.

H4. In systems product industries, the greater the number of in-citations by the patentee to patents in its rival's portfolio, the more likely is settlement of the patent suit.

Counter-suits are more likely in systems product industries since patents can be used defensively in these industries to counteract attempts by other firms to exclude firms from commercial opportunities. Sometimes, the mere threat of a counter-suit may be inadequate to induce a plaintiff to drop its suit. Then, the actual filing of a counter suit acts as a credible signal of the defendant's willingness and ability to enforce its own patents. Counter-suits should therefore make the mutual hold-up threat that much more imminent, and encourage settlement.

H5. Counter-suits are more likely in systems products industries, and when they occur, they will increase the probability of settlement.

In the empirical analysis, patent litigation outcomes are investigated in two industries – research medicines and computers – of which the latter exhibits strong systems products features. The biotechnology and pharmaceutical industries are combined into an encompassing "research medicines" industry because the two share many common features and their activities are increasingly overlapping (Henderson, Orsenigo & Pisano, 1998). Genomic and research tool innovations of the early 1990s imply that new products in the industry increasingly combine large numbers of innovations like systems product industries (Heller & Eisenberg, 1998; Somaya & Teece, 2003), but, due to the time frame of this study (1983–1993), this is a relatively minor concern for the analysis.

In computers, patents typically issue much after products are launched, and the technology they represent is no longer cutting edge by then. Nonetheless, permission to use the technologies covered by these patents is necessary to continue commercialization of products. In true systems product fashion, computer products generally involve a very large number of innovations, and the freedom to design and operate despite competing patent claims is vital to firm strategy. Prior research suggests that, relative to pharmaceuticals, computer firms value patents much more as tools to negotiate licenses with other firms, and to counter patent suits (Cohen, Nelson & Walsh, 2000). The patent counsel at IBM, the firm with the largest patent portfolio in computers, underscored the value of access to other patents provided by IBM's own patents, claiming that it was "many times larger than the [licensing] income [from the patents], perhaps an order of magnitude larger" (Grindley & Teece, 1997, p. 12).

4. DESCRIPTION OF DATA AND INTER-INDUSTRY DIFFERENCES

The hypotheses presented in the previous section are tested using matched USPTO-FJC patent litigation data described earlier. The construction of the two-industry sample, and the observed differences between the industries are discussed below.

4.1. Sample Construction

Only suits filed between the years 1983 and 1993 (inclusive) are used in the empirical analysis, primarily because of the relative stability in the patent regime for the two industries during this time period. In particular, it falls short of the spurt in software and gene fragment/research tool patenting activity in the early-mid 90s, and begins after the creation of the Federal Circuit (1982), which was an important change in the patent regime in the U.S. (Nies, 1993). In addition, as is evident from Fig. 2, it avoids the period of poor reporting of suits to the USPTO in the early years, as well as under-sampling in later years (due to truncated lags in the reporting of litigation data).

The industry samples were built using a technology-based sampling process. First, a small sample of "pure-play" firms was chosen from each industry based on within-industry sales reported in Compustat®. All patents issued to these firms between 1975 and 1992 (approximately 16,000) were then collated and sorted into patent classes, which were arranged by decreasing numbers of patents. Those classes that captured a large fraction of the patents were then identified as "representative technologies" for each industry. The resulting research medicines sample includes biotechnology, drug delivery systems, assays, and dental innovations, while the computers sample includes semiconductors, data storage, computer systems, I/O devices, computer applications and networking technologies. Using both U.S. and international patent classes, this methodology accounted for approximately 96% of the patents issued to firms in the research medicines "pure-play" sample, and 48% in the computers "pure-play" sample. The sampling process yielded 700 and 366 observations from 1983 to 1993 in computers and research medicines respectively. Finally, each litigated patent was manually checked to ensure that it belonged to one of the two industries.[8]

The unit of observation for the sample is a patent litigation event, which is defined as the aggregate of all suits that cover the same patent and involve the same parties. This definition accounts for the skirmishing over forum-choice, validity challenges, and multiple suit filings that sometimes occur early in patent disputes. In all such cases in the sample, the suits were eventually consolidated into one proceeding at the same venue. After manually checking for conformity with industry definitions, and consolidating observations into litigation events, the sample was reduced to 709 litigation events – 449 in computers and 260 in research medicines.

A random matched sample (with replacement) was drawn from the intersection set of patents belonging to the primary 4-digit international and 3 digit U.S. patent class of each litigated patent within a one-month window on either side of its issue date. The matched sample is used to model the likelihood of filing suit, conditional on which the likelihood of settlement is modeled. The filing date of the first suit in the litigation event is treated as the "event date" for the analysis. In addition to the litigation data, patent data from the USPTO (provided by Micropatent®), and the patent portfolios of each litigant[9] are also used in the analyses. As before, when multiple patents are involved in a single suit, the main (first) patent listed in the data is used, and the number of patents in the suit is introduced as a control variable.[10]

4.2. Patenting and Patent Litigation in Computers and Research Medicines

Figure 5 traces the growth of patenting and patent litigation in the two industries since 1975 based on the industry definitions used in this paper.[11] Compared with the overall growth in U.S. patenting, patenting in Computers has grown even faster, increase to more than thrice its level in the 1970s. Research medicines experienced a patenting slump in the eighties, recovering to its mid-1970s levels only in 1989. The patenting trends in the industry samples are consistent with other research that has found sharp increases in patenting propensity (patents obtained per R&D dollar) in computers, and decreases in pharmaceuticals (Hall & Ziedonis, 2001).

Patent litigation trends in the two industries are similar, with suit filings in computers rising very rapidly, and research medicines experiencing a decline through the mid-80s before picking up towards the end of the decade. It is not readily apparent however if the growth in patenting led to more litigation, or if more litigation led to increases in patenting. Yearly ratios of suit filings to patenting show no clear trend (not reported), except to reaffirm the finding (below) that litigation propensity in research medicines is lower than that for computer patents or all patents.

4.3. Features of Litigated Patents in Computers and Research Medicines

For patents issued in 1975–1978, for which the entire litigation histories are available, slightly below two suits were filed per 100 issued patents over their lifetime (Table 4). However, this litigation propensity is significantly lower for research medicine patents, and somewhat (but not significantly) higher for computers. Although the estimated litigation propensity for research medicines is only about a third of the average for these early years, later research medicine patents appear more likely to be litigated, based on their partial litigation histories. Nonetheless, as Fig. 6 illustrates, the litigation propensity for research medicine patents is consistently lower than that for computers or for all patents. Figure 6 also illustrates that litigation propensity is high immediately after the patent is issued, and tapers off over the life of the patent.

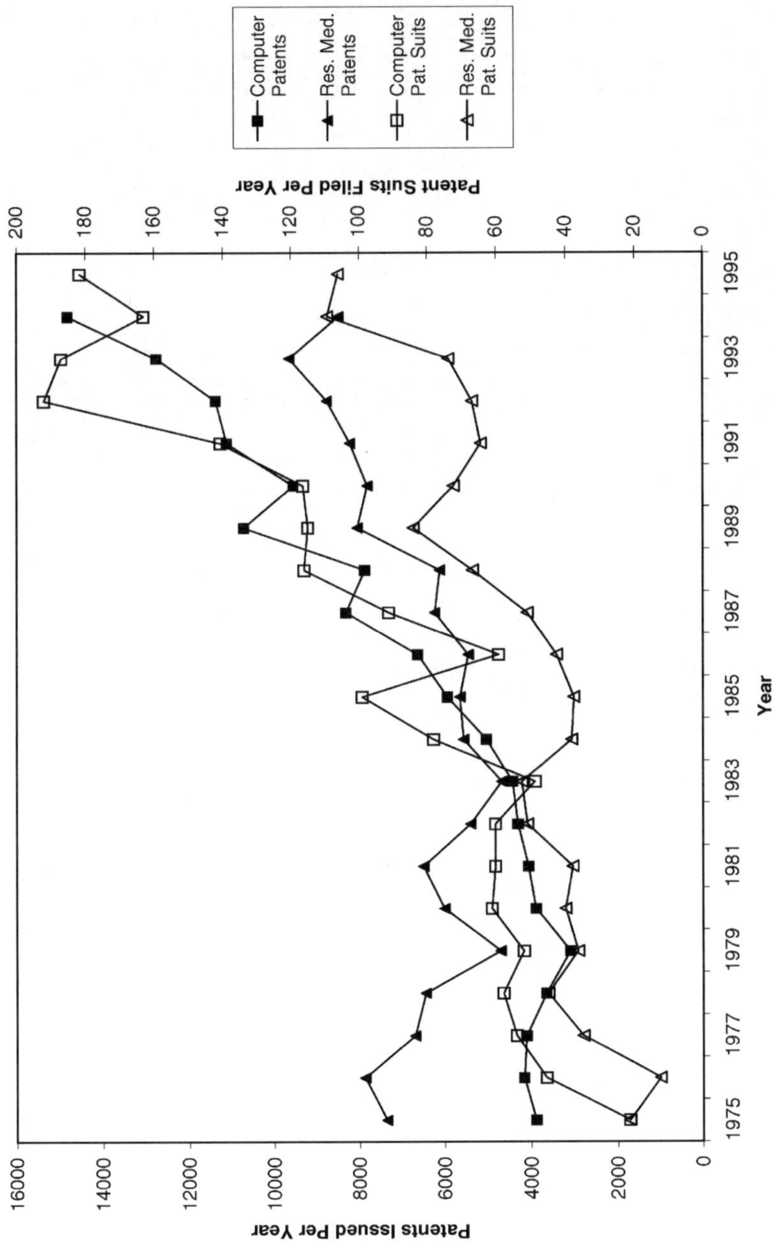

Fig. 5. Patenting and Patent Litigation in Computers and Research Medicines.

Table 4. Sample Characteristics of Computer and Research Medicine Patent Suits.

	Computers	Research Medicines	All Patents/Differences (As Appropriate)
Litigation rate over lifetime (est.)	Suits filed per 100 patents issued		
Patents issued in 1975	2.36	0.70	2.11
Patents issued in 1976	1.44	0.79	1.77
Patents issued in 1977	2.48	0.79	1.98
Patents issued in 1978	3.61	0.37	1.63
Mean (patents issued 1975–1978)	2.47	0.66	1.87
Standard errors for means	0.89	0.20	0.21
No. of patent suits in sample (contains Type I errors)	700	366	
Most Commonly Used Forums	9th circuit (41.2%) 2nd circuit (14.2%)	3rd circuit (31.1%) 9th circuit (22.2%)	
No. of observations (litigation events)	449	260	
No. of litigated patents	263	199	
Right truncated observations	27	9	
Unable to code (no FJC match found)	18	14	
No. of multi-patent obs.	34	6	
Ratios (std. err.)	0.076 (0.012)	0.023 (0.009)	0.053 (0.016)
No. of individual patentee suits	43	11	
Ratios (std. err.)	0.096 (0.014)	0.042 (0.012)	0.053 (0.019)
No. of university patentee suits	2	10	
Ratios (std. err.)	0.004 (0.003)	0.038 (0.012)	−0.034 (0.012)
No. of non-patentee filed suits	43	32	
Ratios (std. err.)	0.096 (0.014)	0.123 (0.020)	−0.027 (0.024)
No. of counter-suit obs.	36	4	
Ratios (std. err.)	0.080 (0.013)	0.015 (0.008)	0.065 (0.015)

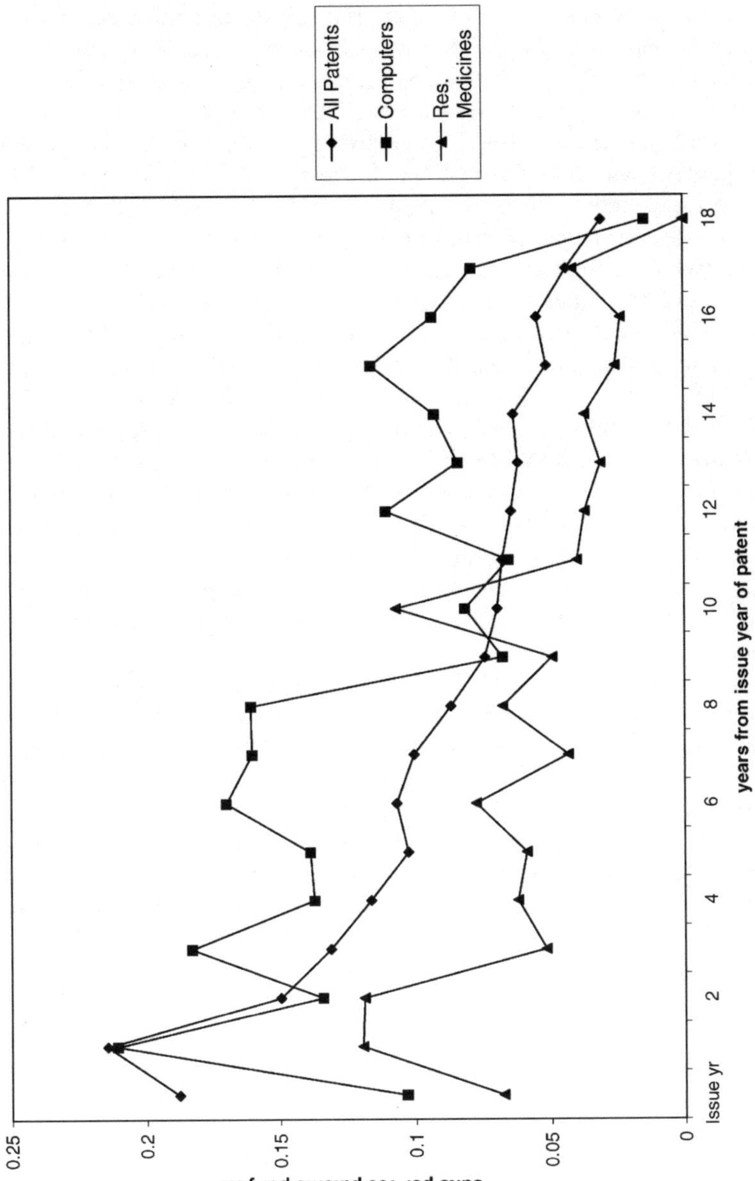

Fig. 6. Average Rates of Litigation Over the Life of Patents.

The lower litigation rate in research medicines seems somewhat surprising since prior survey research has found that firms rely disproportionately on patents for protecting innovations in this industry (Cohen, Nelson & Walsh, 2000; Levin, Klevorick, Nelson & Winter, 1987), and value them highly to recoup investments in research and development (Mansfield, 1986). Studies by Lerner (1995) and Lanjouw and Schankerman (1997) also report higher than average litigation rates in biotechnology and in "drugs and health," respectively. However, the industry samples in these studies are not comparable to this one, with the former being based on a selected sample of firms, and the latter on different technology-class definitions (which excludes many chemistry-based pharmaceutical patent classes).

The greater effectiveness of patents in research medicines (Grabowski & Vernon, 1983) may actually lead to a lower litigation propensity because firms tend to patent a higher fraction of their patentable innovations (Cohen, Nelson & Walsh, 2000). Since the value of patented technologies is known to have a highly skewed distribution (Grabowski & Vernon, 1990, 1994; Scherer & Harhoff, 2000), only a small number of inventions will be worth litigating over anyway, and patenting a higher fraction of inventions could lead to a correspondingly lower propensity for patent litigation. Alternatively, the lower litigation propensity in research medicines may be the result of most disputes being resolved before litigation due to the "comparatively clear standards . . . to assess a . . . patent's validity and to defend against infringement" (Levin, Klevorick, Nelson & Winter, 1987, p. 798) in this industry.

Computer suits are much more likely to include more than one patent, consistent with the perception that individual computer patents are much less effective at blocking rivals (Cohen, Nelson & Walsh, 2000), and that thickets of patents may be required instead. Patent challenges (suits filed by the non-patentee) are somewhat more common in research medicines, but the difference is not significant. Consistent with the settlement-inducing role of mutual hold-up in systems product industries (H4b), computer suits in the sample are more likely (8%) than research medicine suits (1.5%) to be a part of counter suits between rivals in litigation.[12] The four research medicine counter suits all relate to biotechnology patents.

Individual patentees are also much more common in computer litigation (9.6%) than in research medicine litigation (4.2%). In research medicines' litigation, except in biotechnology, large research-intensive firms are often the enforcers of patents against generic and small drug firms. By contrast, even large research oriented computer firms are quite commonly the targets of litigation by other patentees. Universities are more likely to be litigants in research medicine patents, reflecting their active role in innovation and licensing in the life-sciences. In 1997, University licensing revenues from the life-sciences were 87% of their total revenues from licensing (AUTM, 1998).

5. VARIABLES AND EMPIRICAL METHODS

5.1. Empirical Model

The patent litigation process consists of many sequential steps in which a dispute may be resolved, either by a judicial decision or by a settlement between the parties (Fig. 7). Typically, after a suit is filed, one or more preliminary motions follow, of which summary judgment motions are the most important and can result in the disposal of the suit by the court. Other court decisions – like the granting or denial of preliminary injunction – may allow the suit to continue, but with new signal about its "quality." After preliminary motions, the suit goes to trial, with the pre-trial conference stage (after discovery has been completed) being an important intermediate milestone. The parties can choose to settle their dispute at any time during the entire process. The majority of patent suits are dismissed or settled even before motions begin, and only a very small fraction actually make it to trial.

Here, this complex multi-stage litigation process is modeled relatively simply in two-stages, where the first stage is the selection of the dispute for suit (SEL or S), and the second stage is the decision to continue litigation (L), i.e. non-settlement. Two types of litigation outcomes are used as alternate formulations of L – whether the suit experiences adjudication by the court (CRTDEC), and whether the suit survives till summary judgment motions are completed (SMJMOT). The use of SMJMOT is based on estimates from AIPLA's Report of Economic Survey (AIPLA, 1997) and discussions with lawyers, which indicate that the costs of

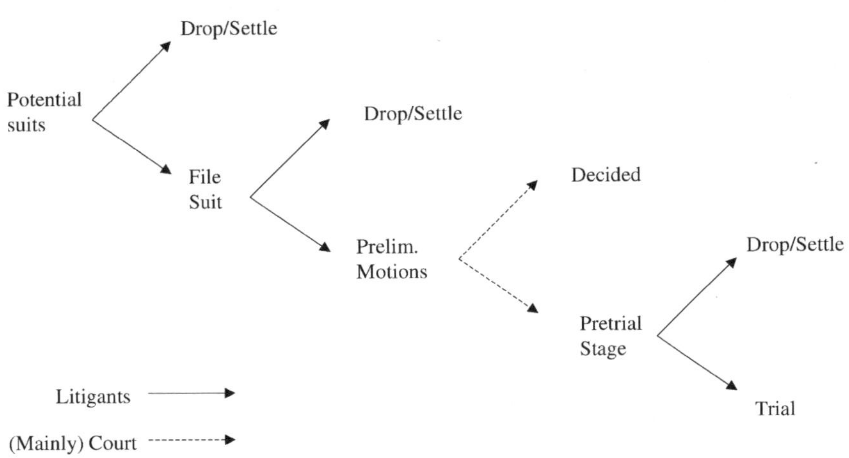

Fig. 7. Description of Sequential Steps in Court Process.

patent litigation begin to increase substantially at the summary judgment stage, suggesting a significant hurdle that should induce settlement before this stage. CRTDEC is used because the theoretical arguments for non-settlement suggest that when stake asymmetries in a patent suit are high, the parties will prefer to obtain a decision from the court rather than reach a settlement.

The likelihood of each of these litigation outcomes is estimated, conditional on the patent being selected for suit. To account for potential unobserved heterogeneity, which can produce biased standard errors, a sample selection Probit model is used (Boyes, Hoffman & Low, 1989; Meng & Schmidt, 1985). The selection and litigation equations can be written in latent variables as:

$$S^* = X_1\beta_1 + \varepsilon_1$$
$$L^* = X_2\beta_2 + \varepsilon_2$$

The possibility that unobserved heterogeneity in the data may drive both selection and litigation choices is accounted for by allowing a correlation ρ between the errors (in the latent variables S^* and L^*), β_1 and β_2. These errors are modeled as being jointly distributed with a bivariate normal distribution, which results in the sample selection Probit specification. The log likelihood function for the model is as follows:

$$\ln L(\beta_1, \beta_2, \rho) = \sum_{n=1}^{N} S_n L_n \ln F(X'_{1n}\beta_1, X'_{2n}\beta_2; \rho) + S_n(1 - L_n)\ln[\phi(X'_{1n}\beta_1) - F(X'_{1n}\beta_1, X'_{2n}\beta_2; \rho)] + (1 - S_n)\ln[1 - \phi(X'_{1n}\beta_1)]$$

where n indexes the observations, and F and Φ are the bivariate and univariate standard cumulative normal distribution functions respectively. The model is estimated by maximum likelihood (FIML).

5.2. Variables

The primary independent variables measure litigant stakes in the suit, and the extent to which mutual hold-up may exist. The patentee's self-citations to the litigated patent, and the non-patentee's citations to the litigated patent are transformed into ratios (PATSTK and NPTSTK) that "orthogonalize" the variables to the overall citation variable, CITVAL. The variables appear in log form, which is appealing because the distribution of citation counts is skew. FILLAG measures the time lag between patent issue and suit filing. The potential for mutual hold up is measured by the patentee's citations to all the patents in the non-patentee's portfolio (PORTCIT), and by a dummy variable if a counter suit is filed (CNTRST). A full list of the

Table 5. List of Variables Used.

Dependent variables	
SMJMOT	= 1 if a suit remained unsettled till the end of summary judgment motions
CRTDEC	= 1 if the suit is disposed from the court process by adjudication (as against a settlement)
SEL	= 1 if a patent is ("selected" to be) sued in a federal district court
Independent variables	
PATSTK[a]	Ratio of patentee self-citations (of the litigated patent) to all citations
NONPSTK[a]	Ratio of defendant's citations to all citations
FILLAG	Lag in years from patent issue date to the date when suit is filed
PORTCIT[a]	The number of citations from patentee's patents to non-patentee's patents (for the 8 years prior to suit)
CNTRST	A dummy variable (= 1 if a counter-suit is simultaneously brought against the patentee on another patent)
Control variables	
CITVAL[a]	Number of citations to the patent (estimated for 17 years from the date of issue)
PORTSIZ[a]	The number of patents in the patentee's portfolio
PORTRAT[a]	The ratio of the number of patents in the patentee's portfolio to those in the non-patentee's portfolio
SUBVAL	Number of IPC subclasses into which the patent has been classified
CLMVAL[a]	Number of (independent) claims made in the patent
BCKVAL[a]	Number of "backward" citations listed in the patent
ISSLAG[a]	Lag in years from application date to issue date for the litigated patent
TIMETR	Time trend variable with 1/1/1983 = 0, and each year = 1 unit, based on suit filing date
FOREN	A dummy variable (= 1 if the patentee is based in a foreign country)
JAPAN	A dummy variable (= 1 if the patentee is based in Japan)
INDIV	A dummy variable (= 1 if the patent is being litigated by an individual)
REPEXT[a]	The number of patent suits involving either litigant in a 10 year window centered around the litigation event
EXPEXT[a]	The total number of suits filed by the patentee over the same patent
NUMPAT[a]	The number of patents included in the lawsuit
MULST	A dummy variable (= 1 if the firms in the suit are involved in another suit in the data-set at a different time)
NUMNPT[a]	The number of commercially distinct defendants included in the lawsuit
PATDEF	A dummy variable (= 1 if the litigation event involves a suit filed by the non-patentee)
RHO	The correlation between errors in the selection and second stage equations

[a]These variables are used in log form. For variables that are measured in levels ("numbers"), this takes the form Log (1 + variable), and for ratios, it takes the form Log (1 + variable 2)/Log (1 + variable 1).

variables used in the empirical analysis, along with brief definitions and the specific log transformations used, appears in Table 5.

A number of controls for overall patent value are introduced into the model based on past research (Hall, Jaffe & Trajtenberg, 2000; Jaffe, Fogarty & Banks, 1998; Lanjouw & Schankerman, 1999; Trajtenberg, 1990a, b). These include citations to the patent (CITVAL), the number of "backward" citations made by the patent (BCKVAL), the number of IPC subclasses assigned to the patent (SUBVAL), and the number of claims it contains (CLMVAL). A suit involving a patent with greater overall value may show an inclination towards non-settlement irrespective of asymmetric stakes, since it can increase differential expectations in the suit, and lead to a greater willingness to play strategic games with asymmetric information.

The use of "forward" citations in the model poses two potential problems. First, the citations are either left or right censored for many observations, and must be estimated using available data. Since citation counts of individual patents are noisy, a citation function is not fitted to each truncated citation series; instead, an unbiased estimate is obtained from the lifetime citation function for the average patent in the industry. This methodology inherently assumes that the citation series for patents within an industry have the same "diffusion and decay coefficients" and differ only in their "levels." Second, citations to the disputed patents can occur after litigation, and their causal effect may therefore be called into question. However, consistent with prior research (Hall, Jaffe & Trajtenberg, 2000), citations are used in this paper as proxies for other underlying features of a patent, which are known at the time of suit. Research has also shown that the impact of litigation on patent citations is an order of magnitude smaller than the citation levels (Lanjouw & Schankerman, 1997).

Control variables are also included for potential externalities created by the patent suit for the litigants. Such externalities, being non contractible, can influence settlement decisions by altering the asymmetric stakes in the suit. Positive externalities, like reputations built from fighting the suit, increase incentives for non-settlement, while negative externalities induce settlement. Reputation externalities (REPEXT) are measured using an estimate of the overall patent litigation activity of the litigants (in a ten year window on either side of the litigation date). A second variable measures expectation externalities (EXPEXT), i.e. the change in expectations of other rivals about the patent resulting from the outcome of the suit. Expectation externalities are measured by the total number of litigation events over the same patent, and may be either positive or negative in character, depending on the patentee's beliefs about the "quality" of its patent.

In addition, controls are introduced to account for the size and relative size of patent portfolios (PORTSIZ and PORTRAT) to account for size effects and potential predation effects in litigation (Lanjouw & Lerner, 2001; Lerner, 1995).

Table 6. Summary Statistics for Variables in Computers and Research Medicines.

	Computers					Research Medicines				
	Mean	Std. Dev.	Min.	Max.	No. Obs.	Mean	Std. Dev.	Min.	Max.	No. Obs.
Litigation stage										
CRTDEC	0.215	0.4116	0	1.00	404	0.376	0.4853	0	1.00	237
SMJMOT	0.238	0.4262	0	1.00	404	0.262	0.4404	0	1.00	237
PATSTK	0.257	0.2817	0	0.96	404	0.347	0.2968	0	1.00	237
NONPSTK	0.091	0.1933	0	0.95	404	0.102	0.2014	0	0.88	237
CITVAL	3.276	1.0778	0	6.01	404	3.407	1.1070	0	7.04	237
PORTSIZ	4.227	2.9359	0	9.19	404	5.736	2.6271	0	9.05	237
ISSLAG	1.219	0.3263	0.4	2.94	404	1.185	0.3432	0.42	2.64	236
PORTRAT	0.107	0.2018	0	0.99	404	0.082	0.1768	0	0.94	237
PORTCIT	0.386	1.0197	0	5.82	404	0.205	0.5275	0	2.77	237
CNTRST	0.082	0.2742	0	1.00	404	0.017	0.1291	0	1.00	237
FILLAG	7.344	5.8147	0	21.31	404	6.731	5.2807	0	20.13	237
REPEXT	3.110	1.0661	0	5.42	404	3.322	0.9688	0.96	4.99	237
EXPEXT	1.031	1.2142	0	3.58	404	0.972	1.3360	0	3.66	237
INDIV	0.228	0.4199	0	1.00	404	0.051	0.2197	0	1.00	237
FOREN	0.052	0.2223	0	1.00	404	0.236	0.4257	0	1.00	237
JAPAN	0.015	0.1211	0	1.00	404	0.021	0.1440	0	1.00	237
SUBVAL	1.706	0.959	1	4.00	327	2.676	1.2565	1	4.00	210
CLMVAL	0.904	0.6101	0.69	4.13	327	0.762	0.3837	0.69	3.50	210
BCKVAL	2.227	0.7257	0.69	4.32	327	1.354	0.8243	0	3.64	210
NUMPAT	0.115	0.4076	0	2.20	404	0.038	0.2759	0	2.94	237
MULST	0.074	0.2625	0	1.00	404	0.097	0.2967	0	1.00	237
NUMNPT	0.095	0.3159	0	2.30	404	0.066	0.2385	0	1.39	237
PATDEF	0.092	0.2888	0	1.00	404	0.114	0.3184	0	1.00	237
TIMETR	6.741	2.9256	0.16	11.06	404	6.100	2.9126	0.09	11.06	237
Selection stage										
SEL	0.500	0.5003	0	1.00	808	0.500	0.5005	0	1.00	474
PATSTK	0.138	0.2325	0	1.00	808	0.208	0.2803	0	1.00	474
CITVAL	2.756	1.1822	0	6.01	808	2.633	1.3947	0	7.04	474
PORTSIZ	4.012	3.3493	0	9.59	808	4.521	2.8957	0	8.89	474
ISSLAG	1.185	0.3081	0	2.94	808	1.162	0.3212	0.42	2.64	472
INDIV	0.177	0.3819	0	1.00	808	0.099	0.2992	0	1.00	474
FOREN	0.219	0.4139	0	1.00	808	0.348	0.4769	0	1.00	474
JAPAN	0.119	0.3238	0	1.00	808	0.063	0.2437	0	1.00	474
SUBVAL	1.598	0.8898	1	4.00	647	2.395	1.2173	1	4.00	418
CLMVAL	0.847	0.5329	0.69	4.13	647	0.766	0.3737	0.69	3.50	418
BCKVAL	1.995	0.7223	0.69	4.32	647	1.385	0.7658	0	4.09	418

Dummy variables were used to code cases in which the patentee was an individual (INDIV), was based outside the U.S. (FOREN) or in Japan (JAPAN), for whether there were multiple (non-contemporaneous) suits between the same parties (MULST), and for litigation events which included a suit filed by the non-patentee (PATDEF). The number of patents in the suit (NUMPAT) and the number of non-patentee rivals in the suit (NUMNPT) were also encoded. Finally, two time controls were used – a time trend (TIMETR), and a variable that measures the amount of time it took the patent office to issue the patent (ISSLAG).

5.3. Sample Statistics

The mean of the dependent variable SMJMOT is similar for the two industries, but a significantly higher fraction of research medicine suits (38%, as against 22%) are disposed of through court decisions (Table 6). This is consistent with our understanding of patent use in the two industries – not only do clearer standards for infringement of research medicine patents make it less costly to obtain a court adjudication, the incentives for non-settlement due to strategic stakes are also higher than a systems products industry like computers.

Other variable means are broadly comparable for the two industries, with the exception of PORTCIT and FOREN, and the differences in CNTRST, NUMPAT and INDIV discussed earlier. The higher mean for PORTCIT in computers is consistent with the industry's systems product character, indicating that firms are often building on each other's technology. Based on the means in the combined (matched plus litigated) sample, it is evident that the lower likelihood of suits involving foreign patentees in computers is more than accounted for by the litigation propensities among Japanese computer patentees. Correlations between variables for the two industries (not reported) suggest few problems with collinearity, except for JAPAN, which is perfectly collinear with the dependent variable in research medicines.

6. RESULTS AND DISCUSSION

Parameter estimates for the sample selection Probit model(s) using a pooled sample (with both industries) are reported in Tables 7 and 8, with the estimates in the first column (within each model) being for the selection stage, and the second column for the litigation ("non-settlement") choice stage. The adjudication of a patent suit by the court (CRTDEC) and the non-settlement of a suit till at least the summary judgment stage (SMJMOT) are the dependent variables for the litigation stages in Tables 7 and 8, respectively. Five models were estimated by adding control

Table 7. Sample Selection Model Estimates for Court Adjudication in Pooled Sample.

Model No.	1		2		3		4	
Dependent Variable	SEL Est. [*p*-Val]	CRTDEC Est. [*p*-Val]	SEL Est. [*p*-Val]	CRTDEC Est. [*p*-Val]	SEL Est. [*p*-Val]	CRTDEC Est. [*p*-Val]	SEL Est. [*p*-Val]	CRTDEC Est. [*p*-Val]
CONSTANT	−2.145***	−0.897	−1.222***	−1.044**	−1.098***	−0.957*	−1.186***	−1.839***
	[0.000]	[0.504]	[0.000]	[0.042]	[0.000]	[0.085]	[0.000]	[0.000]
COMP	−0.091	−0.383***	−0.108	−0.266	−0.158	−0.282	0.032	−0.002
	[0.280]	[0.002]	[0.206]	[0.115]	[0.082]	[0.118]	[0.776]	[0.990]
PATSTK	−0.232	0.782***	0.552***	1.024***	0.542***	1.063***	0.370*	1.009***
	[0.170]	[0.002]	[0.003]	[0.000]	[0.004]	[0.000]	[0.056]	[0.001]
NONPSTK		0.242		0.191		0.187		0.019
		[0.421]		[0.494]		[0.515]		[0.952]
PATSTK × COMP				−0.323		−0.313		−0.333
				[0.388]		[0.426]		[0.393]
CITVAL	0.613***	0.139	0.573***	0.173*	0.542***	0.146	0.531***	0.204**
	[0.000]	[0.531]	[0.000]	[0.088]	[0.000]	[0.179]	[0.000]	[0.018]
PORTSIZ		0.006	−0.161***	−0.020	−0.129***	0.002	−0.146***	−0.056
		[0.819]	[0.000]	[0.555]	[0.000]	[0.961]	[0.000]	[0.106]
ISSLAG	0.483***	−0.188	0.254*	−0.118	0.238*	−0.154	0.045	−0.103
	[0.000]	[0.438]	[0.060]	[0.447]	[0.087]	[0.343]	[0.772]	[0.546]
PORTRAT		−0.072*		−0.077**		−0.079**		−0.044
		[0.058]		[0.041]		[0.041]		[0.245]
PORTCIT		−0.102		−0.159**		−0.157*		−0.110
		[0.157]		[0.041]		[0.053]		[0.178]
CNTRST				0.515**		0.556**		0.453*
				[0.033]		[0.024]		[0.056]
FILLAG				−0.017		−0.022*		−0.019
				[0.161]		[0.095]		[0.195]
REPEXT								0.112
								[0.133]

Table 7. *(Continued)*

Model No.	1		2		3		4	
Dependent Variable	SEL Est. [*p*-Val]	CRTDEC Est. [*p*-Val]	SEL Est. [*p*-Val]	CRTDEC Est. [*p*-Val]	SEL Est. [*p*-Val]	CRTDEC Est. [*p*-Val]	SEL Est. [*p*-Val]	CRTDEC Est. [*p*-Val]
EXPEXT								0.062 [0.328]
INDIV					0.178 [0.187]	0.212 [0.293]	−0.256 [0.120]	0.168 [0.440]
FOREN					−0.450*** [0.000]	−0.061 [0.769]	−0.514*** [0.000]	−0.100 [0.620]
JAPAN					−0.537*** [0.008]	−0.516 [0.317]	−0.494** [0.022]	−0.701 [0.165]
SUBVAL							0.167*** [0.000]	0.157*** [0.004]
CLMVAL							0.135 [0.153]	−0.027 [0.847]
BCKVAL							−0.032 [0.636]	−0.052 [0.534]
TIMETR		−0.009 [0.634]		−0.017 [0.354]		−0.016 [0.394]		−0.018 [0.353]
RHO	0.263 [0.711]		0.582* [0.085]		0.519 [0.193]		0.808*** [0.006]	
No. of obs.		1280		1280		1280		1065
Log likelihood		−1058.57		−982.79		−958.64		−794.17
Schwarz B.I.C.		1115.81		1054.34		1051.65		912.67

Note: Standard errors computed using the BHHH estimator, and *p*-values determined from the corresponding *t* statistics.

Table 8. Sample Selection Estimates for Non-settlement Till Summary Judgment in Pooled Sample.

Model No.	1		2		3		4	
Dependent Variable	SEL SMJMOT Est. [*p*-Val]	SMJMOT Est. [*p*-Val]	SEL Est. [*p*-Val]	SMJMOT Est. [*p*-Val]	SEL Est. [*p*-Val]	SMJMOT Est. [*p*-Val]	SEL Est. [*p*-Val]	SMJMOT Est. [*p*-Val]
CONSTANT	−2.093***	2.160***	−1.238***	1.059*	−1.139***	1.088**	−1.234***	0.767
	[0.000]	[0.000]	[0.000]	[0.054]	[0.000]	[0.045]	[0.000]	[0.323]
SOMP	−0.115	−0.039	−0.119	−0.205	−0.162*	−0.135	0.033	−0.255
	[0.166]	[0.713]	[0.167]	[0.134]	[0.076]	[0.335]	[0.770]	[0.150]
PATSTK	−0.237	0.139	0.539***	−0.044	0.528***	−0.046	0.395*	−0.056
	[0.117]	[0.359]	[0.003]	[0.846]	[0.005]	[0.836]	[0.051]	[0.826]
NONPSTK		0.544**		0.074		0.085		−0.272
		[0.022]		[0.868]		[0.846]		[0.606]
NONPSTK × COMP				1.047*		1.017*		1.121*
				[0.053]		[0.055]		[0.072]
CITVAL	0.613***	−0.451***	0.578***	−0.283**	0.549***	−0.302***	0.543***	−0.199
	[0.000]	[0.000]	[0.000]	[0.010]	[0.000]	[0.005]	[0.000]	[0.182]
PORTSIZ		−0.021	−0.159***	0.011	−0.123***	0.006	−0.139***	−0.016
		[0.206]	[0.000]	[0.748]	[0.000]	[0.857]	[0.000]	[0.723]
ISSLAG	0.452***	−0.384***	0.257*	−0.139	0.235*	−0.138	0.051	−0.165
	[0.000]	[0.006]	[0.055]	[0.419]	[0.090]	[0.422]	[0.739]	[0.430]
PORTRAT		0.009		0.014		0.012		0.017
		[0.635]		[0.636]		[0.683]		[0.941]
PORTCOT		0.053		0.077		0.081		0.010
		[0.256]		[0.259]		[0.237]		[0.911]
CNTRST				0.244		0.254		0.247
				[0.271]		[0.243]		[0.319]
FILLAG				−0.022*		−0.020		0.000
				[0.062]		[0.109]		[0.987]
REPEXT								0.050
								[0.529]

Table 8. *(Continued)*

Model No.	1		2		3		4	
Dependent Variable	SEL SMJMOT Est. [p-Val]	SMJMOT Est. [p-Val]	SEL Est. [p-Val]	SMJMOT Est. [p-Val]	SEL Est. [p-Val]	SMJMOT Est. [p-Val]	SEL Est. [p-Val]	SMJMOT Est. [p-Val]
EXPEXT								−0.165**
								[0.041]
INDIV					0.218	−0.149	−0.143	0.075
					[0.106]	[0.420]	[0.391]	[0.736]
FOREN					−0.441***	0.352**	−0.499***	0.273
					[0.000]	[0.038]	[0.000]	[0.221]
JAPAN					−0.550***	−0.287	−0.505**	−0.301
					[0.007]	[0.576]	[0.020]	[0.614]
SUBVAL							0.157***	−0.081
							[0.001]	[0.204]
CLMVAL							0.096	−0.045
							[0.332]	[0.727]
BCKNAL							−0.030	0.089
							[0.658]	[0.281]
TIMETR		−0.026*		−0.033*		−0.029		−0.038*
		[0.071]		[0.076]		[0.106]		[0.075]
RHO	−0.940***		−0.489*		−0.573**		−0.325	
	[0.000]		[0.060]		[0.038]		[0.471]	
No. of obs.		1280		1280		1280		1065
Log likelihood		−1053.53		−981.30		−955.86		−787.08
Schwarz B.I.C.		1110.76		1052.85		1048.87		905.58

Note: Standard errors computed using the BHHH estimator, and p-values determined from the corresponding t statistics.

Table 9. Comparison of Results from Industry Sample Selection Models.

Dep. Vars.	Computers				Research Medicines			
	SEL = 1		LITIG = 1		SEL =1		LITIG = 1	
	Est.	*p*-Value	Est.	*p*-Value	Est.	*p*-Value	Est.	*p*-Value
LITIG = CRTDEC								
CONSTANT	−0.647	[0.030]**	−0.698	[0.375]	−2.353	[0.000]***	0.107	[0.944]
PATSTK	0.889	[0.002]***	0.796	[0.020]**	0.009	[0.974]	0.847	[0.015]**
NONPSTK			0.319	[0.412]			−0.014	[0.978]
CITVAL	0.459	[0.000]***	0.071	[0.635]	0.668	[0.000]***	−0.091	[0.733]
PORTSIZ	−0.208	[0.000]***	−0.007	[0.916]	0.024	[0.439]	0.032	[0.405]
ISSLAG	0.183	[0.303]	−0.091	[0.721]	0.398	[0.064]*	−0.282	[0.278]
PORTRAT			−0.146	[0.137]			−0.042	[0.340]
PORTCIT			−0.122	[0.256]			−0.171	[0.369]
CNTRST			0.472	[0.104]			1.267	[0.071]*
FILLAG			−0.041	[0.035]**			−0.002	[0.920]
INDIV	−0.026	[0.872]	0.204	[0.449]	0.553	[0.039]**	−0.049	[0.920]
FOREN	−0.870	[0.000]***	0.494	[0.283]	−0.103	[0.520]	−0.133	[0.568]
JAPAN	−0.026	0.934	−0.494	[0.591]	−0.483	[0.142]		
TIMETR			−0.036	[0.165]			0.010	[0.748]
RHO	0.254	[0.661]			−0.345	[0.591]		
No. of obs.				808				472
LITIG = SMJMOT								
CONSTANT	−0.665	[0.027]**	0.489	[0.487]	−2.285	[0.000]***	1.160	[0.494]
PATSTK	0.894	[0.002]***	−0.338	[0.284]	0.005	[0.986]	0.462	[0.235]
NONPSTK			1.054	[0.002]***			0.301	[0.530]
CITVAL	0.467	[0.000]***	−0.156	[0.271]	0.668	[0.000]***	−0.382	[0.196]
PORTSIZ	−0.206	[0.000]***	0.009	[0.877]	0.018	[0.562]	−0.046	[0.281]
ISSLAG	0.177	[0.322]	−0.001	[0.996]	0.364	[0.094]*	−0.213	[0.487]
PORTRAT			−0.013	[0.767]			0.037	[0.410]
PORTCIT			0.127	[0.169]			0.110	[0.538]
CNTRST			0.132	[0.627]			1.706	[0.030]**
FILLAG			−0.029	[0.091]*			−0.015	[0.523]
INDIV	−0.021	[0.895]	−0.050	[0.828]	0.564	[0.036]**	−0.374	[0.476]
FOREN	−0.882	[0.000]***	0.771	[0.017]**	−0.108	[0.502]	−0.006	[0.981]
JAPAN	−0.024	[0.939]	−0.979	[0.323]	−0.458	[0.172]		
TIMETR			−0.055	[0.034]**			−0.012	[0.688]
RHO	−0.448	[0.239]			−0.226	[0.782]		
No. of obs.				808				472

Note: Significance levels: *** = 1%; ** = 5%; * = 10%. Standard errors are computed using the BHHH estimator, and *p*-values determined from the corresponding *t* statistics.

variables, but only models 1–4 are reported in the tables. Model 5, with a full set of control variables, showed no significant improvement in fit over model 4. The parameters were also estimated separately for each industry, and the estimates for model 3, the last model to show a significant improvement in fit in all four cases, are reported in Table 9. Limitations and normative implications are discussed below.

6.1. Strategic Stakes

The results broadly support the hypotheses on strategic stakes. The patentee's strategic stake (PATSTK) has a pronounced and robust positive effect on the probability that the suit goes to a court decision (CRTDEC). In determining whether patent suits progress at least as far as the summary judgment stage (SMJMOT), the coefficient for non-patentee strategic stakes (NONPSTK), and not PATSTK, is significant. Moreover, interaction terms and the separate industry models suggest that this effect is primarily present in the computer industry. Similarly, it appears that the settlement inducing effect of FILLAG, while significant overall (H2), is more pronounced in the computer industry.

The influence of the two strategic stake variables PATSTK and NONPSTK on different aspects of the settlement decision is intuitively appealing – patentees with high stakes would prefer to obtain a court decision that establishes their exclusivity over the technology rather than settle the suit (H1), whereas non-patentees with high stakes would be unwilling to abdicate the patent's domain without fighting the suit into its later stages (H3). In general, there appears to be consistent evidence across a number of studies, including this one, that patentees and their opponents care more about some patents than others when settling disputes. In other research, this has been interpreted as stemming from reputations that arise either out of repeat play by institutional (non-individual) patentees (Waldfogel, 1995; Siegelman & Waldfogel, 1999), or out of concerns about large numbers of other firms operating in the technological neighborhood of the litigated patent (Lanjouw & Schankerman, 2001). In this paper, the emphasis is on patents that firms build on selectively so that they develop non-contractible strategic stakes in them.

The inherent limitation of the data used in this paper however, is that cannot entirely refute the hypothesis that what is measured in follow-on patenting is merely private information that patentees or their rivals have privileged access to, implicating an asymmetric information explanation. But, in addition to the corroborative evidence from other research discussed above, other factors suggest that this may not be the case. Anecdotal reports from patent attorneys indicate that litigants are very well informed about the patent within the very early stages of litigation. In addition, it would be hard to explain why non-patentee citations should matter under an asymmetric information explanation.

The insignificance of FILLAG and NONPSTK in research medicines suggests possible inter-industry differences in the salience of these variables, which we examine below. A more detailed investigation of the research medicine sample also revealed differences in FILLAG and NONPSTK between the pharmaceutical and biotechnology segments of the industry, with the average values of the two variables being significantly lower and higher respectively for biotechnology. But even when the pharmaceutical sub-sample alone was used in the estimation (not reported), the NONPSTK and FILLAG coefficients were not significant.

6.2. Mutual Hold Up and Inter-Industry Differences

The evidence for mutual hold up considerations in litigation is somewhat less encouraging. Of the two variables that test mutual hold-up hypotheses, the coefficient for PORTCIT is not significant except for two models with CRTDEC (although it has the right sign), and CNTRST not only has the wrong sign, it is also significant in many instances. Despite this seemingly dismal verdict, it is clear from the data that mutual hold up does play a role in suit settlement. In almost every case involving counter suits, the two suits were disposed within a day of each other, even though they were typically filed in separate courts. Since there is no legal or administrative reason for this to occur with essentially separate legal proceedings, it seems clear that there is an element of strategic interaction occurring through counter suits.

Viewed in this light, the likely explanation for the negative statistical result is that the sample is analyzed in a cross-sectional framework, which merely implies that counter-suits are more commonly associated with suits that are not easily settled between the parties. This finding can be consistent with the mutual hold-up hypothesis; a non-patentee will not incur the additional costs of filing a counter-suit if it is able to encourage settlement by merely threatening such a suit. Therefore, it is suits that are more difficult to settle in the first place that are likely to invite counter suits. Ultimately, this calls for a longitudinal approach to studying the impact of counter-suits. One such study, using a duration model, finds much stronger evidence than this one for the role of mutual hold-up in inducing patent suit settlement in systems product industries (Somaya, 2002).

Recall that a significantly higher proportion of counter-suits are found in computers, the more systems product oriented of the two industries, which is consistent with the existence of mutual hold-up among computer patents. The differences between the two industries in the salience of NONPSTK and FILLAG may also have similar roots in their different patterns of patent use. In particular, it may relate to the somewhat idiosyncratic role played by patents in research medicines, where they typically protect very narrowly defined but highly valuable products, which in turn have demonstrated an optimal combination of therapeutic efficacy, acceptable side effects and clinical safety. As a result, related ("citing") patents owned by the non-patentee, which lie in the surrounding technology space, probably convey little information about that firm's strategic stake in the litigated patent itself. Similar but non-identical therapeutic compounds are generally not competitors to a pharmaceutical product, especially since they would be required to obtain a separate (and expensive) regulatory clearance.

Since mark-ups over marginal costs are extremely high in research medicines, exclusivity in the specific successful products is very important to patentees in this industry, irrespective of the remaining life of the patent. Moreover, pharmaceutical firms often make a large share of their patent-related profits in the last few years of the patent, when the product is well established in the marketplace. It is therefore not surprising that FILLAG plays a minimal role in the settlement of research medicine suits.

By contrast, litigation strategy in computers, as described by patent lawyers, is often determined by a choice between a royalty/damages gambit and an attempt to obtain exclusivity. An exclusivity strategy, where the patentee is less likely to settle, naturally tends to be used early in the patent's life. On the other hand, there are clear incentives to delay litigation if the patentee is interested in royalties, in view of the possibility that the patentee could switch to substitute technologies if sued earlier in the adoption cycle.

6.3. The Selection of Patents for Suit and Control Variables

Interestingly, CITVAL, which has a major influence on the decision to file suit, and is widely recognized as a proxy for patent value, is inconsistent in explaining non-settlement during patent litigation. Therefore, in the language of the patent valuation literature (Hall, Jaffe & Trajtenberg, 2000; Harhoff, Scherer & Vopel, 1999), raw citations may be a better measure of overall importance (or public value) of patents, whereas self citations may be a more consistent measure of their private or strategic value to firms, which in turn appears to be the operative variable in patent litigation.

In the selection equation, consistent with past research (Lanjouw & Schankerman, 1997), (forward) citation counts (CITVAL) and, where estimated, backward citations (BCKVAL) and number of IPC classes (SUBVAL) increase the probability of suit over a patent. Moreover, the higher litigation propensity for individual patentees found in that research is replicated in the research medicines' sample, but computer patents assigned to individuals (INDIV) display the opposite trend (when importance controls are used). Self-citations (PATSTK) and foreign patentees (FOREN) are only significant for the computer industry, and respectively increase and decrease the probability of suit filing.

Interestingly, ISSLAG also has some predictive power in forecasting the likelihood of suit in both industries. ISSLAG has an intuitive explanation – greater uncertainty about the validity or scope of a patent is likely to increase lags in issuing patents, and is also likely to result in litigation over patents. In computers, although foreign firms are less likely to file suit, they are also less likely to settle suits once they are filed. This lends support to the explanation proposed by Lanjouw and Schankerman (1998) that higher litigation costs may lead foreign firms to avoid litigation in the U.S., except for their most important patents. It appears, however, that once they do file suit on these "important patents," foreign firms pursue litigation very aggressively (at least in computers).

The size variable (PORTSIZ) is significant only for the computer industry, where patents owned by firms with larger portfolios are less likely to end up in suit. This is consistent with the proposition that large firms, which make integrated systems products, are more likely to need reciprocal access to technologies, and are therefore less likely to sue, whereas smaller specialized patentees are more likely to defend their technology niches aggressively. Large firms may also be more aggressive in patenting their innovations, so that they have a smaller percentage of high value patents in their portfolios than small firms. PORTRAT is significant in two models for CRTDEC, lending some credibility to possible predation in the litigation. Finally, the correlation parameter, RHO, is significant in some of the models, suggesting the value of using sample selection models to account for unobserved heterogeneity in studies of litigation.

7. CONCLUSION

The strategic use of intellectual property assets to enhance firm competitive advantage is increasingly becoming a vital aspect of corporate strategy in the knowledge economy. This paper demonstrates how patents function as strategic assets – conveying benefits as isolating mechanisms that are often difficult to contract for – and thus serve as a source of competitive advantage. Although the paper uses patent litigation as its empirical setting, it has implications for patent strategy in general.

In determining their patent licensing strategy, firms must make similar tradeoffs (as in litigation) regarding the contractibility of a patent's potential benefits. Instead of a bargaining surplus, a license may benefit firms through externalities from standards creation or from access to the licensee's valuable complementary technologies or assets. Firms should also actively license low-stake patents that have little strategic value, if they could potentially be used by other firms. In systems products industries, mutual hold-up is a consideration in licensing decisions as well, often leading to cross-licenses or patent pools of broad scope. But individuals and niche specialists with strong patent positions, who do not need reciprocal access, are better able to withstand hold-up threats in systems products industries, and thus gain an upper hand in licensing negotiations.

The patenting decisions of firms are also often driven by similar considerations. When strong, watertight patents are available, as in pharmaceuticals, firms may be able to rely on them to isolate key commercial opportunities (Merges, 1998). On the other hand, in systems products industries, thickets of patents may be necessary to foil attempts to invent around the patent, and obtain a robust patent position. Moreover, defensive patenting – the building of large patent portfolios – may become necessary if rivals, aided by a strong enforcement regime, are able to effectively threaten to hold-up a firm's commercial operations (Hall & Ziedonis, 2001).

Our understanding of patent strategies also suggests a possible explanation for the trend towards the earlier resolution of patent suits discussed at the beginning of the paper. Encouraged by the pro-patent changes in patent enforcement, firms and individuals may be increasingly bringing patents with lower strategic stakes to court with the objective of leveraging licensing revenues from them. In turn, these suits would be less likely to survive into the later stages of litigation and to be adjudicated by the court. While there appears to be some anecdotal support for this conjecture, it should be noted that this trend in patent litigation is also consistent with the idea that less uncertainty in enforcement (as a result of uniformity in case law engendered by the Federal Circuit) may be encouraging parties to settle patent suits earlier. Ultimately, an in-depth investigation of this question awaits future research.

This paper reaffirms that there are substantial differences in the ways in which different industries use patents. Survey evidence suggests that some industries like chemicals and petroleum are similar to research medicines in their use of patents, and others like electronics and communications equipment are similar to computers (Cohen, Nelson & Walsh, 2000; Levin, Klevorick, Nelson & Winter, 1987). To the extent that these similarities are accurate, the understanding about patent litigation behavior gained through this paper can be extended to these industries as well. In addition, it is also clear that – at least in industries like computers – the strategies pursued by firms with their patents probably varies over the life-cycle of the technology, with exclusivity being emphasized in the early phase, and licensing when the technology becomes mature.

Finally, the strategic concerns faced by firms when enforcing their patents also have implications for public policy in the patent arena. Both courts and the patent office need to be cognizant of the inter-industry and technology life-cycle differences in patent use highlighted above when making decisions to award patents, find infringement, and determine damages. In some contexts, strong patents may be valuable as isolating mechanisms, and even encourage entry from new innovative firms who may not possess strong complementary organizational assets. In other contexts, system product considerations create somewhat complex incentive structures for innovation than in the simple independent innovation case (Somaya & Teece, 2003).

NOTES

1. Baechler, Mary. "My life without lawyers," Inc. v18, n18 (Dec 1996, pp. 37–38). Baechler is the CEO of Racing Strollers Inc., a Yakima-WA based company that was involved in patent litigation over its stroller designs.
2. Federal Judicial Center. Federal Court Cases Integrated Database (various versions, 1970–2000), Ann Arbor, MI: Inter-university Consortium for Political and Social Research (distributor).
3. The data was kindly provided by Derwent Information®. Help from David Mowery in obtaining the data is gratefully acknowledged.
4. Since the matching techniques used were extremely thorough, these missing suits were likely classified under some other "nature of suit" category in the FJC data. This is probable because the same suit can involve a multiplicity of issues, and court clerks are required to pick only one "nature of suit" category for the suit.
5. Since the durations have a skew distribution, the geometric means or medians are reported in Table 2, and later.
6. Mathematically, the expected value from trial for the two parties are as follows:

$$\text{Patentee: } R + WT - C \quad \text{Non-patentee: } r - wt - c$$

where R and r are the respective rent streams to the parties absent litigation, T and t are the respective transfers imposed by a patentee win, C and c are the respective costs of litigation, and W and w are the respective estimates of a patentee win. Settlement incentives are provided by $-(C + c)$, which is the total cost of litigation for the two parties, taken as a unit. A settlement solution exists if $(wt - WT) + (C + c) \geq 0$.

7. Mathematically, this means that $T \gg t$, such that $(WT - wt) > (C + c)$. Then, no bargaining solution is available.
8. Lerner (1995) and Kortum and Lerner (1999) use a similar approach with only the IPC classes, but do not attempt to weed out "false positives." Since manual elimination of such type 2 errors is feasible for a smaller data set like the one used here, type 1errors could be lowered by using IPC as well as USPC patent classes, and also by including more patent classes, which would otherwise have increased type 2 errors. Thus, the samples that were eventually assembled are highly representative of the two industries.
9. Help from Bronwyn Hall is gratefully acknowledged for access to the Case Western database (Hall, Jaffe & Trajtenberg, 2000), which contains a match of roughly 3000 manufacturing firms from the Compustat® database to over 30,000 of their patenting subsidiaries using a 1989 "Who Owns Whom" directory. In addition, all name variations for firms not listed in that database were tracked to identify patents assigned to each firm in the dataset.
10. This approach seems reasonable, given the finding (in validity suits) that such sets of patents usually stand or fall together (Allison & Lemley, 1998).

11. The industry definition is static over time, and does not capture technology or patent classification changes. In addition, the patent classes used may systematically introduce both type 1 and type 2 errors, which cannot be mitigated for such a large sample.

12. As estimates of counter suing, these fractions are biased downward because counter-suits may exist that are not in the USPTO sample or in the specific patent classes used. Correcting for the first bias alone (assuming a random process of selection into the sample) results in estimates of 14.3 and 2.7% respectively.

ACKNOWLEDGMENTS

The author wishes to thank Bronwyn Hall, Mark Lemley, Greg Linden, Alan Marco, David Mowery, Carl Shapiro, Edward Sherry, David Teece, Brian Wright, and two anonymous referees for their comments on earlier drafts. Research support from the University of California President's Initiative for University-Industry Research, and the Ryoichi Sasakawa Young Leaders Foundation is also gratefully acknowledged. Customary disclaimers apply.

REFERENCES

AIPLA (1997). *Report of economic survey*. Washington, DC: American Intellectual Property Law Association.

Allison, J., & Lemley, M. A. (1998). *Empirical evidence on the validity of litigated patents*. Stanford Workshop on Intellectual Property and Industry Competitive Standards, Stanford, CA.

AUTM (1998). *Licensing survey: Fiscal year 1997*. Norwalk, CT: Association of University Technology Managers.

Babcock, L., Loewenstein, G., Issacharoff, S., & Camerer, C. (1995). Biased judgments of fairness in bargaining. *American Economic Review*, *85*(5), 1337–1343.

Bebchuk, L. A. (1984). Litigation and settlement under imperfect information. *Rand Journal of Economics*, *15*(3), 404–415.

Bhagat, S., Brickley, J. A., & Coles, J. L. (1994). The costs of inefficient bargaining and financial distress: Evidence from corporate lawsuits. *Journal of Financial Economics*, *35*(2), 221–247.

Bittlingmayer, G. (1988). Property rights, progress, and the aircraft patent agreement. *Journal of Law & Economics*, *31*(1), 227–248.

Boyes, W. J., Hoffman, D. L., & Low, S. A. (1989). An econometric analysis of the bank credit scoring problem. *Journal of Econometrics*, *40*(1), 3–14.

Cohen, W. M., Nelson, R. R., & Walsh, J. (2000). Protecting their intellectual assets: Appropriability conditions and why U.S. manufacturing firms patent (or not). NBER Working Paper No. 7552.

Grabowski, H. G., & Vernon, J. M. (1983). The pharmaceutical industry. In: R. R. Nelson (Ed.), *Government and Technical Progress* (pp. 283–360). New York: Pergamon.

Grabowski, H. G., & Vernon, J. M. (1990). A new look at the returns and risks to pharmaceutical R&D. *Management Science*, *36*, 804–821.

Grabowski, H. G., & Vernon, J. M. (1994). Returns on new drug introductions in the 1980s. *Journal of Health Economics*, *13*, 383–406.

Grindley, P. C., & Teece, D. J. (1997). Managing intellectual capital: Licensing and cross-licensing in semiconductors and electronics. *California Management Review*, *39*(2), 8–41.

Hall, B. H., Jaffe, A. B., & Trajtenberg, M. (2000). Market value and patent citations: A first look. NBER Working Paper No. 7741 (June).

Hall, B. H., & Ziedonis, R. M. (2001). The patent paradox revisited: An empirical study of patenting in the semiconductor industry, 1979–1995. *RAND Journal of Economics*, *32*(1), 101–128.

Harhoff, D., Scherer, F. M., & Vopel, K. (1999). Citations, family size, opposition, and the value of patent rights. Working Paper (September).

Heller, M. A., & Eisenberg, R. S. (1998). Can patents deter innovation? The anticommons in biomedical research. *Science*, *280*(5364), 698–701.

Henderson, R., Orsenigo, L., & Pisano, G. (1999). The pharmaceutical industry and the revolution in molecular biology: Interactions among scientific, institutional and organizational change. In: D. C. Mowery & R. R. Nelson (Eds), *The Sources of Industrial Leadership: Studies of Seven Industries* (pp. 267–311). Cambridge, NY.

Hicks, D., Breitzman, T., Olivastro, D., & Hamilton, K. (2001). The changing composition of innovative activity in the U.S. – a portrait based on patent analysis. *Research Policy*, *30*(4), 681–703.

Jaffe, A. B., Fogarty, M. S., & Banks, B. A. (1998). Evidence from patents and patent citations on the impact of NASA and other federal labs on commercial innovation. *Journal of Industrial Economics*, *46*(2), 183–205.

Kitch, E. W. (1977). The nature and function of the patent system. *Journal of Law and Economics*, *20*, 265–290.

Kortum, S., & Lerner, J. (1999). What is behind the recent surge in patenting? *Research Policy*, *28*(1), 1–22.

Lanjouw, J. O., & Lerner, J. (1998). The enforcement of intellectual property rights: A survey of the empirical literature. *Annales d'Economie et de Statistique*, (49–50), 223–246.

Lanjouw, J. O., & Lerner, J. (2001). Tilting the table? The use of preliminary injunctions. *Journal of Law and Economics*, *44*(2), 573–603.

Lanjouw, J. O., & Schankerman, M. (1997). Stylized facts of patent litigation: Value, scope and ownership. NBER Working Paper No. 6297.

Lanjouw, J. O., & Schankerman, M. (1998). Patent suits: Do they distort research incentives? CEPR Working Paper No. 2042.

Lanjouw, J. O., & Schankerman, M. (1999). The quality of ideas: Measuring innovation with multiple indicators. NBER Working Paper No. 7345.

Lanjouw, J. O., & Schankerman, M. (2001). Characteristics of patent litigation: A window on competition. *RAND Journal of Economics*, *32*(1), 129–151.

Lerner, J. (1995). Patenting in the shadow of competitors. *Journal of Law & Economics*, *38*(2), 463–495.

Levin, R. C., Klevorick, A. K., Nelson, R. R., & Winter, S. G. (1987). Appropriating the returns from industrial research and development. *Brookings Papers on Economic Activity* (3), 783–820.

Linden, G., & Somaya, D. (2003). System-on-a-chip integration in the semiconductor industry: Industry structure and firm strategies. *Industrial and Corporate Change* (forthcoming).

Lippman, S. A., & Rumelt, R. P. (1982). Uncertain imitability: An analysis of interfirm differences in efficiency under competition. *Bell Journal of Economics*, *13*(2), 418–438.

Loewenstein, G., Issacharoff, S., Camerer, C., & Babcock, L. (1993). Self-serving assessments of fairness and pretrial bargaining. *Journal of Legal Studies*, *22*(1), 135–159.

Mansfield, E. (1986). Patents and innovation: An empirical study. *Management Science*, *32*(2), 173–181.

Meng, C.-L., & Schmidt, P. (1985). On the cost of partial observability in the bivariate probit model. *International Economic Review*, *26*(1), 71–85.

Merges, R. P. (1996). Contracting into liability rules: Intellectual property rights and collective rights organizations. *California Law Review*, *84*(5), 1293–1393.

Merges, R. P. (1998). Intellectual property rights, input markets, and the value of intangible assets. Working Paper, Boalt School of Law, University of California Berkeley.

Merges, R. P., & Nelson, R. R. (1990). On the complex economics of patent scope. *Columbia Law Journal*, *90*(4), 839–916.

Merz, J. F., & Pace, N. (1994). Trends in patent litigation: The apparent influence of strengthened patents attributable to the court of appeals for the federal circuit. *Journal of Patent and Trademark Office Society*, *76*, 579–590.

Meurer, M. J. (1989). The settlement of patent litigation. *Rand Journal of Economics*, *20*(1), 77–91.

Nies, H. W. (1993). Ten years of patent law development under the U.S. court of appeals for the federal circuit. *IIC: International Review of Industrial Property and Copyright Law*, *24*(6), 797–802.

P'ng, I. P. L. (1983). Strategic behavior in suit, settlement, and trial. *Bell Journal of Economics*, *14*(2), 539–550.

Priest, G. L., & Klein, B. (1984). The selection of disputes for litigation. *Journal of Legal Studies*, *13*, 1–55.

Rivette, K. G., & Kline, D. (2000). Discovering hidden value in intellectual property. *Harvard Business Review*, *78*(1), 54–66.

Rumelt, R. P. (1984). *Towards a strategic theory of the firm. Competitive strategic management*. R. B. Lamb (pp. 556–570). Englewood Cliffs, NJ: Prentice-Hall.

Scherer, F. M., & Harhoff, D. (2000). Technology policy for a world of skew-distributed outcomes. *Research Policy*, *29*, 559–566.

Schweizer, U. (1989). Litigation and settlement under two-sided incomplete information. *Review of Economic Studies*, *56*, 163–178.

Scotchmer, S. (1991). Standing on the shoulders of giants: Cumulative research and the patent law. *Journal of Economic Perspectives*, *3*, 29–41.

Siegelman, P., & Waldfogel, J. (1999). Toward a taxonomy of disputes: New evidence through the prism of the Priest/Klein model. *Journal of Legal Studies*, *28*(1), 101–130.

Somaya, D. (2002). The duration of patent litigation: Firm strategies and litigation tactics in computers and research medicines. Working Paper, R. H. Smith School of Business, University of Maryland.

Somaya, D., & Teece, D. J. (2003). Combining inventions in multi-invention products: Patents, organizational performance, and public policy. Working Paper, R. H. Smith School of Business (University of Maryland), earlier version (2000) available online at: www.ssrn.com.

Spier, K., & Spulber, D. (1993). Suit settlement and trial: A theoretical analysis under alternative methods for the allocation of legal costs. *Journal of Legal Studies*, *11*, 55–81.

Stefanadis, C. (1997). Downstream vertical foreclosure and upstream innovation. *Journal of Industrial Economics*, *45*(4), 445–456.

Taylor, D. S. (1992). The sinking of the United States electronics industry within Japanese patent pools. *George Washington Journal of International Law & Economics*, *26*(1), 181–212.

Teece, D. J. (1986). Profiting from technological innovation: Implications for integration, collaboration, licensing and public policy. *Research Policy*, *15*(6), 285–305.

Teece, D. J. (1998). Capturing value from knowledge assets: The new economy, markets for know-how, and intangible assets. *California Management Review*, *40*(3), 55–79.

Teece, D. J., Pisano, G., & Shuen, A. (1997). Dynamic capabilities and strategic management. *Strategic Management Journal*, *18*(7), 509–533.

Trajtenberg, M. (1990a). *Economic analysis of product innovation*. Cambridge, MA: Harvard University Press.

Trajtenberg, M. (1990b). A penny for your quotes: Patent citations and the value of innovations. *Rand Journal of Economics*, *21*(1), 172–187.

Von Hippel, E. (1988). *The sources of innovation*. New York: Oxford University Press.

Waldfogel, J. (1995). The selection hypothesis and the relationship between trial and plaintiff victory. *Journal of Political Economy*, *103*(2), 229–260.

Waldfogel, J. (1998). Reconciling asymmetric information and divergent expectations theories of litigation. *Journal of Law and Economics*, *41*(2), 451–476.

CROSSING THE GREAT DIVIDE: USING ADVERSE POSSESSION TO RESOLVE CONFLICTS BETWEEN THE ANTITRUST AND INTELLECTUAL PROPERTY REGIMES

Constance E. Bagley and Gavin Clarkson*

ABSTRACT

This paper focuses on two related questions at the intersection of antitrust and intellectual property law. First, under what circumstances must the holder of a patent or a copyright or the owner of a trade secret allow others to use that intellectual property? Second, under what circumstances can the holder of an intellectual property right use that right to make it difficult for another party to succeed in a related market? These questions have vexed antitrust and intellectual property scholars alike ever since the Federal Circuit ruled in 2000 that patent holders "may enforce the statutory right to exclude others from making, using, or selling the claimed invention free from liability under the antitrust laws," a ruling that directly contradicted the

*Certain ideas and arguments presented in this paper were first discussed in "*Adverse Possession for Intellectual Property: Adapting an Ancient Concept to Resolve Conflicts Between Antitrust and Intellectual Property Laws in the Information Age*," 16 HARV. J.L. & TECH. 327 (2003), which contains an expanded legal analysis (including several hundred additional footnotes).

Intellectual Property and Entrepreneurship
Advances in the Study of Entrepreneurship, Innovation and Economic Growth, Volume 15, 149–199

ISSN: 1048-4736/doi:10.1016/S1048-4736(04)01505-X

Ninth Circuit ruling that antitrust liability could be imposed for almost identical conduct, depending on the motivations of the patent holder. The various proceedings in United States v. Microsoft only added fuel to the firestorm of controversy.

After briefly retracing the jurisprudential path to see how this situation arose, we propose a solution that primarily involves a variation on the real property concept of adverse possession for the intellectual property space along with a slight extension of the Essential Facilities Doctrine for industries that exhibit network effects. We examine, both for firms with and without market power, how our proposal would resolve the situations presented by large fixed asset purchases, the introduction of entirely new products, and operating systems with network effects. We also demonstrate how our proposal could be applied in the European antitrust enforcement context.

1. INTRODUCTION

This paper focuses on two related questions. First, under what circumstances must the holder of a patent or a copyright or the owner of a trade secret allow others to use that intellectual property? Second, under what circumstances can the holder of an intellectual property right unilaterally[1] use that right to make it difficult for another party to succeed in a related market? Although the Supreme Court addressed these issues somewhat obliquely in 1992 in its decision in *Kodak v. Image Technical Services* ("*Kodak I*"),[2] these questions assumed new urgency in 2000 after the Federal Circuit created a split in the circuits by ruling in *CSU v. Xerox* that patent holders "may enforce the statutory right to exclude others from making, using, or selling the claimed invention free from liability under the antitrust laws."[3] The Federal Circuit expressly declined to follow a Ninth Circuit ruling that held antitrust liability could be imposed for almost identical conduct, depending on the motivation of the patent holder.[4] The District of Columbia Circuit's statement in *United States v. Microsoft* ("*Microsoft 2001*")[5] that Microsoft's assertion that its copyrights on its software programs entitled it to impose restrictions on its original equipment manufacturer ("OEM") licensees was "no more correct than the proposition that use of one's personal property, such as a baseball bat, cannot give rise to tort liability"[6] only added fuel to the firestorm of controversy.

The depth of this quandary became even more evident when, presented with a seemingly obvious split in the circuits, the Solicitor General suggested that the U.S. Supreme Court deny certiorari in the *Xerox* case to "allow these difficult issues to percolate further in the courts of appeals."[7] Indeed, the Antitrust Division of the Justice Department and the Federal Trade Commission held an entire day of hearings on exactly these questions as part of their series of hearings on competition and intellectual property policy.[8] The hearings produced a substantial amount of thought and discussion, but no clear solution.

Given that determining antitrust liability for refusing to deal is already one of the most challenging questions in the antitrust field, the additional requirement of determining the scope of the intellectual property grant and the concomitant right to exclude makes this problem that much more complex.[9] In order to arrive at a potential solution, it is necessary to retrace briefly the jurisprudential path to see how this issue arose. Section 2 of this paper discusses some of the intellectual property and antitrust principles that should guide the development of any potential solution. Section 3 then reviews the origin and propagation of the problems created when intellectual property rights ("IPR") holders unilaterally refuse to deal. In Section 4 we present our proposed solution, which calls for the application of an adapted version of the real property concept of adverse possession in the intellectual property arena, along with a slight extension of the Essential Facilities Doctrine for industries that exhibit network effects. Our proposed solution is designed to allow these cases to be resolved via summary judgment, and we demonstrate this potential judicial efficiency by applying our framework to various potential situations involving refusals to deal. In Section 5 we attempt to answer some potential critiques of our solution, and in Section 6 we conclude with suggestions for further exploration.

2. IP AND ANTITRUST BASICS

2.1. Patents, Copyrights, and Trade Secrets

Intellectual property is any product or result of a mental process that is given legal protection against unauthorized use (Bagley, 2002a). Intellectual property law identifies four categories of innovation – patents, copyrights, trade secrets, and trademarks – and provides separate protection for each.

Although the present controversy focuses on patents and copyrights because they are the most likely forms of IP to be at issue in a refusal to license, refusal to disclose trade secrets has also been a problem in the antitrust context. As part of the consent decree[10] settling the suit brought by the United States and a number of individual states, Microsoft Corporation is obligated to reveal certain trade secrets to competitors, for instance, providing early disclosure of its application programming interfaces ("APIs").

A copyright[11] includes the government-granted right to prevent others from copying the original expression embodied in an original work of authorship. The exclusivity of copyright protection has certain limits. "Fair use" is permitted, such as copying portions of the copyrighted work for criticism, comment, news reporting, teaching, scholarship, or research.

A patent is the government-granted right that protects any "useful process, machine, manufacture, or composition of matter, or any new and useful improvement thereof"[12] that is both novel and non-obvious. After a period of time the patent expires, and the invention is dedicated to the public. The patent holder need not personally make use of the invention during the life of the patent to enjoy protection. Patents do not entitle the patent holder to make or sell the patented invention. Instead, what is granted is a right to exclude others from making, using, or selling the same invention. As with copyrights, the exclusivity of the patent grant is not unlimited. The doctrine of patent misuse keeps the patent owner from overreaching and attempting to do more than is legitimately authorized under the patent grant. Unlike copyright, there is no "fair use" defense to patent infringement. Experimental use of a patented invention is permitted but is strictly limited to instances where the actions performed are "for amusement, to satisfy idle curiosity, or for strictly philosophical inquiry."[13]

Unlike copyrights and patents, which are based on federal law, trade secrets are protected by state laws, most of which are based on the Uniform Trade Secrets Act.[14] A trade secret is any process, innovation, or information that gives its owner a competitive advantage by reason of its being kept secret. The owner of a trade secret has the right to prohibit anyone from engaging in the unauthorized use or disclosure of the trade secret for as long as the owner takes reasonable steps to keep it confidential.

Given that the intellectual property laws grant exclusive rights to the holders of patents, copyrights, and trade secrets, it is not surprising that a regime with a predisposed hostility to monopolies, such as the antitrust system, would find aspects of the intellectual property regime problematic.

2.2. Unreasonable Restraints on Trade

Section 1 of the Sherman Act provides that "[e]very contract, combination in the form of trust or otherwise, or conspiracy, in restraint of trade or commerce among the several states, or with foreign nations, is declared to be illegal."[15] Although the language in the statute is unequivocal, it has been interpreted to bar only "unreasonable" restraints on trade. As then Judge Breyer explained: "Virtually *every* contract to buy 'forecloses' or 'excludes' alternative sellers from *some* portion of the market, namely the portion consisting of what was bought."[16] Under this judicial construction of the antitrust laws, the "Rule of Reason," the courts compare the potential harms to competition against the benefits or procompetitive justifications relating to the restraint. If the procompetitive effects of the challenged conduct outweigh any anticompetitive harm, then under the Rule of Reason, the practice is not unlawful.

2.3. Monopolization

Section 2 of the Sherman Act makes it a felony for a firm to monopolize, or attempt to monopolize, any part of trade or commerce. The offense of monopolization has two elements: (1) monopoly power in the relevant market; and (2) willful acquisition or maintenance of that power through means other than superior product development, business acumen, or historic accident.

The U.S. Supreme Court defines monopoly power as "the power to control prices or exclude competition" in the relevant market.[17] The relevant market includes all products that consumers can reasonably interchange for the same purposes. As the ability of consumers to switch to other suppliers restrains a given firm from raising prices above the competitive level, precise definition of the relevant product market is critical. The broader the definition, the less dominant one firm's position will be. In *Microsoft 2001*, Microsoft argued, unsuccessfully, that the relevant market was not just Intel-compatible PC operating systems but also middleware (such as Java or Internet browsers), hand-held computers, and the Apple Computer operating system.

Having a monopoly does not by itself violate Section 2. A Section 2 violation occurs only when a firm acquires or maintains, or attempts to acquire or maintain, a monopoly by engaging in anticompetitive conduct. As Judge Learned Hand explained, "The successful competitor, having been urged to compete, must not be turned upon when he wins."[18] Anticompetitive conduct includes refusal to deal and other exclusionary practices. In many cases, it is difficult to determine whether any particular act is exclusionary or merely a form of vigorous competition.

2.4. Duty to Deal

In general even a monopolist does not have a duty to deal with others. Indeed, the Patent Act expressly provides that refusing to license one's patented property or imposing conditions on licenses of such property is not itself unlawful in most circumstances. Unilateral refusals to deal are most likely to be problematic when they are used to "punish" others for doing business with a competitor, such as when Microsoft charged OEMs higher prices for Microsoft's Windows operating system if those OEMs installed Netscape Navigator. The instances where the courts have found a duty to deal fall into two categories.

2.4.1. Essential Facilities Doctrine

A monopolist has a duty to deal with its rivals when it controls an essential facility – a resource necessary to its rivals' survival that they cannot feasibly duplicate. To prevail on an essential facilities claim, a plaintiff must establish that (1) access to the facility is essential to other firms' ability to compete; (2) the facility is controlled by a monopolist; (3) the plaintiff cannot practically or reasonably duplicate the facility; (4) the plaintiff has been denied the use of the facility; and (5) it is feasible to provide the facility to the plaintiff.[19]

The earliest application of the Essential Facilities Doctrine involved a group of railroads that used the same bridge over the Mississippi River to transport freight and passengers to St. Louis. Several of the railroads bought the company that controlled the bridge and terminal at St. Louis, then refused to permit the other railroads to use the bridge even though it was the only bridge in the area. The U.S. Supreme Court held in *Terminal Railroad*[20] that the bridge constituted an essential facility that had to be made available, on reasonable terms, to all of the railroads.

The Essential Facilities Doctrine does not require a firm to share resources that are merely useful. A company need not share technology that would allow competitors to compete more effectively nor resources that competitors could duplicate on their own. In practice, courts are reluctant to characterize products or facilities as "essential," particularly when there is no horizontal element involving competitors. This reluctance stems in part from the fact that there is no "free lunch" in regulation of essential facilities or natural monopolies (Lipsky & Sidak, 1999). While there is an existing pricing literature (Caillaud & Tirole, 2001; Fels, 2001; Gans, 2001; McAndrews, 1995), courts are generally ill-equipped to regulate the prices, terms, and conditions for access to the essential facility.

Reiffen and Kleit (1989) argue that *Terminal Railroad* was simply a case of horizontal monopoly, and thus to the extent the Essential Facilities Doctrine is based upon this case, the doctrine should be reexamined. Although Carlton posits that "no economic concept of essentiality exists as distinct from market power in an analysis of competition [and thus the doctrine] has only the power to confuse" (2001, p. 4, n. 4), most of the doctrine's remaining critics agree that it is appropriate in certain circumstances. The late Professor Areeda referred to the Essential Facilities Doctrine as "an epithet in need of limiting principles," arguing that it should be limited to the regulation of natural monopolies (1989, p. 841).

Certain scholars have analogized the Essential Facilities Doctrine to the law and economics of common carriers (Epstein, 1998; Langlois, 2001). The local cable television provider is one example of such a natural monopoly, as it would not be efficient to connect each home to more than one physical network. Because the local cable monopoly is usually government granted, however, cable companies are highly regulated. Accordingly, the government may impose certain requirements on the cable companies because they are operating what amounts to an essential facility. For example, one of the requirements imposed by the federal government under 47 U.S.C. § 534 is a "must carry" provision, which stipulates that cable companies must carry the signals of at least three local broadcast stations.

While most applications of the doctrine have been limited to tangible assets, a few courts have applied the doctrine to intangible assets, including those protected by intellectual property (Pitofsky, Patterson & Hooks, 2002), despite strong arguments to the contrary (Cotter, 1999; Lipsky et al., 1999). We examine several of those cases in Section 3.

2.4.2. Changes in Preexisting Business Practices

In certain instances courts have found violations of Section 2 when a firm with market power acted to change a pre-existing distribution pattern. In *Aspen Skiing Co. v. Aspen Highlands Skiing Corp.*, the U.S. Supreme Court closely scrutinized an instance where a "monopolist elected to make an important change in a pattern of distribution that had originated in a competitive market and had persisted for several years."[21] The three original Aspen skiing facilities had previously offered a combination lift ticket that allowed access to all three mountains. When one of the original facilities, Aspen Skiing, purchased one of the other facilities and opened a fourth mountain, the practice was continued and extended to the fourth mountain. At a certain point, however, Aspen Skiing terminated the program, which severely damaged the lone independent facility, Aspen Highlands. The Court found this sudden refusal to deal exclusionary because Aspen Skiing was not motivated by efficiency concerns and was willing to sacrifice short-run benefits and consumer goodwill in exchange for a perceived long-run impact on its smaller rival.

Like the Essential Facilities Doctrine, the *Aspen Skiing* decision has also been criticized. Lipsky and Sidak (1999) point out that the Supreme Court ignored any price competition between the operators or adverse consumer welfare effects of the joint marketing agreement. Carlton (2001) characterizes the situation in *Aspen Skiing* as a contract dispute without any antitrust consequence. He argues that if the joint venture produced a superior product than either company could produce independently, then there would be an incentive to produce that product. As long as Aspen Highlands would have obtained more by participating than by not participating, it should have been willing to participate. In the absence of court intervention, the parties have an incentive to arrive at an efficient bargain.

Following *Aspen Skiing*, other courts have also imposed antitrust liability when a change in longstanding business practices has occurred.[22] Carlton suggests that courts may be more willing to impose an antitrust duty to deal on parties that have successfully bargained previously because their prior behavior demonstrates that the transaction costs are not prohibitive. If there have been no prior agreements, courts would probably be more reluctant to find a duty to deal.

2.5. Monopoly Leveraging

Ordinarily, Section 2 liability is restricted to monopolistic behavior within the specific market in which the firm has monopoly power. Through leveraging, however, a firm with monopoly power in one market can gain an advantage in a separate market. A firm clearly violates Section 2 when it uses that advantage to attain monopoly power in the second market. It is less clear whether a firm can legally use monopoly power in one market to gain a competitive advantage, short of actual monopolization, in another market.

2.6. Tying

In a "tying arrangement," the seller will sell product A (the "tying," or desired, product) to the customer only if the customer agrees to purchase product B (the "tied," or undesired, product). A tying arrangement forces a buyer to purchase a product or service that it would not buy based solely on the product's or service's own merits.

There are four elements to a per se tying violation: (1) the tying and tied goods must be two separate products; (2) the defendant must have market power in the tying product market; (3) the defendant must afford consumers no choice but to purchase the tied product from it; and (4) the tying arrangement must foreclose a not insubstantial volume of commerce in the tied product.[23]

The primary concern behind the prohibition on tying "is that tying prevents goods from competing directly for consumer choice on their merits, i.e. being selected as a result of buyers' independent judgment."[24] Direct competition on the merits of the tied product is foreclosed when the tying product either is sold only in a bundle with the tied product or, though offered separately, is sold at a bundled price, so that the buyer pays substantially less for the bundled product than he or she would pay for the tying product and tied products if purchased separately (an "economic tie"). In either case the consumer becomes less willing to buy a competitor's version of the tied product even if the competitor's version is superior in terms of quality and price.

3. ORIGINS OF THE INTELLECTUAL PROPERTY RIGHT REFUSAL TO DEAL QUANDARY

While a durables manufacturing firm generally has only a single price lever in the market for its product (the foremarket), the firm may have additional levers if that firm also manufactures spare parts for a separate service aftermarket. Enterprising equipment service firms ("Independent Service Organizations" or "ISOs") may then arise to compete with the firm in that aftermarket, separate and apart from the competition the firm faces in the foremarket from other manufacturers. In a price competition model developed by MacKie-Mason and Metzler (2003), the firm considers both prices ($p^{\mathrm{f}}, p^{\mathrm{a}}$) when maximizing profits.[25] Assuming constant costs ($c^{\mathrm{f}}, c^{\mathrm{a}}$) and perfectly competitive fore- and aftermarkets ($p^{\mathrm{f}} = c^{\mathrm{f}}$ and $p^{\mathrm{a}} = c^{\mathrm{a}}$), the firm picks these prices to maximize profits (π):

$$\pi = \pi^{\mathrm{f}} + \pi^{\mathrm{a}} = (p^{\mathrm{f}} - c^{\mathrm{f}})q^{\mathrm{f}}(p^{\mathrm{f}}, p^{\mathrm{a}}) + (p^{\mathrm{a}} - c^{\mathrm{a}})q^{\mathrm{a}}(p^{\mathrm{f}}, p^{\mathrm{a}})$$

If the firm could acquire market power in the aftermarket by restricting the supply of spare parts available to competing ISOs, it could then raise p^{a} above c^{a} and increase profits, $\partial\pi^{\mathrm{a}}/\partial p^{\mathrm{a}} > 0$. Such an increase, however, would prompt a simultaneous decrease in foremarket profits, $\partial\pi^{\mathrm{f}}/\partial p^{\mathrm{a}} < 0$, because increasing p^{a} increases total cost of ownership, and buyers would switch to less costly alternatives available in the competitive foremarket. The question for the firm would then be whether $\partial\pi^{\mathrm{f}}/\partial p^{\mathrm{a}} + \partial\pi^{\mathrm{a}}/\partial p^{\mathrm{a}}$ is positive or negative. If $\partial\pi^{\mathrm{f}}/\partial p^{\mathrm{a}}$ is sufficiently negative to offset the aftermarket monopoly profits, then aftermarket monopolization will be unprofitable overall. The assertion that a competitive foremarket precludes anticompetitive aftermarket actions, which MacKie-Mason and Metzler describe as the "systems" theory, was a central question in the Supreme Court decision in *Kodak I*, a case that spawned several generations of ISO litigation. Much has been written about the *Kodak* line of cases (Joskow, 2002; Klein, 1996; MacKie-Mason et al., 2003; Shapiro & Teece, 1994), so we will only briefly summarize the essential elements of each case while highlighting those aspects that relate to intellectual property.

3.1. Kodak I

Kodak manufactures and sells photocopiers and micrographic equipment and also provides service and spare parts to its customers. Kodak directly produces some of these parts, and it uses OEMs to produce the rest. During the early 1980s ISOs began repairing and servicing Kodak photocopier equipment at prices substantially lower than those charged by Kodak. Some customers purchased their own parts from Kodak and hired the ISOs to provide the service, while others used the ISOs for both parts and service. The ISOs either obtained parts from Kodak or the OEMs, or they scavenged used parts and refurbished them.

In 1985 Kodak implemented a policy of selling replacement copier parts only to customers who purchased service from Kodak or who repaired their own machines. Kodak implemented a similar policy for its micrographic equipment products in 1986. In both instances Kodak also mandated that the OEMs follow a similar policy of not selling to ISOs. When their sources of spare parts began to evaporate, many ISOs were forced out of business; those that managed to survive lost substantial revenue.

A number of ISOs filed an antitrust suit in 1987 alleging that Kodak had violated Section 1 of the Sherman Act by tying the sale of copier service to the sale of parts. The ISOs also alleged that Kodak unlawfully monopolized and attempted to monopolize the sale of service for Kodak equipment in violation of Section 2 of the Act.

After only limited discovery, and before trial, the district court granted Kodak's motion for summary judgment, which the Ninth Circuit overturned. In order to prevail on a motion for summary judgment, Kodak had to prove that, even if all questions of fact were resolved in favor of the ISOs, Kodak would still win at trial as a matter of law. Because Kodak was appealing the Ninth Circuit's determination that a summary judgment in favor of Kodak was improperly granted, the sole issue before the Supreme Court was whether Kodak should prevail as a matter of law even if all of the facts were viewed in the light most favorable to the ISOs. To that end, Kodak argued that, as a matter of law, the existence of a competitive foremarket precluded the exercise of market power for antitrust purposes in the aftermarkets for parts and service, and thus $\partial\pi/\partial p^{\mathrm{a}}$ would always be less than or equal to zero when the foremarket is competitive.

The Supreme Court rejected Kodak's assertion and found instead that there was no fundamental economic principle that competition in the foremarket eliminates the possibility of market power in an aftermarket. The Court also held that products manufactured by a single firm – in this case, Kodak copier parts – could constitute the relevant market, particularly when the costs of switching to another supplier's product are very high and the lack of availability of the product from other sources is not known in advance. Although Kodak did not have monopoly power in the copier market, the costs of replacing the copier were high relative to the costs of parts and service and consumers were unlikely to discard their Kodak copier when offered replacement parts only as part of a service contract.

Because of these high switching costs, the Court held that the relevant product market for the purposes of summary judgment was the market for Kodak copier replacement parts. As sole provider of Kodak parts, Kodak clearly had monopoly power in that aftermarket. As a result, the Supreme Court concluded that Kodak's policy of selling replacement parts only to buyers of Kodak equipment who used Kodak repair service could conceivably be anticompetitive and illegal unless the policy had a valid business justification.

Although the issue of whether Kodak's conduct was within the scope of its intellectual property rights was not before the Court, the Court included in its opinion a footnote that became quite important in later cases. Footnote 29 states, in part:

> [We have] held many times that power gained through some natural and legal advantage such as a patent, copyright, or business acumen can give rise to liability if a seller exploits his dominant position in one market to expand his empire into the next.[26]

Because the summary judgment motion was decided early in the discovery phase, the factual record was incomplete. The Supreme Court remanded it back to the district court for a trial on the merits of the case. As guidance for remand, the Court stated:

> In the end, of course, Kodak's arguments may prove to be correct. It may be that its parts, service, and equipment are components of one unified market, or that the equipment market does discipline the aftermarkets so that all three are priced competitively overall, or that any anticompetitive effects of Kodak's behavior are outweighed by its competitive effects.[27]

Thus the court left open the possibility that $\partial\pi/\partial p^{a}$ could be less than or equal to zero, but such a determination would require a trial before the district court.

3.2. Kodak II

At trial, Kodak asserted for the first time that a valid business justification for its refusal to sell parts was its desire to exploit its copyrighted and patented products. Although the Supreme Court had stated in *Kodak I* that a refusal to deal is justified if legitimate competitive reasons exist for the refusal, the district court concluded that Kodak's proffered business justification was pretextual. Kodak's parts manager "testified that patents 'did not cross [his] mind' at the time Kodak began the parts policy. Further, no distinction was made by Kodak between 'proprietary' parts covered by tooling or engineering clauses and patented or copyrighted products."[28] In addition, the district court noted that, although Kodak equipment "requires thousands of parts, ... only 65 were patented. [Thus] this case concern[ed] a blanket refusal that included protected and unprotected products."[29]

The jury awarded treble damages of $71.8 million, and the court imposed a ten-year injunction requiring Kodak to sell "all parts" to ISOs. On appeal, the Ninth Circuit held that "a monopolist's 'desire to exclude others from its [intellectual property] is a presumptively valid business justification for any immediate harm to consumers' "[30] resulting from that monopolist's unilateral refusal to license a patent or copyright or to sell its patented or copyrighted work. Nonetheless, the Ninth Circuit also held that such a presumption may be rebutted by evidence that the intellectual property protection was acquired unlawfully or that its use as a business justification was merely pretext.

The Ninth Circuit recognized the tension between antitrust law's concerns about monopoly as a threat to competition and the public policy underlying the copyright and patent laws granting limited monopolies as an incentive to innovate. The court acknowledged that patent and copyright holders might refuse to sell or license their protected work, even if the result would give them a monopoly on the patented or copyrighted products. The court noted that it could "find no reported case in which a court has imposed antitrust liability for a unilateral refusal to sell or license a patent or copyright."[31]

The Ninth Circuit cited *Kodak I*, however, for the proposition that a monopolist who acquires a dominant position in one market through patents and copyrights may violate Section 2 if it exploits that dominant position to extend the lawful monopoly into a separate market. According to the Ninth Circuit, Kodak did not have a valid business justification for its policies and therefore violated Section 2 when it used its monopoly power in the Kodak copier parts market to obtain a monopoly in the Kodak copier service market.

3.3. Xerox

During the same period as the Kodak litigation, Xerox also manufactured, sold, and provided service for photocopiers. In 1984 Xerox implemented a policy of not selling parts unique to its new series 10 copiers to ISOs unless they were also end-users of the copiers. Xerox expanded the policy in 1987 to include all new products as well as its older series 9 copiers.

In contrast to the policy implemented by Kodak, however, Xerox end-users "remained free to supply the parts they purchased to any service provider, including ISOs," as long as the parts were ultimately installed in end-users' copiers (Gleklen, 2002a, p. 2). Additionally, all ISOs could still purchase parts at the same prices with the same quantity discounts as any other end-user as long as the parts were installed in their own machines. Although Xerox limited the quantity of parts to any given end-user to the quantity reasonably necessary to repair the end-user's equipment, end-users were free to resell the parts they purchased to ISOs.

At issue was a change Xerox made in its policy in 1989. Xerox tightened its policy and implemented an "on-site end-user verification" procedure to confirm that the parts ordered for end-users were actually installed in end-users' copiers. CSU filed an antitrust lawsuit alleging that Xerox's refusal to sell its patented parts or to license its copyrighted manuals and software violated Section 2 of the Sherman Act. Unlike the plaintiff in *Kodak I*, however, CSU did not make any allegations of tying.

Xerox responded by arguing that its unilateral refusal to sell patented parts and to license patented and copyrighted diagnostic software was a lawful unilateral exercise of its intellectual property rights and hence could not violate Section 2 of the Sherman Act. In an attempt to circumvent the issue of intent highlighted by the Ninth Circuit in *Kodak II*, Xerox did *not* claim that its refusal to deal was in fact motivated by its intellectual property rights; instead it asserted that the existence of such IP rights immunized its conduct.

In essence, Xerox argued that even if it could generate monopoly profits through aftermarket activities (i.e. if $\partial\pi/\partial p^{a} > 0$), there could be no antitrust violation as long as a lawful exercise of intellectual property rights was the source of the monopoly profits. Xerox also argued that CSU could not show that it was actually injured by Xerox's refusal to sell unpatented parts. Thus any injury was caused solely by "Xerox's lawful refusal to sell or license patented parts and copyrighted software."[32]

As in *Kodak I*, the district court granted summary judgment and dismissed CSU's antitrust claims, holding that if a patent or copyright is lawfully acquired, the unilateral refusal to sell or license is not unlawful exclusionary conduct under the antitrust laws, even if such a refusal impacts competition in more than one market. The appeal was taken by the Federal Circuit because, at the time, antitrust issues involving the scope of the patent grant were viewed as being within the Federal Circuit's exclusive jurisdiction.[33]

In its brief to the Federal Circuit, Xerox argued that its patents claim inventions that made *servicing* copiers and printers easier and more effective. For example, Xerox's '797 pressure roll patent notes that the invention provides the advantage of "less down-time required for replacing pressure rolls," i.e. the need for less frequent copier and printer service. Additionally, Xerox's '256 fuser roll patent states that the object of this invention is to provide a new fuser roll with a longer lasting coating, a goal that is desirable because fuser rolls are expensive. Xerox's '006 patent for a document handler belt disclosed a belt with "good document handling performance without requiring the intermittent or periodic adjustment required" by servicemen for proper operation of the belt.

The Federal Circuit held that Xerox could refuse to sell its patented replacement parts to independent service providers such as CSU. In doing so, the Federal

Circuit expressly declined to follow the Ninth Circuit's position in *Kodak II* that determination of the defendant's subjective motivation is necessary to determine whether the defendant's actions really were motivated by a desire to exploit its intellectual property.

The Federal Circuit acknowledged that intellectual property rights do not confer a privilege to violate antitrust laws, and the D.C. Circuit cited *Xerox* with approval for this proposition in *Microsoft 2001*. Unlike the D.C. Circuit, however, the Federal Circuit went on to state that, as a general rule, the antitrust laws do not prevent the owner of intellectual property rights from excluding others from use of its patented property. A patent owner who sues or counter-sues to exclude others from making, using, or selling the claimed invention is exempt from antitrust laws unless: (1) the "patent was obtained through knowing and willful fraud" on the Patent and Trademark Office; or (2) the patent infringement suit or counter-claim is a "mere sham" to disguise an attempt "to interfere directly with the business relationships of a competitor."[34] Neither of these two exceptions was at issue in the case.

The Federal Circuit rejected CSU's reliance on the Supreme Court's suggestion in *Kodak I* that a seller cannot exploit its dominant position in one market to expand into another market. The Federal Circuit characterized *Kodak I* as a tying case and suggested that when properly viewed within the framework of a tying case, footnote 29 from *Kodak I* can be interpreted as restating the undisputed premise that the patent holder cannot use the statutory right to refuse to sell patented parts to gain a monopoly in a market beyond the scope of the patent. Unlike the plaintiff in *Kodak I*, CSU had made no claims that Xerox had tied the sale of its patented parts to its unpatented products or services.

The Federal Circuit also declined to inquire into the subjective motivation behind Xerox's refusal to sell or license its patented works. Instead, the court reiterated that unless the patent infringement claims are objectively baseless, "an antitrust defendant's subjective motivation is immaterial."[35] As long as the antitrust defendant did not engage in illegal tying, Patent and Trademark Office fraud, or sham litigation, it could enforce its statutory intellectual property rights without violating the antitrust laws.

With respect to Xerox's refusal to license its copyrighted manuals and software, the court cited Supreme Court precedent emphasizing that property rights "granted by copyright law cannot be used with impunity to extend power into the marketplace beyond what Congress intended."[36] As the Ninth Circuit had done in *Kodak II*, the Federal Circuit held that "an author's desire to exclude others from use of its copyrighted work is a presumptively valid business justification for any immediate harm to consumers" resulting from the exclusionary conduct.[37] Unlike the Ninth Circuit, the Federal Circuit declined to examine Xerox's subjective motivation for relying on the copyright laws to refuse to sell its copyrighted materials. As it had held with respect to patented property, the Federal Circuit held that the antitrust plaintiff can rebut the presumption of validity only by proving that the antitrust defendant had obtained the copyrights by unlawful means or used them to gain monopoly power beyond the statutory copyright granted by Congress.

3.4. Microsoft

While this debate over unilateral refusals to deal was raging, Microsoft Corporation was embroiled in defending itself against sweeping allegations that it had maintained a monopoly in the market for Intel-compatible personal computer ("PC") operating systems through anticompetitive means. The cases arose, in large part, out of Microsoft's various attempts to unseat Netscape Navigator as the dominant Internet browser. The first case was brought by the U.S. Justice Department and nineteen states, but there were also relevant private antitrust actions as well. We will first discuss the antitrust enforcement cases, and then we will discuss a relevant private antitrust action.

3.4.1. Public Antitrust Enforcement

Less than a year and a half after the Federal Circuit decision in *Xerox*, the D.C. Circuit rendered its unanimous decision in *Microsoft 2001*. In rendering its opinion, the court acknowledged that it was deciding this case

> against a backdrop of significant debate amongst academics and practitioners over the extent to which "old economy" § 2 monopolization doctrines should apply to firms competing in dynamic technological markets characterized by network effects. In markets characterized by network effects, one product or standard tends towards dominance, because "the utility that a user derives from consumption of the good increases with the number of other agents consuming the good." For example, "an individual consumer's demand to use (and hence her benefit from) the telephone network . . . increases with the number of other users on the network whom she can call or from whom she can receive calls." Once a product or standard achieves wide acceptance, it becomes more or less entrenched. Competition in such industries is "for the field" rather than "within the field."
>
> In technologically dynamic markets, however, such entrenchment may be temporary, because innovation may alter the field altogether. Rapid technological change leads to markets in which "firms compete through innovation for temporary market dominance, from which they may be displaced by the next wave of product advancements."[38]

In particular, the court noted the lack of consensus among commentators as to whether, and to what extent, current monopolization doctrine should be modified to account for competition in technologically dynamic markets characterized by network effects. Evans and Schmalensee (2000), economists retained by Microsoft, argued that even with network effects, Microsoft's operating system could have been supplanted by a superior alternative. Fisher and Rubinfeld (2000), economists retained by the Justice Department, argued that Microsoft's anticompetitive actions have deterred innovation and hurt consumers.

In addition to articles written by the expert witnesses, much has been written about the various *Microsoft* cases (Bagley, 2002b; Bresnahan, 2002; Economides, 2001; Gilbert & Katz, 2001; Klein, 2001; Whinston, 2001). Although the D.C. Circuit's opinion covered a wide range of alleged transgressions and at times exonerated Microsoft, we focus on the aspects of the opinion addressing the OEM licensing restrictions and the alleged tying of Windows and Internet Explorer.

The D.C. Circuit upheld the district court's finding of monopoly power in a relevant market, defined as Intel-compatible PC operating systems, rejecting Microsoft's attempt to define the market more broadly. Microsoft had a greater than 95% share of that market, and its market position was protected by a substantial entry barrier, namely the applications barrier to entry. That barrier stemmed from the fact that consumers generally prefer operating systems for which a large number of applications have already been written, and developers generally prefer to write for operating systems that already have a substantial consumer base.

Microsoft viewed Sun Microsystems' Java as a threat to this applications barrier to entry because Java ran on top of Windows as well as other operating systems and thus applications written for Java could easily be ported to those other operating systems. The Netscape Navigator browser included Java, and was therefore also a threat to the applications barrier to entry, as versions of Navigator were available for multiple operating systems. In addition to its assault on Netscape, Microsoft also attempted to either wrest control of Java away from Sun or at least turn Java into merely another method of writing Windows applications. To further this effort, Microsoft developed a faster but incompatible version of the Java Virtual Machine to run on top of Windows.

Although the appeals court concluded that Microsoft did not violate the antitrust laws when it developed and promoted its own Java Virtual Machine, the court upheld the finding that the provisions requiring OEMs to install Microsoft's Java Virtual Machine as the default violated Section 2 because Microsoft had offered no procompetitive justification. Similarly, the court upheld the finding that Microsoft's campaign to deceive developers and its threats to Intel to dissuade Intel from supporting Java both violated Section 2.

The appeals court also held that "with the exception of the one restriction prohibiting automatically launched alternative interfaces, all the OEM license restrictions at issue represent uses of Microsoft's market power to protect its monopoly, unredeemed by any legitimate justification" in violation of Section 2.[39] Microsoft had attempted to defend its OEM license restrictions as lawful conditions on the rights it gave others to use its intellectual property. The court rejected Microsoft's "bold and incorrect" position that if intellectual property rights have been lawfully acquired, then their subsequent exercise cannot give rise to antitrust liability. The court characterized this assertion as "no more correct than the proposition that use of one's personal property, such as a baseball bat, cannot give rise to tort liability."[40] As noted earlier, the D.C. Circuit cited with approval language from the Federal Circuit's opinion in *Xerox*: "Intellectual property rights do not confer a privilege to violate the antitrust laws."[41]

The D.C. Circuit did, however, accept Microsoft's argument that even a copyright holder with monopoly power may legally limit a licensee's ability to engage in significant and deleterious modifications of a copyrighted work. Even so, the court concluded that the only license restriction necessary to prevent a substantial alteration of Microsoft's copyrighted work was a restriction on OEMs' ability to automatically launch a substitute user interface upon completion of the boot process. None of the other elements of the ruling regarding product design specifically implicated intellectual property rights.

In perhaps the most groundbreaking part of the opinion, the D.C. Circuit deviated from the Supreme Court's holding that tying is per se illegal if there are separate products and the defendant has market power in the tying product. Notwithstanding its earlier ruling[42] that Windows and Internet Explorer were not separate products for purposes of construing the 1995 consent decree,[43] the *Microsoft 2001* court determined that Windows and Internet Explorer were, in fact, two separate products. Nonetheless, the court concluded that applying per se analysis to software platform products creates undue risk of error and of thwarting welfare-enhancing innovation. Accordingly, it remanded the case to the district court for its evaluation of whether, under the Rule of Reason, the procompetitive advantages of integration outweighed the costs to consumers of impairing consumers' ability to make direct price/quality tradeoffs in the tied market.

After remand, on November 1, 2002, the District Judge Kollar-Kotelly entered a final consent decree. The consent decree not only included a variety of provisions intended to address the allegations that were at the core of the prior litigation, but also included a provision that had received scant attention during most of the litigation. Whereas much of the litigation had focused on Microsoft's efforts to keep Sun Microsystem's Java from developing as a viable cross-platform threat to Windows, in the midst of a tremendous amount of discussion of the applications barrier to entry, the issue of whether Microsoft withheld access to the Windows operating system APIs in order to provide an advantage to its own applications developers was never proven. Both Judge Jackson's final judgment, which was overturned, and Judge Kotar-Kotelly's final judgment, which seems to have ended the case, specifically mention access to the Windows APIs. Although Judge Jackson's emphasis on API disclosure was part of his subsequently overturned divestiture remedy, the consent decree approved by Judge Kotar-Kotelly requires Microsoft to disclose all of the APIs that its own products use to interface with the Windows operating system. This requirement not only gives developers of competing middleware access to the necessary APIs to be fully compatible with Windows but also gives other applications developers writing directly for the Windows platform simultaneous access to the same information, in terms of APIs, as Microsoft developers.

3.4.2. Private Antitrust Actions

There are at least two private antitrust actions that questioned whether Microsoft Windows could be categorized as an essential facility. In *David L. Aldridge Co. v. Microsoft Corp*,[44] a small software manufacturer argued that Microsoft's Windows 95 was an essential facility. The district court held that an operating system such as Windows could be an essential facility for application software, such as a word processing program, but not for a software product whose sole purpose is to improve on the facility at issue, such as a disk cache accelerator improving the performance of the underlying operating system. Therefore, the application of the Essential Facilities Doctrine turned on whether the software product ran on top of the operating system or inside of it. If the latter, the Essential Facilities Doctrine would have been inapplicable. If the former, then the Essential Facilities Doctrine might have applied.

In more a recent case, *Sun Microsystems v. Microsoft*,[45] which was decided after *Microsoft 2001* and the subsequent consent decree, the district court had ruled that Microsoft's prior anticompetitive behavior had fragmented Sun's Java platform and had destroyed its channels of distribution for that platform. In a remedy somewhat analogous to common carrier regulation, the court imposed a preliminary injunction that required Microsoft to include Sun's most current, unmodified version of Java in every copy of Windows or Internet Explorer. Although Microsoft had other obligations as part of the injunction, the "must-carry" provision seemed to treat Microsoft Windows as an essential facility, although not quite in the same fashion as in the *Aldridge* case.

On appeal, the Court of Appeals for the Fourth Circuit emphasized the requirement that in order for a court to impose a preliminary injunction before trial, the court must find that there would be immediate irreparable harm in the absence of such an injunction such that the change in the status quo prior to trial would significantly hinder the court's ability to craft a remedy.[46] Although the Fourth Circuit disagreed with the district court as to whether such circumstances existed, it did not criticize the "must carry" remedy itself when it partially overturned the preliminary injunction and remanded the case back to the district court.

This remedy is similar to that sought by the plaintiffs in the public antitrust enforcement action, where the court was being asked to force Microsoft to include the latest version of Netscape's browser. In discussing that proposed remedy, which was ultimately rejected by the D.C. Circuit in *Microsoft 2001*, Lipsky and Sidak (1999) point out that the analogy to a common carrier is not entirely appropriate, as Microsoft would not have been required to carry browsers from other companies, just the Netscape browser.

3.5. Other Relevant Cases

Discussion of several additional patent and copyright cases is necessary to demonstrate the full range of decisions in this area. As above, the courts struggle in these cases to clearly and consistently define the extent of power that intellectual property law gives patent and copyright holders in the face of antitrust law. Also as above, each court approaches this dilemma uniquely, in both interpretation and application of intellectual property and antitrust law; thus courts come to distinct and sometimes conflicting conclusions. This confusion, of course, only adds to the complexity in this field and increases the need for a simpler, comprehensive solution.

3.5.1. Other Patent Cases

3.5.1.1. Intergraph v. Intel. Intergraph Corp. v. Intel Corp.[47] involved Intel's threat to discontinue providing advanced technical information to a microprocessor customer that had sued Intel for patent infringement. Intel manufactured high performance microprocessors that were sold to a variety of OEMs. One such OEM, Intergraph, made workstations for computer-aided graphics. From 1987 to 1993, Intergraph based its workstations on its own microprocessor technology called "Clipper," but in 1993 Intergraph changed over to Intel microprocessors. In 1994 Intel designated Intergraph a "strategic customer" and began to provide special benefits such as early access to technology under non-disclosure agreements.

In 1996, Intergraph determined that certain OEMs were infringing several of its Clipper patents through their use of Intel microprocessors. Intel attempted to negotiate access to the Clipper technology, but unsuccessfully. Intergraph sued Intel for patent infringement in 1997. Intel responded by threatening to cut off the special strategic customer benefits, and Intergraph sought to enjoin Intel from doing so. The district court characterized Intel's refusal to deal as denying access to an essential facility and found that this refusal violated Section 2 of the Sherman Act.

In an opinion issued two years before *Xerox*, the Federal Circuit reversed, holding that unilateral conduct permitted under patent and copyright laws is not subject to antitrust scrutiny.[48] The court also held that "Intel's conduct with respect to Intergraph does not constitute the offense of monopolization or the threat thereof in any market relevant to competition with Intergraph."[49] Thus, absent direct competition between Intel and Intergraph, no essential facility for antitrust purposes existed.[50]

Although Intel prevailed before the Federal Circuit, a concurrent FTC action ultimately forced Intel to acquiesce. Per the consent decree, Intel agreed to continue providing advanced technical information for ten years to microprocessor customers even if one of those customers initiated an infringement action against Intel.

3.5.1.2. SCM Corp. v. Xerox Corp. Another relevant case involving antitrust liability for actions within the scope of the patent grant is *SCM Corp. v. Xerox Corp.*,[51] which the Ninth Circuit cited several times in *Kodak II*. As part of its strategy to develop the plain-paper copier, "Xerox had more than 1,000 patents, mostly related to xerography and plain-paper copying."[52] Xerox only used 35–40% of those patents in actual Xerox products. In essence, Xerox had accumulated most of those patents to create a defensive thicket around its photocopier technology. Having made substantial R&D investments, Xerox wanted to exploit its copier technology on its own. As a result, during the 1960s and early 1970s, Xerox refused to grant licenses for plain paper copying under its patents but did grant licenses under its patents for other fields, including coated paper copiers. SCM had obtained a license for coated paper copying patents but then attempted unsuccessfully to obtain a license for plain paper copying patents.

SCM filed an antitrust claim against Xerox alleging that Xerox's acquisition of its patents and subsequent exclusion of their use by others violated the antitrust laws and injured SCM. SCM asserted that Xerox acquired most of its patents to block others from making plain paper copiers rather than for protecting improved Xerox products and processes, and that Xerox's failure to use the majority of its patents reflected this intent. Although Xerox's patents were so numerous and complex that they created a "thicket" that prevented SCM from designing around the patents, the Second Circuit ultimately held "that where a patent has been lawfully acquired, subsequent conduct permissible under the patent laws cannot trigger any liability under the antitrust laws."[53]

3.5.2. Other Copyright Cases

3.5.2.1. MAI Systems v. Peak Computer. In *MAI Systems Corp. v. Peak Computer Inc.*,[54] MAI was a computer manufacturer that designed software to run on its computers. MAI also provided service for its customers' computers and the software necessary to operate its computers, including the operating system. Peak Computer maintained computer systems for its clients, including more than one hundred clients that used MAI computers. Peak competed with MAI in servicing MAI equipment, which comprised between 50 and 70% of Peak's business. As part of its service program, Peak technicians would often operate the computer and its operating system software in order to service the machine. In 1992, MAI sued Peak for copyright infringement, alleging, among other things, that every time a Peak technician operated the equipment or loaded a diagnostic program, an infringing copy of the software was created in the computer's RAM. The Ninth Circuit held that "the loading of copyrighted software into RAM creates a 'copy' of that software in violation of the Copyright Act"[55] and that such copying by Peak was unauthorized, even if for purely diagnostic purposes as part of servicing or repairing the customer's equipment.

3.5.2.2. Data General v. Grumman. The last case we discuss in this section is *Data General Corp. v. Grumman Corp.*,[56] which the courts in both *Kodak II* and *Xerox* cited. Data General and Grumman competed in the service market for computers manufactured by Data General. Although Data General controlled approximately 5% of the highly competitive primary market for mini-computers, it controlled approximately 90% of the secondary market for service of Data General computers. Grumman was the leading ISO[57] with approximately 3% of the available service business.

The first ISOs appeared in the early 1970s. Data General fostered the growth of these ISOs with relatively liberal policies concerning ISO access to service tools from 1976 until some point in the mid-1980s. Much as Kodak and Xerox changed their policies to maximize revenue from their service business, so too did Data General. In the mid-1980s, Data General stopped providing many service tools directly to ISOs or allowing them to purchase certain repair items. Additionally, Data General severely restricted the licensing of ADEX, the new software diagnostic for its MV computers, its most advanced computer systems and an increasingly important source of sales and service revenue for Data General.

While the ISOs could obtain parts, service tools, and most required software from other sources, ADEX was only available to Data General technicians or those equipment owners that performed their own service. ISOs were also unable to obtain ADEX equivalents from sources other than Data General. ISOs such as

Grumman found various ways to circumvent Data General's ADEX restrictions and acquire copies of ADEX in order to maintain libraries of diagnostics, so that their technicians could more easily provide service to customers with Data General's MV computers.

When Data General sued Grumman for copyright infringement in 1988, Grumman asserted that Data General could not maintain a copyright infringement action because Data General had used its ADEX copyrights to violate Sections 1 and 2 of the Sherman Act The district court granted Data General's motions for summary judgment with respect to Grumman's tying claim under Section 1 as well as its monopolization claim under Section 2,[58] and the First Circuit affirmed both rulings. The First Circuit held that Grumman failed to prove that ADEX and Data General support were two distinct products, and thus no tying arrangement could exist. As to the Section 2 claims, the First Circuit held that "while exclusionary conduct can include a monopolist's unilateral refusal to license a copyright, an author's desire to exclude others from use of its copyrighted work is a presumptively valid business justification for any immediate harm to consumers."[59]

4. MAKING SENSE OF *KODAK*, *XEROX*, AND *MICROSOFT*

The situation that confronts the intellectual property marketplace is one fraught with confusion and conflicting directives. The Ninth Circuit's approach in *Kodak II* seems deeply flawed.[60] By making liability turn on the defendant's subjective intent, the Ninth Circuit would make it difficult or impossible for defendants to have the case decided on a motion for summary judgment. It basically rewards the company that has best educated its managers on the need to constantly repeat in their e-mails and elsewhere that the reason they are not dealing with others is their desire to exploit their intellectual property rights.[61] Another company engaging in essentially the same conduct would be treated differently, not because its conduct was more or less harmful to competition, but because it had done a poorer job schooling its managers. Instead, the test should be whether the practice in fact harms competition.

The D.C. Circuit's position in *Microsoft 2001* seems equally problematic. Following the court's baseball bat analogy, it would seem that if a fan ran onto the field and approached a player warming up inside the on-deck circle from behind, the player should not be liable if the fan were struck by the bat during the normal course of swinging the bat. Yet if one followed the logic of *Microsoft 2001*, liability would have to be imposed no matter how legitimate the bat swinging may have been. Had the D.C. Circuit correctly followed the logic of the Federal Circuit's *Intergraph* and *Xerox* opinions, it would have focused on whether Microsoft's activity exceeded the scope of its grant of intellectual property protection. While its baseball bat analogy carries rhetorical heft, absent an examination of the context of batsmanship, the analogy ends up like a corked bat that has shattered at the plate, exposed as hollow for all the world to see.

Although *Microsoft 2001* dealt only with copyrighted products, *Kodak II* dealt with both patented and copyrighted products. Since *Nobelpharma*, which was decided less than one year after *Kodak II*, all cases where patent law conflicts with other causes of action, including antitrust, have been handled by the Federal Circuit.[62]

In *Xerox*, the Federal Circuit explicitly rejected the intent test of *Kodak II*; therefore, for patents at least, one could argue that *Kodak II* is no longer valid law. Instead, *Xerox* holds that any activity within the scope of the patent grant is fair game. We see no reason why the ability to refuse to license intellectual property should turn on whether the asset is protected by copyright or by patent. Such an approach would be particularly troublesome as applied to software code and other products eligible for both forms of protection.

In the context of refusals to deal, some have argued that the Supreme Court position in *Kodak I* was correct but may have been subsequently misinterpreted by the circuit courts (MacKie-Mason, 2002; MacKie-Mason et al., 2003; Pitofsky, 2000). While not addressing the intellectual property issues, Joskow suggests that after almost ten years of post-*Kodak* litigation,

> the end result appears to be moving toward a legal rule which effectively says that postcontractual holdups are not an antitrust problem if the ex ante market is reasonably competitive, buyers (franchisees) have been adequately informed about the suppliers' (franchisers) intentions, and the supplier (franchiser) does not implement ex post changes in policies that were not revealed or reasonably anticipated ex ante. If these conditions are met then the buyer (franchisee) may get a trial on its postcontractual holdup claims, but it must show that a holdup that exploits specific investments has actually occurred and that there is no countervailing business or efficiency justification for the changes in behavior (Joskow, 2002, pp. 111–112).

Others argue either that *Kodak* was wrongly decided (Carlton, 2001) or that *Xerox* represents a better approach when the intellectual property issues are fully considered (Gleklen, 2002b).

As for the *Microsoft* case, Economides (2001) considers the sunk costs and network effects to be so strong that Microsoft might be impossible to displace. Others suggest that although network effects may indeed allow monopolies such as Windows to erect high barriers to entry, those barriers can be overcome in certain circumstances, such as when there is divided technical leadership (Bresnahan, 2001) or the introduction of disruptive innovations (Bresnahan, 2001; Christensen, 1997; Liebowitz & Margolis, 1999). The existence of significant network effects was the lynchpin for much of the Government's position in the *Microsoft* case, yet the empirical verification of such effects in the software industry has been elusive. Certain studies claim to demonstrate network effects in the software industry (Brynjolfsson & Kemerer, 1996; Gandal, 1994), while others question whether network effects have any significant consequence in the software industry (Liebowitz & Margolis, 2001). Bresnehan documents that executives at Microsoft took the issue of network effects quite seriously, however, and responded accordingly (Bresnahan, 2001).

Although we are most troubled by the treatment of intellectual property in *Kodak II* and *Microsoft 2001*, we also feel that *Xerox* goes too far in that it allows Xerox to change its policies retroactively. Had the Federal Circuit limited *Xerox* to the implementation of new policies for new products rather than also allowing retroactive policy changes for existing products, the resulting opinion would have avoided the anticompetitive problems that the Supreme Court identified in *Aspen Skiing* and which courts have admonished against in franchise tying cases.

While granting absolute immunity within the scope of the patent grant allows for quick disposal of cases at summary judgment, the subjective intent test in *Kodak II* will almost always have to be decided by a jury. An objective test that can be handled at summary judgment is desirable, but the *Xerox* position goes too far. Additionally, what is needed is an objective test based on market realities rather than "[l]egal presumptions that rest on formalistic distinctions."[63] We propose such a test in the following section.

4.1. Our Proposal

Our proposal involves a refinement of the Federal Circuit's position that any activity within the scope of the intellectual property grant is permissible. We accomplish that refinement by incorporating a variation on the concept of adverse possession for the intellectual property space. Just as adverse possession has potential benefits in the arena of real property, similar benefits could be obtained by applying a variant of this principle to intellectual property. Of course, analogizing a real property concept in the intellectual property space is never a perfect fit. The nature of intellectual property is such that almost any real property analogy will require modification. Unlike real property, intellectual property is an intangible good and therefore non-rivalrous (Romer, 1990). Intellectual property is also non-excludable and thus difficult to monitor and control in terms of dissemination and use without additional legal protections.

Alternative analogies may also exist that are preferable on certain dimensions yet are also inadequate on other dimensions. An obvious alternative to the adverse possession analogy would be an easement for intellectual property, but we reject the easement analogy for reasons discussed in Section 5 and proceed with the adverse possession analogy.

Given that intellectual property owners should be rightfully entitled to claim the full scope of their property grant, we agree with the Federal Circuit's position that they should be able to exclude other firms from developing aftermarkets within the scope of the original intellectual property grant, via unilateral refusals to license their intellectual property. If other firms are not excluded, however, and the aftermarkets develop and provide a service that is separately demanded by consumers, then we propose that after a sufficient period of time, adverse possession would preclude the original intellectual property owners from excluding the other firms from the aftermarket for existing products.

In essence, we are proposing that a firm's ability to increase profits based on activities in the aftermarket, $\partial\pi/\partial p^{a} > 0$, should not raise antitrust concerns as long as the activities are grounded in a lawful exercise of intellectual property rights, unless: (1) the intellectual property owner has failed to claim the entire scope of the intellectual property grant relative to potential aftermarkets; or (2) the intellectual property owner has market power in the foremarket and the underlying product is a platform technology characterized by network effects, such as an operating system. In the first situation, we propose applying our variation of adverse possession. In the latter (and certainly rarer) situation, we propose an application of the Rule of Reason if the underlying product is found to be an essential facility and the holder of the essential facility competes in the aftermarket. Before we explore the specific application of our proposal, however, it is instructive to review some of the justifications for adverse possession as applied to real property.

4.1.1. Relevant Justifications for Adverse Possession

The most relevant economic justification for adverse possession is that "adverse possessors who have occupied a piece of land for a long period may have developed considerable reliance interests that would be lost if the true owner could reclaim title at any time" (Miceli & Sirmans, 1995, p. 161). Judge Posner characterized the rationale as follows:

> Over time, a person becomes attached to property that he regards as his own, and the deprivation of the property would be wrenching. Over the same time, a person loses attachment to property that he regards as no longer his own, and the restoration of the property would cause only moderate pleasure. This is a point about diminishing marginal utility of income. The adverse possessor would experience the deprivation of the property as a diminution of his wealth; the original owner would experience the restoration of the property as an increase in his wealth. If they have the same wealth, then probably their combined utility will be greater if the adverse possessor is allowed to keep the property (Posner, 1998, pp. 89–90).

Other scholars have characterized this justification as moral in nature (Merrill, 1984; Singer, 2001). Professor Singer described the situation this way:

> The true owner and the adverse possessor [develop] a kind of relationship. The possessor comes to expect and may have come to rely on the fact that the true owner will not interfere with the possessor's use of the property. If the adverse possessor were to be ousted from the property, she would experience a loss. The adverse possessor's interests grow stronger over time as she develops legitimate expectations that the true owner will continue to allow her to control the property. In other words, she may rely on continued access, both in the sense of relying on use of the land itself and relying on the relationship that makes such access possible.
>
> ... The possessor has come to expect continued access to the property, and the true owner has fed those expectations by her actions (or her failure to act). It is morally wrong for the true owner to allow a relationship of dependence to be established and then to cut off the dependent party.
>
> ... After the true owner has acquiesced in the adverse use for a long enough period of time, ... [t]he rules protect the more vulnerable party to the relationship by shifting ownership from the true owner to the adverse possessor (1988, pp. 666–667).

This situation seems quite analogous to the fact patterns in both *Kodak* and *Xerox*. If we view the maximum scope of the patent grant as a piece of property, the servicing of products based on the patent grant is initially within the metes and bounds of the property grant. Kodak and Xerox initially left part of that space vacant, and the ISOs subsequently occupied that space.

Given that antitrust is designed to enhance consumer welfare by protecting competition rather than competitors, a reliance interest by the ISOs alone would be insufficient. A social welfare analysis that is agnostic as to the intent of the antitrust regime, however, would also examine the investment of the ISOs in terms of such costs, separate and apart from any customer-based reliance interest. Allowing ISOs to invest heavily in product specific training or equipment and then forcing them to abandon those investments would seem deleterious to overall economic welfare.

Kodak and Xerox customers also have a reliance interest. When they were purchasing their equipment, the existence of ISO service may have been factored into their purchase decision. It is the reliance interest of the consumers that provides the justification for the application of adverse possession in the intellectual property space.[64] As a result, the right to limit the IP owner's right to exclude rests with the existing customers and is not specific to any one ISO. Once that customer reliance interest is created, any ISO, regardless of when it entered the market, can assert it on behalf of the existing customers.

A different economic justification for adverse possession "is that it tends to prevent valuable resources from being left idle for long periods of time by specifying procedures for a productive user to take title from an unproductive user" (Cooter & Ulen, 1988, p. 156). This Lockean notion of highest and best use of the land, which rewards "the useful laborer at the expense of the sluggard" (Rose, 1985, p. 79), would suggest that if the ISOs are able to deliver a superior level of service, they should be allowed to continue to do so if Kodak and Xerox did not exclude them from the service market from the beginning.

Our proposal of adverse possession for intellectual property only applies if the adverse possession occurs in a marketplace in which the adverse possessor is already established. If the ISO service were so inferior that consumers did not demand it, then under the separate products test of *Jefferson Parish*,[65] no separate market and thus no anticompetitive conduct would exist. In *Jefferson Parish* the Court held that no tying arrangement can exist unless there is a sufficient demand for the purchase of service A, separate from product B, to identify a distinct product market in which it is efficient to offer service A separately from product B. Thus, adverse possession for a portion of the intellectual property space can be justified based on the reliance interest of the customers of the patented products as well as the increased customer welfare of allowing the competitive service to continue to be offered.

Adverse possession for real property is far from instantaneous and is difficult to accomplish. To prevail on a claim for adverse possession, the adverse possessor must demonstrate: (a) actual possession that is; (b) open and notorious; (c) exclusive; (d) continuous; (e) adverse or hostile; and (f) for the statutory period (Singer, 2001, p. 137). Any application of adverse possession to intellectual property should be similarly constrained; however, the nature of intellectual property necessitates some modification to the doctrine. Actual possession would be established by one or more firms occupying the aftermarket that falls within the scope of the patent grant, so long as those firms are not infringing on the patent itself. This modification would also satisfy the open and notorious requirement, given that the ISO-type firms would be purchasing or licensing with the full knowledge of the intellectual property owner.

Exclusive occupation would differ from the real property concept because of inherent differences between real property and intellectual property. It is physically impossible for a sufficiently small piece of real property to be simultaneously occupied by more than one person; yet those same two people could simultaneously make use of the same piece of intellectual property, a distinction that highlights the difference between the rivalrous nature of real property and the non-rivalrous nature of intellectual property. In the intellectual property arena, the determining factor would be whether the intellectual property owner (or an exclusive licensee) occupies the aftermarket exclusively. In essence, non-exclusivity on the part of the intellectual property owner would be the required criteria for a successful claim for adverse possession, as long as the additional occupiers were non-infringing. We should make it clear that infringement of the intellectual property owner's IP rights would never qualify for adverse possession.

While the concept of continuity would be similar, the requirement that the possession be adverse or hostile would be different. Given that the intellectual property owner is the source for patented spare parts, one could argue that the occupation of the aftermarket is permissive. Since permissive use of real property generally defeats a claim for adverse possession, we would therefore propose that the occupation of the aftermarket must only be competitive to qualify as adverse or hostile. If the intellectual property owner chooses not to compete, the potential for competition would also be sufficient for a claim of adverse possession. Again, while infringement might be viewed as an adverse or hostile activity, we do not suggest that such infringement ever qualify for adverse possession.

In terms of hostile or adverse occupation, our proposal for the intellectual property space clearly differs from traditional adverse possession for real property. Our construction, however, closely follows Singer's reinterpretation of adverse possession, which suggests that the doctrine requires only longstanding use rather than requiring non-permissive longstanding use. We propose that as long as the use is non-infringing, the adverse possession doctrine for the intellectual property space should not necessarily distinguish between longstanding permissive use and longstanding non-permissive use.

The last requirement for a successful adverse possession claim is that the occupation be for a sufficiently long period of time. Given that intellectual property grants are limited in duration, it would seem to follow that its required timeframe would be shorter than the required timeframe for real property. Although we do not suggest what the appropriate period of time should be for intellectual property, the *Data General* case discussed earlier might provide a basis for determining a sufficient period. As Hovenkamp, Janis and Lemley point out, the First Circuit held that the withdrawal of Data General's support could not be proof of anticompetitive effect because the old policy of permitting ISO competition had never led to a competitive market. Thus, a sufficient period of time might be stated in terms of market competitiveness rather than in months or years, which would also be consistent with our requirement that the occupation be competitive to qualify as hostile or adverse. Alternatively, the sufficient time period could be stated in terms of the time necessary to make sufficient investment to compete in the aftermarket. Such a determination would take into account such factors as the level of investment required or the asset specificity of such investment.

In addition to a theoretical justification for our proposal, cases from several circuits provide additional support for the idea that a change in longstanding business practices should be the threshold question from an antitrust standpoint. Both the First[66] and the Seventh[67] Circuits have limited *Kodak I* to situations in which the manufacturer's policy was not generally known. The situation before the Sixth Circuit in *PSI Repair Services v. Honeywell, Inc.*[68] seems the closest to both the situations in *Kodak* and *Xerox*, and the court's reasoning seems quite consistent with our proposal.

Honeywell manufactured and sold industrial control equipment that contained and depended heavily on printed circuit boards. Honeywell did not repair the boards when they failed, but instead replaced them and charged the customer for the replacement so long as the customer returned the defective board. Although companies such as PSI offered circuit-board repair services to owners of industrial control equipment, Honeywell's stated policy was to not sell its components to anyone. Thus, there was no ISO market for Honeywell equipment because Honeywell's own actions essentially limited the existence of a separate market for components. In distinguishing *Kodak I*, the Sixth Circuit found merit in the argument that unlike Kodak, which changed its service and parts policy to include the tie after many Kodak customers purchased their equipment, Honeywell consistently maintained its policy of board replacement. In essence, the definition of the relevant market in *Kodak I* resulted from Kodak's change in policy. Although *Kodak I* did not explicitly state the extent that the change in Kodak's policy was determinative, the Court nonetheless suggested that the outcome might have been different had Kodak presented evidence that its restrictive parts policy was consistently maintained and generally known beforehand. The Sixth Circuit stated "that the change in policy in *Kodak I* was the crucial factor in the Court's decision. By changing its policy after its customers were 'locked in,' Kodak took advantage of the fact that its customers lacked the information to anticipate this change."[69] Thus, it was Kodak's own actions that increased customer information costs.

One of the Sixth Circuit's main holdings, which strongly supports our proposal, declared that

> [i]n light of our reading of *Jefferson Parish* and *Kodak*, we thus hold that an antitrust plaintiff cannot succeed on a *Kodak*-type theory when the defendant has not changed its policy after locking-in some of its customers, and the defendant has been otherwise forthcoming about its pricing structure and service policies.[70]

Having found judicial precedent to bolster the theoretical justification basis for subjecting part of the intellectual property grant to adverse possession, we extend this concept in the next two sections first to firms without market power and then to firms with market power.

4.1.2. Firms Without Market Power

If a firm without market power in any relevant product market introduces a new product based in whole or in part on a collection of intellectual property rights, it would have three options for an aftermarket that would implicate its intellectual property rights (such as the sale of patented spare parts). First, it could prohibit the development of a secondary market at the outset by refusing to license or sell to anyone. For example, Apple Computer pursued this strategy when it refused to license its Mac operating system to hardware manufacturers interested in selling Apple clones (except for a brief period when it gave short-term licenses to select hardware manufacturers). Because Apple did not have market power in either the personal computer operating system or hardware markets, this tying of the Mac OS to Apple hardware was permissible under standard tying analysis, even though the OS and hardware are separate products. Our proposal would lead to the same result.

A second alternative would be to grant limited rights to the aftermarket using contracts of relatively short duration. This second situation would be analogous to a lease of real property, with ISO investment being analogous to leasehold improvements, whereby the issue of sunk costs could then be dealt with contractually. The third alternative would be to abstain from excluding other firms in the aftermarket, thus allowing that market to arise naturally.

It is in this third scenario that our concept of adverse possession of the intellectual property space comes into play. Just as real property owners have a right to exclude others from their property, intellectual property rights should also confer a right to exclude. Open and notorious occupation of real property, however, can confer ownership to the trespasser after a given period of time. As discussed earlier, we propose that the same principle be extended to the intellectual property space. Just as the original property owner loses the right to eject those who adversely possess the real property in question successfully, a firm allowing an aftermarket to develop should not be able to extinguish that market later by changing its policy of intellectual property right enforcement. This adverse possession concept is consistent with *Kodak I* in that even if the firm does not have market power in the primary product market, it may have market power in the secondary market if there are high switching costs and it permits a secondary market to develop. Note that in this third scenario, if access to protected intellectual assets is initially required for such aftermarkets to arise, the cooperation of the firm is still required. If the aftermarket arises due to infringement, our proposal would not apply.

The prohibition against making an important change in a pattern of distribution that had originated in a competitive market and had persisted for a given period of time, as in *Aspen Skiing*, only applies retroactively. The firm would retain the right to exclude the development of an aftermarket for future models of existing products or new products. Just as Ski Co. could have opened a new mountain without including it in the All-Aspen ski pass program, the firm could sell future primary products and maintain exclusive control of any potential aftermarket for those products.

Under this proposal, customers who purchased products when a competitive aftermarket existed would not be disadvantaged. Because our proposal prevents the firm from retroactively foreclosing a competitive aftermarket, the firm is unable to increase aftermarket prices above the competitive level without sacrificing aftermarket profits, $\partial\pi^{a}/\partial p^{a} < 0$. Customers who purchased new products after the firm announced a policy of not selling replacement parts to ISOs would do so with the full knowledge of the unavailability of the aftermarket for their specific purchase. These new customers would be able to factor the lack of an aftermarket into their purchase decision. Relative to a hypothetical competitive aftermarket, the firm could increase prices in the aftermarket but still might suffer a loss in overall profits because the total cost of ownership for its product would increase relative to other products in the competitive foremarket, $\partial\pi/\partial p^{a} < 0$, as all of the necessary conditions for the systems theory to hold (MacKie-Mason et al., 2003) would seem to be present.

Despite its environmental unpalatability, an apt analogy is the creation of new coastal land through the process of dredging, such as occurred with Boston's Back Bay. Once the land was created, it theoretically would have been possible for someone, through open and notorious occupation of the landfill areas, to obtain rights in that land through adverse possession. That right of occupancy would not extend, however, to lands subsequently created unless there was similar open and notorious occupation of that new land. If the property creator excluded the prior adverse possessor from the new land, no adverse possession of that new land would take place.

Continuing with our expanded real property analogy, just as there may be public policy reasons to grant easements or rights-of-way across this newly created coastal landfill property for development purposes, it may be necessary to grant easements in the intellectual property space. Although this situation seems more likely in the copyright context given the fair use doctrine, and less likely in the patent context given the Federal Circuit's narrow interpretation of experimental use, our expanded notion of essential facilities discussed in detail in the next section could be considered an easement by necessity.

An open question remains as to whether any refusal to deal could include a refusal to sell unpatented parts. If the firm in this instance does not have market power, such a refusal would not seem to run afoul of antitrust principles. If the firm had allowed an aftermarket to arise, however, it is possible that it could be found to have market power, as in *Kodak I*. In that event, the issue of refusing to deal in unpatented parts might be more problematic, and we would propose a Rule of Reason analysis.

4.1.3. Firms with Market Power

If a firm has market power in any market, the same three options would still be available for the introduction of a new product: no ISOs, contracted ISOs, or selling parts to anyone. The firm could act to prevent any aftermarket from arising by refusing to sell or license products embodying its intellectual property, or it could grant short term contracts to provide service. If the firm allowed the aftermarket to develop, however, it would be subject to the same adverse possession rule as the firm without market power.

Because the firm would also have market power in the primary market, its ability to exclude other firms from the aftermarket ex ante is subject to an additional constraint, a variation of the Essential Facilities Doctrine. Historically, an essential facility has been found only in instances involving direct competitors. In fact, one of the core requirements for a facility to be deemed essential is a direct or horizontal competitor's inability, practically or reasonably, to duplicate the essential facility. In this instance we propose that the doctrine be extended to downstream competitors but only in industries that exhibit network effects and only for new generations of technology where the prior technology could have been characterized as an essential facility. In this instance, the Rule of Reason would apply. Part of the rationale for this extension is to maintain and enhance the incentives for interoperability.

An example of a situation involving downstream competitors would be Microsoft, as developer of an operating system, relative to an applications developer that develops a competing product to one also offered by Microsoft (such as a word processor or spreadsheet). Relative to Microsoft, both Corel and Lotus would be examples of downstream competitors in the applications market that rely on Microsoft's operating system as an essential facility as would Sun Microsystems in the Internet-enabled distributed computing platform market. We will further examine our application of essential facilities to downstream competitors in industries that exhibit network effects later in Section 4.2.3.

Our extension of the Essential Facilities Doctrine is also based in part on a civil antitrust case mentioned earlier in Section 3.4 in which one of the authors testified as an expert witness against Microsoft. In that case, *Aldridge v. Microsoft*,[71] the court stated that an operating system such as Windows could be an essential facility for application software, such as a word processing program, that ran on top of the operating system. Of course, our proposed extension of essential facilities is not limited to Microsoft operating systems.

4.2. Application to Unilateral Refusal to Deal

Having developed our proposal establishing the constraints on a firm's ability to refuse to license or sell items embodying its intellectual property, we now proceed to evaluate several relevant fact patterns to show how our proposal would be applied.

4.2.1. Large Fixed Assets (the Kodak and Xerox Scenarios)

Under our proposal, a firm that offers a large fixed asset product to the market has three choices:

- Announce a policy that it will not sell replacement parts to anyone, so that consumers buying the product will not expect ISO service to be available.
- Enter into exclusive ISO contracts of relatively short duration that will subsequently be put up to bid so that there is competition in the contract (a situation akin to a lease of real property).
- Sell replacement parts to anyone.

Kodak initially chose the third option, but then later changed to the first option. Under our proposal, the ruling in *Kodak I* would remain the same. While the outcome of *Kodak II* would also be the same, the determining factor would be entirely different. Whereas the Ninth Circuit focused on the subjective intent of Kodak's management, our proposal eschews the examination of intent altogether. Unlike the intent examination, which would almost always require that the matter go to a jury, the adverse possession determination could be made by a judge as a matter of law pursuant to a motion for summary judgment.

If Kodak introduced a new type of copier, given that Kodak does not have market power in the copier market, it would be permitted to sell its machines under a policy announced at the time of sale that there will be no replacement parts sold to ISOs. Even if Kodak initially were willing to sell replacement parts to ISOs, Kodak would be permitted to change that policy for new models going forward but not retroactively for existing models. In other words, if Kodak permitted the ISOs to occupy part of the space that Kodak could have occupied initially, then Kodak could not now extinguish that market. If this were not the case, Kodak could impose costs on consumers not contemplated at the time of purchase. This restriction means that when consumers are deciding which copier to buy, they need to take into account the effect of Kodak's service contract fees when deciding whether to purchase the Kodak copier or a copier produced by another manufacturer.

Our analysis of the facts of the *Xerox* case produces an outcome similar, but not identical, to that of the Federal Circuit's decision. Our proposal would allow Xerox to continue its policy of refusing to sell parts to the ISOs, but it would differ significantly from the Federal Circuit's decision in its treatment of Xerox's retroactive application of the policy. Xerox would be able to refuse to sell parts to ISOs for service on new models sold after the policy change, but it could not refuse to sell parts for copiers sold during the time of the previous policy. Although the ISOs would be foreclosed from servicing a part of the future market, they would still be able to service existing customers; thus those customers would also be protected. In either instance, the copier industry clearly does not exhibit network effects, so our modified Essential Facilities Doctrine would not apply.

4.2.2. Creation of a New Product Market: Segway

The Segway Human Transporter ("HT") is one of inventor Dean Kamen's latest creations. The Segway HT is a self-balancing, electric powered personal transportation machine that emulates human balance, which is controlled by subtle shifts in body weight. Prior to its unveiling in December 2002, the product was codenamed "Ginger" in order to keep the identity of the device secret. Since then, the Segway HT has gained widespread interest from consumers and businesses around the globe.

The Segway HT is protected by a number of patents, and the device is arguably sufficiently unique as to define a market in and of itself, although it could also arguably be viewed as a substitute for bicycles or motorized scooters. For our purposes here, we assume that the market for self-balancing transporters is sufficiently distinct to be a separate market. In that market, Segway LLC clearly has market power, but its initial three options would remain the same as for a firm without market power: no ISOs, contracted ISOs, or selling parts to anyone.

Currently it appears that Segway has chosen option number one, although it is offering to provide training for service technicians. Although Segway clearly has market power in the market for self-balancing, electric-powered personal transportation machines, such a market does not exhibit network effects, and thus our modification to the Essential Facilities Doctrine would not apply.

4.2.3. Operating Systems and APIs (Microsoft)

Copiers and other capital equipment have long service lives and are often replaced when they wear out. Operating systems, however, never wear out and are instead replaced when a new version is released, often with a relatively brief time between releases. Switching costs for copiers are based primarily on the cost of the new copier with the labor costs being a small fraction of total costs. Conversely, the cost of a software upgrade is usually fairly small; most of the switching costs are a function of the labor cost associated with both installing the new software and training the user how to operate it.

The switching costs for operating systems are increased dramatically, however, if all of the applications software must be replaced. To reduce those costs, most software development in the last decade or so has been done using connections to the operating system called Application Programming Interfaces ("APIs"). As the D.C. Circuit pointed out:

> Operating systems . . . function as platforms for software applications . . . by "exposing" – i.e. making available to software developers – routines or protocols that perform certain widely-used functions [called APIs]. For example, Windows contains an API that enables users to draw a box on the screen. Software developers wishing to include that function in an application need not duplicate it in their own code. Instead, they can "call" – i.e. use – the Windows API. Windows contains thousands of APIs, controlling everything from data storage to font display.
>
> Every operating system has different APIs. Accordingly, a developer who writes an application for one operating system and wishes to sell the application to users of another must modify, or "port," the application to the second operating system. This process is both time consuming and expensive.[72]

A potential problem arises when the operating system programmers work for a different company from some, but not all, of the applications programmers. If the operating system company provides its own applications programmers with earlier access to the APIs than it provides to applications programmers at other companies, then those other companies would be at a tremendous competitive disadvantage.

The Justice Department alleged a variation of this fact pattern in its first suit against Microsoft where Microsoft purportedly conditioned access to APIs on agreements by software manufacturers not to work with competing technologies. The allegation was neither part of the first consent decree approved in 1995, however, nor raised as an issue in the subsequent litigation. Yet it seems that Judge Jackson's remedy of breaking up Microsoft into an operating systems company and an applications company evidenced his concern that Microsoft was leveraging its dominance in the OS market to obtain monopoly power in the applications market. Similarly, the requirement imposed by the Final Judgment in *United States v. Microsoft Corp.* that Microsoft provide rival applications firms earlier access to its Windows APIs is intended to eliminate at least some of the advantages enjoyed by the Microsoft applications group.

Operating systems clearly demonstrate network effects, given that the more users there are for a particular operating system, the more applications are likely to be written to run on top of that operating system (Abramson, 2002; Gilbert et al., 2001; Katz & Shapiro, 1994; Lemley & McGowan, 1998; Shapiro, 1996). In his findings of fact, Judge Jackson described the situation as follows:

> Consumer demand for Windows enjoys positive network effects. A positive network effect is a phenomenon by which the attractiveness of a product increases with the number of people

> using it. The fact that there is a multitude of people using Windows makes the product more attractive to consumers. The large installed base attracts corporate customers who want to use an operating system that new employees are already likely to know how to use, and it attracts academic consumers who want to use software that will allow them to share files easily with colleagues at other institutions.[73]

The D.C. Circuit referred to this phenomenon as a "chicken-and-egg" situation that ensures applications will continue to be written for the already dominant Windows platform, which in turn ensures that consumers will continue to prefer Windows over other operating systems. Therefore, if an operating system manufacturer develops an upgrade to its operating system and refuses to give outside developers of applications software the same access to the new APIs as it gave to its internal applications software developers, application of our framework would likely result in the operating system manufacturer being compelled to grant access to the APIs on a prompt, reasonable, and non-discriminatory basis.[74]

Whereas we might not be able to achieve such a result based on monopoly leveraging, application of essential facilities in the case of a product exhibiting network effects produces the appropriate result. Our proposal would thus lead to the same result as the Final Judgment in *United States v. Microsoft Corp.*, although the Final Judgment does not rely on essential facilities as a rationale.

While Microsoft is the most obvious scenario for the application of our proposed extension of essential facilities, it is by no means the only one. If Intel had manufactured graphical workstations at the time of the *Intergraph* case, our proposal would have suggested that access to the "special strategic customer benefits" would have in fact been an essential facility. The district court found an essential facility in *Intergraph*, but incorrectly so, as there was no competition between Intel and Intergraph in any market.

4.2.4. European Application

The issues addressed in this paper are not unique to the United States, and the European Union is developing its own jurisprudence regarding such matters. Although our framework seems to be more specific than the European approach, the two can be reconciled. In the European Union, the European Commission held that it was unacceptable for IBM to tie sales of its dominant CPU to sales of its memory and service. In contrast, the Commission said that it was acceptable for Kyocera, a Japanese computer printer company, to offer discounted bundles of Kyocera-made printer toner cartridges and other Kyocera-made printer hardware even though the bundle might otherwise have been characterized as an impermissible economic tie. Pelikan, a German manufacturer of toner cartridges for use with Kyocera printers, had challenged both the bundles and Kyocera's policy of not honoring warranty claims for printers damaged by non-Kyocera toner cartridges as an abuse of Kyocera's dominant position. The Commission found: (1) that Kyocera did not have a dominant position in the market for toner cartridges despite its large share of the toner market; (2) that there was fierce competition in the primary market for printers that restrained Kyocera's behavior in the secondary market for toner; (3) that purchasers of printers were well informed about the price charged for toner and appeared to take this into account when selecting a printer; and (4) that the cost of toner was a significant part of the cost of the entire product. As a result, if Kyocera had attempted to raise the price of its toner cartridges, then the low switching costs would lead customers to junk the printer and buy another brand for which cheaper toner cartridges were available. Unlike the situation in *Kodak I* where Kodak changed its policy after customers were locked into a high cost capital asset with high switching costs, in the Kyocera case, all of the necessary conditions for the systems theory to hold (MacKie-Mason et al., 2003) were present, so $\partial\pi/\partial p^{\mathrm{a}}$ would likely be less than or equal to zero.

A case before the European Court of the First Instance involved facts similar to those in *Data General*. In *IMS Health Inc. v. Commission of the European Communities*, IMS Health developed a sophisticated method of tracking pharmaceutical sales involving a new data structure. In November and December of 2000, IMS successfully sued two other competitors for copyright infringement before the Landgericht Frankfurt and obtained an injunction prohibiting them from using the new data structure. The Oberlandesgericht Frankfurt subsequently overturned the injunction. The Commission of the European Communities then imposed a compulsory licensing requirement, concluding both that IMS's data structure was an essential facility and that IMS's refusal to license the use of its copyright constituted an abuse of the dominant position it enjoyed in the relevant market. IMS then appealed to the Court of the First Instance, which suspended the injunction. IMS then successfully argued that it enjoyed copyright protection in its data structure, ultimately prompting the Commission to abandon its efforts to require compulsory licensing of the IMS data structure.

For IMS to be found liable for refusing to license new technology under our test, IMS would have to be in a position of market power, and its copyrighted data structure would have to be both an essential facility and a type of network effects technology. While IMS seems to be in a position of market power, given that the two other competitors already have a significant number of customers and have both developed alternative (and supposedly non-infringing) data structures that they can continue to use and improve, it would appear unlikely that the new data structure would be found to be both an essential facility and to exhibit significant network effects.

4.3. Application to Unilateral but Conditional Refusal to Deal

At first glance it would seem logical that if an intellectual property right holder has the right to refuse to deal with someone unilaterally, that right would include the right to conditionally refuse to deal with that same someone, particularly given that conditional dealing might be more accommodating. As economist Carl Shapiro has testified, however, it is possible to "create incentives through conditional refusals that you can't create through unconditional refusals."[75]

While our framework was designed to answer questions regarding unilateral (or unconditional) refusals to deal, it may still provide some insight for conditional refusals to deal when we reexamine the various scenarios examined above.

4.3.1. Large Fixed Assets

The possibility of adverse possession of certain elements of the intellectual property grant would still apply in this instance, and the analysis would turn on whether incentive modifications resulting from the conditional refusal to deal result in an impermissible recapture of the intellectual property right space that was previously lost to adverse possession. If the firm has market power, its attempt to recapture that space through the imposition of conditions is more likely to be categorized as either tying or monopoly leveraging.

4.3.2. Segway

The analysis for this scenario remains the same, as long as Segway continues to foreclose the possibility of ISOs. As with the prior example, however, if Segway allows a secondary market to develop, it runs the risk of liability for either tying or monopoly leveraging if it attempts to recapture that secondary market through conditional refusals to deal.

4.3.3. Operating Systems

It is hard to discuss this scenario without being heavily influenced by the D.C. Circuit's decision in *Microsoft 2001*. We would argue that in *Microsoft 2001*, the court misinterpreted the *Xerox* opinion. Although the court correctly cites *Xerox* for its statement that "intellectual property rights do not confer a privilege to violate the antitrust laws,"[76] it seems to ignore the very next sentence in the *Xerox* opinion: "it is also correct that the antitrust laws do not negate the patentee's right to exclude others from patent[ed] property."[77]

Many of the allegations against Microsoft that could be characterized as conditional refusals to deal, however, were "garden variety" tying and monopoly leveraging in areas in which Microsoft was not asserting intellectual property protection. Given that our framework only attempts to address instances where intellectual property rights form the basis for refusal to deal, either unconditional or conditional, we have not addressed those other issues in this paper.

To the extent that Microsoft sought to justify limiting its OEMs' ability to change the user interface or its policy of selling Windows at a lower price to OEMs who agreed not to install competing software, the D.C. Circuit correctly characterized those activities as willful attempts to maintain Microsoft's monopoly on Intel-compatible PC operating systems. The primary evil the Final Judgment sought to address by requiring uniform licenses was Microsoft's attempt to deter OEMs from dealing with Microsoft's competitors. Such conduct would appear to be impermissible even in the Federal Circuit, given the Federal Circuit's reasoning in *Xerox* that a patent infringement suit or counter-claim is not immune from antitrust scrutiny when it is a mere sham to disguise an attempt to interfere directly with the business relationship of a competitor.

5. POSSIBLE CRITIQUES

Critiques of our proposal are likely to come from two different perspectives. The first concerns the appropriateness of the adverse possession analogy itself. The second involves the practicability of our proposal. We discuss each in more detail in the following two sections.

5.1. Real Property Critique – Why Not Easements for Intellectual Property?

Given that we are proposing the application of adverse possession to intellectual property, one might ask why is not easement a better analogy for what we are trying to accomplish? We respond fully to this critique elsewhere (Bagley & Clarkson, 2003). The analogy of easement has also found support relevant to our extension of the Essential Facilities Doctrine. Some scholars have suggested that if application of the Essential Facilities Doctrine required the license of an intellectual property right (O'Rourke, 1998), then the result would be an easement across the monopolist's intellectual property (Donahey, 1997). In the applications software context, for example, the software developers own their own intellectual property space, but if they are denied access to the APIs for the next generation of the relevant operating system, they could be viewed as being landlocked in their intellectual property space with no way to reach the customers who upgrade to the new operating system. We concede that an application of the Essential Facilities Doctrine that would allow the applications programmers to access the APIs could appropriately be characterized as an easement, as could the fair use doctrine applicable to copyrights.

5.2. Intellectual Property and Antitrust Critiques

Although we feel that our adverse possession analogy is preferable to an easement analogy, several potential critiques still exist relative to the applicability of our proposal. One critique might point out that the Federal Circuit has been reluctant to extend the Essential Facilities Doctrine to downstream competitors, noting in *Intergraph* that

> no court has taken it beyond the situation of competition with the controller of the facility, whether the competition is in the field of the facility itself or in a vertically related market that is controlled by the facility. That is, there must be a market in which plaintiff and defendant compete . . . . Absent such a relevant market and competitive relationship, the essential facility theory does not support a Sherman Act violation.[78]

At the time of the opinion, Intel and Intergraph were not competitors[79] and thus the Federal Circuit reversed the district court's ruling that Intergraph's dependency upon Intel's superior microprocessor product converted Intel's special customer benefits into an essential facility. Unlike the situation in *Intergraph*, however, we limit our extension of essential facilities to the case where the holder of the essential facility does compete in the downstream market, as would be the case with Microsoft and other desktop applications developers.

The district court in *Data General* also cautioned against an expansive application of essential facilities:

> The "bottleneck" of [Data General's] superior knowledge in the design of [its own] computers is insufficient to invoke the essential facilities doctrine; a better mousetrap is not necessarily an essential facility. The Sherman Act has not been interpreted to require manufacturers to abandon their advantage in creating accessories to their systems. If manufacturers of complex and innovative systems were required to share with competitors the development of accessories, because they had a possibly absolute advantage through producing the system, the incentives of copyright and patent laws would be severely undermined. Not only would the manufacturer, who is in the best position to create these accessories, have less incentive to do so, but also the impetus for competitors to reverse engineer and produce competing solutions would be reduced.[80]

We agree wholeheartedly and thus limit our extension of the Essential Facilities Doctrine to situations where the essential facility owner competes directly in the downstream markets, and only then in industries exhibiting network effects. We would not characterize copier spare parts or software diagnostics as essential facilities, but we would characterize operating system APIs necessary for developing software applications as essential facilities. Thus the application of our proposed extension is quite narrow, although not so narrow as to be limited only to the facts of the Microsoft litigation.

One could also argue that our adverse possession argument could be applied to the problem of patent non-use or patent suppression and thus be used to deprive firms of strategic options regarding their intellectual property portfolios. Our proposal is limited, however, to arenas where the patents in question are already in use either by their owners or by their licensees, and the issue is whether other firms should also be allowed to make use of the patented technology. The idea of compulsory licensing to combat patent non-use or suppression (Saunders, 2002) is thus not relevant in this instance.

Such a distinction should sufficiently insulate our proposal from the concern that it contravenes established Supreme Court jurisprudence that "[t]here is nothing in the patent law that requires working of a patented invention and the general rule is that a patentee need not use its inventions" (Sobel, 1984, p. 691). In *Continental Paper Bag Co. v. Eastern Paper Bag Co.*, the Court upheld a patentee's right to "accumulate patents merely for the purpose of protecting [its] general [industry] and shutting out competitors" in order to make more money with prior patented technology.[81] Numerous cases have subsequently upheld the right not to use a patent.[82]

6. CONCLUSION

Other scholars have attempted to address the conflict between *Xerox* and *Kodak II* from various perspectives (Gleklen, 2002b; MacKie-Mason, 2002; Patterson, 2000), but we believe that our proposal provides a more elegant model for addressing the broader question of how to evaluate refusals to license not just patents but also copyrights and trade secrets. Our proposal triangulates between the conflicting positions on the treatment of intellectual property in the shadow of the antitrust regime contained in the *Kodak II*, *Xerox*, and *Microsoft* opinions, and it stakes out a position that appropriately protects competition while still respecting the rights of the intellectual property owner.

While philosophically closest to the Federal Circuit's position in *Xerox*, our proposal starts with the premise that intellectual property owners should be rightfully entitled to claim the full scope of their property grant, and that any activity within the scope of that grant is permissible. We qualify that original premise, however, by introducing a concept of adverse possession for intellectual property as a restriction on certain activities that would otherwise be within the scope of the intellectual property grant. We justify our application of adverse possession to intellectual property using many of the same arguments that are used for real property, including reliance, moral, and Lockean justifications.

Our proposal dictates that intellectual property owners should be entitled *ab initio* to exclude other firms from developing aftermarkets within the scope of the original intellectual property grant via unilateral refusals to license the intellectual property. If, however, the other firms are not excluded and aftermarkets develop and provide a product or service that is separately demanded by consumers, then, after a sufficient period of time, adverse possession would preclude the original intellectual property owners from excluding other firms from the aftermarkets as they pertain to existing products. Consistent with our desire to protect both competition and consumers, the prohibition against excluding other firms would include both current occupants as well as future entrants. The intellectual property owner would, however, be entitled to foreclose the development of new aftermarkets relating to new products unless: (1) the intellectual property owner again fails to claim the entire scope of the intellectual property grant relative to potential aftermarkets; or (2) the intellectual property owner has market power and the underlying product is one that is characterized by network effects. In the latter case, we propose an application of the Rule of Reason if the underlying product is found to be an essential facility.

One of the requirements that we imposed on ourselves in crafting our proposal was that the resulting test be sufficiently objective so as to be appropriately applied at summary judgment. Resolving such cases at summary judgment is a more efficient use of judicial resources than the intent-based inquiry proposed in *Kodak II*, which requires a jury to subjectively evaluate the intent of the intellectual property owner. Given that the Sixth Circuit upheld the district court's summary judgment ruling in *PSI Repair Services v. Honeywell, Inc.* using a rationale that is quite consistent with our proposal, we are confident that our test can be applied at summary judgment.

Rather than merely stating a theoretical case for our proposal, we have also demonstrated how our proposal would operate in various market scenarios. We examined, both for firms with and without market power, how our proposal would resolve the situations presented by large fixed asset purchases, the introduction of entirely new products, and operating systems with network effects. We have even demonstrated how our proposal could be applied in the European antitrust enforcement context. Thus, our proposal adheres to the admonishment in *Kodak I* that any test must be objectively based on market realities rather than "[l]egal presumptions that rest on formalistic distinctions."[83]

Questions to be considered later include whether a company that would otherwise like to integrate its products should be required to provide unbundled versions so that consumers wanting to buy one product but not another would have the ability and incentive to do so. In other words, again taking the Microsoft example, should Microsoft be required to sell a version of Windows without an Internet browser or other piece of application software? One way to analyze this question might be to use the approach the D.C. Circuit took in *Microsoft 1998* when it asked whether or not the overall benefit of integrating the products exceeded the cost to consumers of not being able to choose the individual component parts. Or, building on the D.C. Circuit's rejection in *Microsoft 2001* of per se treatment of tying as it pertains to operating systems, we would urge the U.S. Supreme Court to abandon the facade of per se treatment of tying arrangements and replace the tortured analysis currently used to determine whether there are two separate products with a straight forward application of the Rule of Reason.

NOTES

1. By unilateral, we mean non-collusive. *See* BLACK'S LAW DICTIONARY 1532 (7th ed. 1999); *see also* HERBERT HOVENKAMP, MARK D. JANIS & MARK A. LEMLEY, IP AND ANTITRUST: AN ANALYSIS OF ANTITRUST PRINCIPLES APPLIED TO INTELLECTUAL PROPERTY LAW (2002), § 13.4 (distinguishing unilateral refusals from conditional and/or concerted refusals) [hereinafter IP AND ANTITRUST].

2. Eastman Kodak Co. v. Image Technical Servs., Inc., 504 U.S. 451 (1992) [hereinafter Kodak I]. Within the Court's discussion of tying, footnote 29 of that opinion states that "power gained through some natural and legal advantage such as a patent, copyright, or business acumen can give rise to liability if 'a seller exploits his dominant position in one market to expand his empire into the next.' " *Id.* at 480, n. 29 (citations omitted). At the time, however, intellectual property rights were not an issue before the court. *See* Jonathan Gleklen, "Antitrust Liability for Unilateral Refusal to License Intellectual Property: Xerox and its Critics," 6, unpublished manuscript *available at* http://www.ftc.gov/opp/intellect/020501gleklen.pdf (stating that "[F]ootnote 29 can be characterized as *dicta* because the Kodak case did not involve the rights of intellectual property owners – the only evidence before the Court was that none of Kodak's parts were patented.").

3. CSU, LLC v. Xerox Corp. (*In re* Independent Service Organizations Antitrust Litigation), 203 F.3d 1322, 1327 (Fed. Cir. 2000), *cert. denied*, 531 U.S. 1143 (2001) [hereinafter Xerox].

4. *See* Image Technical Servs., Inc. v. Eastman Kodak Co., 125 F.3d 1195 (9th Cir. 1997), *cert. denied*, 523 U.S. 1094 (1998) [hereinafter Kodak II].

5. United States v. Microsoft Corp., 253 F.3d 34, 62 (D.C. Cir. 2001) [hereinafter Microsoft 2001].

6. *Id.* at 63.

7. Brief for the United States as Amicus Curiae at 16, Xerox, 531 U.S. 1143 (2001) (No. 00–62), *available at* http://www.usdoj.gov/osg/briefs/2000/2pet/6invit/2000–0062.pet.ami.inv.html.

8. Department of Justice Antitrust Division & Federal Trade Commission, Hearing on Competition and Intellectual Property Law and Policy in the Knowledge Based Economy,

"Strategic Use of Licensing: Is There Cause For Concern about Unilateral Refusals to Deal?," *available at* http://www.ftc.gov/opp/intellect/020501xscript.pdf (May 1, 2002).

9. Determinations of patent scope have huge economic implications. Josh Lerner has empirically demonstrated that patent scope significantly affects firm value. *See* Joshua Lerner, *The Importance of Patent Scope: An Empirical Analysis*, 25 RAND J. ECON. 319 (1994); *see also* Robert P. Merges & Richard R. Nelson, *On the Complex Economics of Patent Scope*, 90 COLUM. L. REV. 839 (1990).

10. United States v. Microsoft Corp., 2002 U.S. Dist. LEXIS 22864 (Nov. 12, 2002).

11. *See* 17 U.S.C. § 101 *et. seq*. (2000).

12. *See* 35 U.S.C. § 101 (2000).

13. Madey v. Duke Univ., 307 F.3d 1351, 1362 (Fed. Cir. 2002) (quoting Embrex Inc. v. Service Eng'g Corp., 216 F.3d 1343, 1349 (Fed. Cir. 2000)).

14. *See*, e.g. CAL. CIV. CODE § 3426 (2003). Note that this and many other state statutes are based on the Uniform Trade Secrets Act (1985).

15. 15 U.S.C. § 1 (2000).

16. Barry Wright Corp. v. ITT Grinnell Corp., 724 F.2d 227, 236 (1st Cir. 1983) (emphasis in original).

17. United States v. E. I. du Pont de Nemours & Co., 351 U.S. 377, 391 (1956).

18. United States v. Aluminum Co. of Am., 148 F.2d 416, 430 (2d Cir. 1945).

19. *See* MCI Communications Corp. v. Am. Tel. & Tel. Co., 708 F.2d 1081, 1132–1133 (7th Cir. 1983), *cert. denied*, 440 U.S. 971 (1983).

20. United States v. Terminal R. R. Ass'n of St. Louis, 224 U.S. 383 (1912).

21. Aspen Skiing Co. v. Aspen Highlands Skiing Corp. 472 U.S. 585, 603 (1985).

22. *See*, e.g. SMS Sys. Maint. Servs., Inc. v. Digital Equip. Corp., 11 F. Supp. 2d 166, 168 (D. Mass. 1998) (finding no liability without change in conduct).

23. *See*, e.g. Indep. Ink, Inc. v. Trident, Inc., 210 F. Supp. 2d 1155, 1162 (C. D. Cal. 2002).

24. Microsoft 2001, 253 F.3d 52, 87 (D.C. Cir. 2001) (internal quotation marks omitted).

25. As MacKie-Mason and Metzler note, aftermarket sales associated with a given foremarket sale will typically occur over a period of time. Therefore, the aftermarket prices and profits are defined as the discounted streams of future expenditures and profits, respectively.

26. *Id*. at 479 n. 29 (internal citations omitted).

27. Kodak I, 504 U.S. at 486.

28. Kodak II, 125 F.3d at 1219.

29. *Id*.

30. *Id*. at 1218 (quoting Data Gen. Corp. v. Grumman Sys. Support, 36 F.3d 1147, 1197 (1st Cir. 1994)).

31. *Id*. at 1216.

32. *Xerox*, 203 F.3d at 1324.

33. The Federal Circuit's exclusivity at the intersection of patent and antitrust law was established in Nobelpharma AB v. Implant Innovations, Inc., 141 F.3d 1059 (Fed. Cir. 1998); *see also* Midwest Indus., Inc. v. Karavan Trailers, Inc., 175 F.3d 1356, 1360 (Fed. Cir. 1999) (en banc in relevant part) ("Pro-Mold and Nobelpharma make clear that our responsibility as the tribunal having sole appellate responsibility for the development of patent law requires that we do more than simply apply our law to questions of substantive patent law. In order to fulfill our obligation of promoting uniformity in the field of patent law, it is equally important to apply our construction of patent law to the questions whether

and to what extent patent law preempts or conflicts with other causes of action."). After Holmes Group v. Vornado Air Circulation Sys., 535 U.S. 826 (2002), however, it is unclear whether such a case would automatically continue to fall within the Federal Circuit's exclusive jurisdiction. In Holmes, the Supreme Court held that the well-pleaded complaint rule requires that for a case to arise under patent law, the plaintiff's complaint must establish either that federal patent law itself created the cause of action or that "the plaintiff's right to relief necessarily depends on resolution of a substantial question of federal patent law." *Id.* at 1893, quoting Christianson v. Colt Indus. Operating Corp., 486 U.S. 800, 809 (1988). If cases involving antitrust immunity for activities within the scope of the patent grant are to remain the exclusive jurisdiction of Federal Circuit, it will likely have to be under this second rationale.

34. Xerox at 1326.

35. *Id.* (quoting Nobelpharma AB v. Implant Innovations, Inc., 141 F.3d 1059, 1072 (Fed. Cir. 1998) (internal quotation marks omitted)).

36. *Id.* at 1328 (citing United States v. Loew's, Inc., 371 U.S. 38, 47–48 (1962)).

37. *Id.* at 1329 (quoting Data Gen. Corp. v. Grumman Sys. Support Corp., 36 F.3d 1147, 1187 (1st Cir. 1994) (internal quotation marks omitted)).

38. Microsoft 2001, 253 F.3d 34, 49 (D.C. Cir. 2001) (citations omitted); *see also* Bruce Abramson, *From Investor Fantasy to Regulatory Nightmare: Bad Network Economics and the Internet's Inevitable Monopolists*, 16 HARV. J. L. & TECH. 159, 173 (2002). ("The first and best-known defining characteristic of a network industry is that the value of a network grows with its size. This phenomenon is known as 'positive feedback' (or alternatively, increasing returns to scale"). The phenomenon of positive feedback "follows standard microeconomic principles describing consumer choice. The rational choice for a consumer . . . is to select the option that will connect her to the most valuable of the competing networks . . .. [Because of the higher utility associated with joining a larger network,] the consumer's rational decision to join it will make the largest network larger – and thus correspondingly more valuable to the next consumer who considers joining.")

39. *Id.* at 64.

40. *Id.* at 62.

41. *Id.* at 63 (quoting Xerox, 203 F.3d 1322, 1325 (Fed. Cir. 2000)). Note, however, that the D.C. Circuit declined to include the subsequent statement from the Xerox opinion ("But it is also correct that the antitrust laws do not negate the patentee's right to exclude others from patent property."). Perhaps this is because the intellectual property at issue was based on copyright rather than a patent. Whether such a divergent outcome should depend on whether the asset in question is a copyright or a patent is discussed in Section 4, *infra*.

42. United States v. Microsoft Corp., 147 F.3d 935 (D.C. Cir. 1998) [hereinafter Microsoft 1998].

43. The court conceded that "we believed our interpretation of the term 'integrated product' [in Microsoft 1998] was consistent with the test for separate products under tying law," but pointed out that the Microsoft 1998 opinion had made it clear that the "antitrust question is of course distinct" and that the conclusion that Internet Explorer and Windows 95 were integrated was subject to reexamination on a more complete record. *Id.* at 92 (internal quotation marks omitted). The Microsoft 2001 opinion did, however, indicate that any disclaimer in Microsoft 1998 of judicial competence to evaluate product design did not conform to prevailing antitrust doctrine. *Id.*

44. 995 F. Supp. 728 (S. D. Tex. 1998). (Mr. Clarkson testified as an expert witness on behalf of the plaintiff in this case.)

45. 237 F. Supp. 2d 639 (D. Md. 2002).

46. Sun Microsystems v. Microsoft, 333 F.3d 517 (4th Cir. 2003).

47. 195 F.3d 1346 (Fed. Cir. 1999).

48. *Intergraph*, 195 F.3d at 1354–1355. ("Unilateral conduct that may adversely affect another's business situation, but is not intended to monopolize that business, does not violate the Sherman Act.")

49. *Id.* at 1356.

50. *Id.* at 1357–1358. ("The notion that withholding of technical information and samples of pre-release chips violates the Sherman Act, based on essential facility jurisprudence, is an unwarranted extension of precedent and can not be supported on the premises presented. The district court erred in holding that Intel's superior microprocessor product and Intergraph's dependency thereon converted Intel's special customer benefits into an 'essential facility' under the Sherman Act.")

51. 645 F.2d 1195 (2d Cir. 1981).

52. Gerald Sobel, *The Antitrust Interface with Patents and Innovation: Acquisition of Patents, Improvement Patents and Grant-Backs, Non-Use, Fraud on the Patent Office, Development of New Products and Joint Research*, 53 Antitrust L. J. 681, 685 (1984).

53. SCM v. Xerox, 645 F.2d at 1206.

54. 991 F.2d 511 (9th Cir. 1993).

55. *Id.* at 518.

56. 36 F.3d 1147 (1st Cir. 1994).

57. The term used in the Data General opinion is "TPM" (Third Party Maintainer), but we will use ISO for the sake of consistency with Xerox and both Kodak opinions.

58. The district court also disagreed with Grumman's characterization of ADEX as an essential facility. Data Gen. Corp. v. Grumman Sys. Support Corp., 761 F. Supp. 185, 192 (D. Mass. 1991).

59. Data General 36 F.3d at 1187.

60. *See*, e.g. Michael A. Carrier, *Unraveling the Patent-Antitrust Paradox*, 150 U. Pa. L. Rev. 761, 794 (2002). ("There are manifest difficulties in centering analysis on the defendant's intent.")

61. *See id.* ("[T]here will always be the concern that documentation of beneficent intent will be made with an eye toward future litigation.")

62. As noted earlier, it is still unclear what effect the Supreme Court's decision in Holmes v. Vornado, 535 U.S. 826 (2002), will have on the Federal Circuit's jurisdiction when an application of antitrust law would reduce the scope of the patent grant.

63. Kodak I, 504 U.S. 451, 466 (1992).

64. We might also make a similar argument under a theory of detrimental reliance or breach of the implied covenant of good faith and fair dealing from a contract law perspective. However, the adverse possession concept seems to be less of a conceptual stretch. Neither Kodak nor Xerox ever explicitly promised that ISO service would be available to their customers, nor did they agree to supply parts to ISOs in perpetuity. We are also unaware of an instance where a reliance interest for one party (the customer) was sustainable based on a separate reliance interest of a third party (the ISOs). Thus, we do not pursue detrimental reliance under contract law any further.

65. Jefferson Parish Hosp. Dist. No. 2 v. Hyde, 466 U.S. 2, 21 (1984) ("a tying arrangement cannot exist unless two separate product markets have been linked.").

66. *See* Lee v. Life Ins. Co. of N. Am., 23 F.3d 14, 20 (1st Cir. 1994), *cert. denied*, 513 U.S. 964 (1994). ("The timing of the 'lock in' at issue in Kodak was central to the Supreme Court's decision.... Had previous customers known, at the time they bought their Kodak copiers, that Kodak would implement its restrictive parts-servicing policy, Kodak's 'market power,' i.e. its leverage to induce customers to purchase Kodak servicing, could only have been as significant as its [market power] in the copier market, which was stipulated to be inconsequential or non-existent.")

67. *See* Digital Equip. Corp. v. Uniq Digital Techs., Inc., 73 F.3d 756, 763 (7th Cir. 1996). ("The Court did not doubt in *Kodak* that if spare parts had been bundled with Kodak's copiers from the outset, or Kodak had informed customers about its policies before they bought its machines, purchasers could have shopped around for competitive life-cycle prices. The material dispute that called for a trial was whether the change in policy enabled Kodak to extract supra-competitive prices from customers who had already purchased its machines.")

68. 104 F.3d 811 (6th Cir. 1997).

69. PSI Repair Servs., 104 F.3d at 820.

70. *Id.*

71. David L. Aldridge Co. v. Microsoft Corp., 995 F. Supp. 728 (S. D. Tex. 1998).

72. Microsoft 2001, 253 F.3d 34, 53 (D.C. Cir. 2001).

73. United States v. Microsoft Corp., 84 F. Supp. 2d 9, 20 (D.D.C. 1999).

74. The question of what is a reasonable royalty is non-trivial. The Federal Circuit has stated that the determination of "a fair and reasonable royalty is often . . . a difficult judicial chore, seeming often to involve more the talents of a conjurer than those of a judge." *Fromson v. Western Litho Plate & Supply Co.*, 853 F.2d 1568, 1574 (Fed. Cir. 1988). As Judge Learned Hand put it, "the whole notion of a reasonable royalty is a device in aid of justice, by which that which is really incalculable shall be approximated." *Cincinnati Car Co. v. New York Rapid Transit Corp.*, 66 F.2d 592, 595 (2d Cir. 1933).

75. *See* Department of Justice Antitrust Division and Federal Trade Commission, Hearing on Competition and Intellectual Property Law and Policy in the Knowledge Based Economy, "Strategic Use of Licensing: Is There Cause for Concern about Unilateral Refusals to Deal?" 156 (May 1, 2002), *available at* http://www.ftc.gov/opp/intellect/020501xscript.pdf.

76. *Id.* at 63 (quoting Xerox, 203 F.3d 1322, 1325 (Fed. Cir. 2000)).

77. Xerox, 203 F.3d at 1325 (quoting Intergraph Corp v. Intel Corp, 195 F.3d 1346, 1362 (Fed. Cir. 1999)).

78. Intergraph Corp. v. Intel Corp., 195 F.3d 1346, 1357 (Fed. Cir. 1999).

79. Hovenkamp, Janis and Lemley suggest, however, that the Federal Circuit was "arguably incorrect to conclude that the parties were not in the same market, [given that] Intergraph's *intellectual property rights*, the assertion of which triggered the dispute, were in the same technology market as Intel's primary line of business (microprocessors)" (emphasis added). The authors do point out that the Federal Circuit's "failure to recognize this does not affect the essential facilities analysis, however, because any such competition would exist in the market for the essential facility itself, not the downstream market Intel was allegedly trying to control." IP AND ANTITRUST, *supra* Note 1, § 13.3c2, n. 23.

80. Data Gen. Corp v. Grumman Corp., 761 F. Supp. 185, 192 (D. Mass. 1991).

81. Continental Paper Bag Co., v. Eastern Paper Bag Co., 210 U.S. 405, 428 (1908).

82. *See*, e.g. Dawson Chem. Co. v. Rohm & Haas Co. 448 U.S. 176, 215 (1980) (discussing the "long-settled view that the essence of a patent grant is the right to exclude others from profiting by the patented invention"); Special Equip. Co. v. Coe, 324 U.S. 370 (1945); United Shoe Mach. Corp. v. O'Donnell Rubber Prods. Co., 84 F.2d 383, 386 (6th Cir. 1936) ("Whatever may be the policy of patent laws elsewhere, it has long been settled that, in the United States, exclusion of competitors is the very essence of the right conferred by a patent, and it is the privilege of any owner of property to use or not to use it without question of motive."); Crown Die & Tool Co. v. Nye Tool & Mach. Works, 261 U.S. 24, 35 (1923) (A patentee may refuse "to make, use or vend his patented invention . . .." The "essence" of "the patent was the right to exclude others . . ."); Motion Picture Patents Co. v. Universal Film Mfg. Co., 243 U.S. 502, 514 (1917) (Under the patent system, a patent owner clearly had the right to "withhold his patent[ed] machine from public use . . .."); Lewis Blind Stitch Mach. Co. v. Premium Mfg. Co., 163 F. 950, 954 (8th Cir. 1908). ("[D]uring the life of his monopoly, a patentee is under no obligation to use or place upon the market . . . his invention.")

83. *Kodak I*, 504 U.S. 451, 466–467 (1992).

ACKNOWLEDGMENTS

The authors wish to thank John Allison, Kristen Carpenter, Rob Huckman, Joe Kalt, Mark Lemley, Josh Lerner, Jeff MacKie-Mason, Lynda Oswald, Joe Singer, Robert Williams, and the peer reviewers at the 2003 Huber Hurst Seminar on Legal Research at the University of Florida for helpful comments and suggestions. Any errors or omissions are solely the fault of the authors.

REFERENCES

Abramson, B. (2002). From investor fantasy to regulatory nightmare: Bad network economics and the internet's inevitable monopolists. *Harvard Journal of Law & Technology*, *16*(Fall), 159.

Areeda, P. (1989). Essential facilities: An epithet in need of limiting principles. *Antitrust Law Journal*, *58*, 841.

Bagley, C. E. (2002a). *Managers and the legal environment: Strategies for the 21st century* (4th ed.). Cincinnati, OH: West/Thomson Learning.

Bagley, C. E. (2002b). *United States v. Microsoft: Application of the antitrust laws to the new economy*. Harvard Business School.

Bagley, C. E., & Clarkson, G. (2003). Adverse possession for intellectual property: Adapting an ancient concept to resolve conflicts between antitrust and intellectual property laws in the information age. *Harvard Journal of Law & Technology*, *16*(2), 327–393.

Bresnahan, T. F. (2001). *Network effects and Microsoft*. Stanford Institute for Economic Policy Research.

Bresnahan, T. F. (2002). *The economics of the Microsoft case*. Stanford Law School.

Brynjolfsson, E., & Kemerer, C. F. (1996). Network externalities in microcomputer software: An econometric analysis of the spreadsheet market. *Management Science*, *42*(12), 1627.

Caillaud, B., & Tirole, J. (2001). *Essential facility financing and market structure*. C.E.P.R. Discussion Papers.

Carlton, D. W. (2001). *A general analysis of exclusionary conduct and refusal to deal: Why Aspen and Kodak are misguided*. NBER.

Carrier, M. A. (2002). Unraveling the patent-antitrust paradox. *University of Pennsylvania Law Review*, *150*, 761.

Cooter, R., & Ulen, T. (1988). *Law and economics*. Glenview, IL: Scott, Foresman.

Cotter, T. F. (1999). Intellectual property and the essential facilities doctrine. *The Antitrust Bulletin*, *44*(1), 211–250.

Donahey, T. I. (1997). Terminal railroad revisited: Using the essential facilities doctrine to ensure accessibility to internet software standards. *American Intellectual Property Law Association Quarterly Journal*, *25*, 277.

Economides, N. (2001). The Microsoft antitrust case. *Journal of Industry, Competition and Trade*, *1*(1).

Epstein, R. A. (1998). *Principles for a free society: Reconciling individual liberty with the common good*. Reading, MA: Perseus Books.

Evans, D. S., & Schmalensee, R. L. (2000). Be nice to your rivals: How the government is selling an antitrust case without consumer harm. Did Microsoft harm consumers?: Two opposing views. Washington, DC: AEI-Brookings Joint Center for Regulatory Studies.

Fels, A. (2001). Regulating access to essential facilities. *Agenda*, *8*(3), 195–207.

Fisher, F. M., & Rubinfeld, D. L. (2000). United States v. Microsoft: An economic analysis. *Did Microsoft Harm Consumers? Two Opposing Views*. Washington, DC: AEI-Brookings Joint Center for Regulatory Studies.

Gandal, N. (1994). Hedonic price indexes for spreadsheets and an empirical test for network externalities. *Rand Journal of Economics*, *25*(1), 160–170.

Gans, J. S. (2001). Regulating private infrastructure investment: Optimal pricing for access to essential facilities. *Journal of Regulatory Economics*, *20*(2), 167–189.

Gilbert, R. J., & Katz, M. L. (2001). An economist's guide to U.S. v. Microsoft. *Journal of Economic Perspectives*, *15*(2), 25–44.

Gleklen, J. (2002a). *Antitrust liability for unilateral refusals to license intellectual property: Xerox and its critics*. Washington, DC: FTC Hearings on Competition and Intellectual Property Law and Policy in the Knowledge-Based Economy.

Gleklen, J. I. (2002b). Per se legality for unilateral refusals to license IP is correct as a matter of law and policy. *The Antitrust Source*, *1*(5).

Hovenkamp, H., Janis, M. D., & Lemley, M. A. (2002). *IP and antitrust: An analysis of antitrust principles applied to intellectual property law*. New York: Aspen Law & Business.

Joskow, P. L. (2002). Transaction cost economics, antitrust rules, and remedies. *Journal of Law, Economics, & Organization*, *18*(1), 95–116.

Katz, M. L., & Shapiro, C. (1994). Systems competition and network effects. *Journal of Economic Perspectives*, *8*(2), 93–115.

Klein, B. (1996). Market power in aftermarkets. *Managerial & Decision Economics*, *17*(2), 143–164.

Klein, B. (2001). The Microsoft case: What can a dominant firm do to defend its market position? *Journal of Economic Perspectives*, *15*(2), 45–62.

Langlois, R. N. (2001). Technological standards, innovation, and essential facilities: Toward a Schumpeterian post-Chicago approach. In: J. Ellig (Ed.), *Dynamic Competition and Public*

Policy: Technology, Innovation, and Antitrust Issues. Cambridge, UK; New York: Cambridge University Press.

Lemley, M. A., & McGowan, D. (1998). Legal implications of network economic effects. *California Law Review*, *86*, 479.

Lerner, J. (1994). The importance of patent scope: An empirical analysis. *Rand Journal of Economics*, *25*(2), 319–333.

Liebowitz, S. J., & Margolis, S. E. (1999). *Winners, losers and Microsoft: Competition and antitrust in high technology*. Oakland, CA: Independent Institute.

Liebowitz, S., & Margolis, S. E. (2001). Network effects and the Microsoft case. In: J. Ellig (Ed.), *Dynamic Competition and Public Policy: Technology, Innovation, and Antitrust Issues*. Cambridge, UK; New York: Cambridge University Press.

Lipsky, A. B., & Sidak, J. G. (1999). Essential facilities. *Stanford Law Review*, *51*, 1187.

MacKie-Mason, J. K. (2002). Antitrust immunity for refusals to deal in (intellectual) property is a slippery slope. *The Antitrust Source*, *1*(5).

MacKie-Mason, J. K., & Metzler, J. (2003). Links between vertically related markets: ITS v Kodak. In: J. E. Kwoka & L. J. White (Eds), *The Antitrust Revolution: Economics, Competition, and Policy* (4th ed.). New York: Oxford University Press.

McAndrews, J. (1995). Antitrust issues in payment systems: Bottlenecks, access, and essential facilities. *Federal Reserve Bank of Philadelphia Business Review*, 3–12.

Merges, R. P., & Nelson, R. R. (1990). On the complex economics of patent scope. *Columbia Law Review*, *90*, 839–916.

Merrill, T. W. (1984). Property rules, liability rules, and adverse possession. *Northwestern University Law Review*, *79*, 1122.

Miceli, T. J., & Sirmans, C. F. (1995). An economic theory of adverse possession. *International Review of Law and Economics*, *15*, 161.

O'Rourke, M. A. (1998). Striking a delicate balance: Intellectual property, antitrust, contract, and standardization in the computer industry. *Harvard Journal of Law & Technology*, *12*, 1.

Patterson, M. R. (2000). When is property intellectual? The leveraging problem. *Southern California Law Review*, *73*(5), 1133–1160.

Pitofsky, R. (2000). *Challenges of the new economy: Issues at the intersection of antitrust and intellectual property*. Washington, DC: American Antitrust Institute Conference: An Agenda for Antitrust in the 21st Century.

Pitofsky, R., Patterson, D., & Hooks, J. (2002). The essential facilities doctrine under United States antitrust law. *Antitrust Law Journal*, *70*, 443.

Posner, R. A. (1998). *Economic analysis of law* (5th ed.). New York: Aspen Publishers.

Reiffen, D., & Kleit, A. N. (1989). *Terminal railroad revisited: Foreclosure of an essential facility or simple horizontal monopoly?* 30.

Romer, P. (1990). Endogenous technological change. *Journal of Political Economy*, *98*(5), S71–S102.

Rose, C. M. (1985). Possession as the origin of property. *University of Chicago Law Review*, *52*, 73.

Saunders, K. M. (2002). Patent nonuse and the role of public interest as a deterrent to technology suppression. *Harvard Journal of Law & Technology*, *15*, 389.

Shapiro, C. (1996). Antitrust in network industries. *Antitrust/Intellectual Property Claims in High Technology Markets*. San Francisco, CA: American Law Institute and American Bar Association.

Shapiro, C., & Teece, D. J. (1994). Systems competition and aftermarkets: An economic analysis of Kodak. *The Antitrust Bulletin*, *39*(1), 135–162.

Singer, J. W. (1988). The reliance interest in property. *Stanford Law Review*, *40*, 611.

Singer, J. W. (2001). *Introduction to property*. Gaithersburg, MD: Aspen Law & Business.

Sobel, G. (1984). The antitrust interface with patents and innovation: Acquisition of patents, improvement patents and grant-backs, non-use, fraud on the patent office, development of new products and joint research. *Antitrust Law Journal*, *53*, 681.

Whinston, M. D. (2001). Exclusivity and tying in U.S. v. Microsoft: What we know, and don't know. *Journal of Economic Perspectives*, *15*(2), 63–80.

INCOMPLETE CONTRACTING AND THE STRUCTURE OF R&D JOINT VENTURE CONTRACTS

Suzanne E. Majewski and Dean V. Williamson

ABSTRACT

There is a tension between the literatures on incomplete contracting and transactions cost economics regarding the importance of ex post governance and the extent to which formal theories of incomplete contracting capture salient aspects of exchange relations. In this paper, we empirically examine how firms structure joint R&D agreements to illuminate how contracts can be incomplete and how governance can matter. We employ a dataset of 96 contracts to construct a taxonomy of the types of mechanisms firms use in organizing collaborative R&D, and indicate how groups of mechanisms line up with various types of contracting hazards. The results suggest that the allocation of property rights over innovations at the time of contracting between R&D partners is an important aspect of contract design. But they also suggest that weak property rights admit scope for other dimensions of contract. In particular, the research indicates that while knowledge spillovers may give rise to appropriability hazards, efforts to contain or channel knowledge spillovers may enable joint venture members to strategically block other members' follow-on commercialization or research. Firms

Intellectual Property and Entrepreneurship
Advances in the Study of Entrepreneurship, Innovation and Economic Growth,
Volume 15, 201–228

ISSN: 1048-4736/doi:10.1016/S1048-4736(04)01506-1

design joint R&D governance mechanisms to balance spillover hazards and strategic blocking.

1. INTRODUCTION

How do parties to complex exchange craft their exchange relations to enable adaptations to unforeseen contingencies? This question might seem innocuous, but it is loaded with concepts that have motivated diverse lines of research relating to the design of governance structures. What do we mean by "exchange relations" and "unforeseen contingencies," and how might firms design contracts to accommodate the problem of adapting exchange relations to unforeseen contingencies?

These are questions that have been taken up in literatures that relate to the more general issue of *incomplete contracting*. These literatures have made progress toward characterizing how contracts can be incomplete, and they have made progress toward sorting out how the incompleteness of contracts, both exogenous and endogenous, can inform the design of governance structures. But tricky issues remain, only one of which we will focus on here. We focus on how weak property rights can motivate incompleteness and how weak property rights can frustrate solutions to hold-up that theories of incomplete contracting suggest. Specifically, literatures on incomplete contracting have suggested that much of the action in contract design relates to hold-up problems, and these literatures have suggested ways in which judiciously allocating property rights can remedy hold-up. Contracting problems involving collaborative R&D may also feature hold-up, but a salient aspect of R&D is that it involves environments in which property rights – in this case intellectual property rights – can be weak. Weak property rights, it turns out, provide scope for strategic behaviors that have some of the flavor of hold-up but yet remain beyond the scope of formal theories of incomplete contracting. And that is not all. Weak property rights provide scope for a class of remedies not anticipated by formal theory.

Weak property rights manifest themselves in at least two ways. First, weak property rights can frustrate the efforts of parties to appropriate rents. For example, informational externalities or knowledge spillovers might prevent firms from fully appropriating the returns to R&D. Appropriability can be an interesting problem in contexts, such as those involving R&D, in which parties must sink investments, because difficulties in appropriating rents may diminish incentives to invest in the first place (Spence, 1984). Second, weak property rights are conspicuously manifested in the ways parties manipulate legal processes to extract rents. The most obvious example is patent litigation. A patent endows a patent holder with the right to assert claims of patent infringement against other parties, and patent holders might assert claims to compel other parties to submit to licensing agreements, cease production, or engage in bargaining beyond the scope of the patent (Cohen, Nelson & Walsh, 2000; Hall & Ham-Ziedonis, 2001; Lanjouw & Lerner, 2001).[1] Such claims might not have much merit, but demonstrating the merits of a claim to a third party (the court) can be costly (Williamson, 2000a). More generally, intellectual property rights may be costly to delineate, because court ordering of claims may be neither timely nor perfect. Knowing this, a patent holder may exploit the litigation process by leveraging the high costs of legal defense for more favorable settlement terms. A patent holder might, for example, exploit legal processes to effectively block another party's efforts to conduct particular R&D or to commercialize a particular product.

In this chapter we explore the intersection of incomplete contracting and weak property rights in the context of collaborative R&D. Parties might collaborate on R&D because they have complementary capabilities or complementary intellectual property to contribute, and they might collaborate to achieve various objectives. Objectives might extend beyond innovation per se and may include establishing de facto technology standards or transferring selected technologies to selected parties. Pursuing these objectives invites the design of structures to govern how parties contribute to R&D as it unfolds, but that is not all. Weak property rights can complicate the pursuit of objectives, and, as is well understood, weak intellectual property rights can frustrate the efforts of would-be innovators to appropriate the returns to R&D. Firms may design governance structures to accommodate problems associated with pursuing objectives and with appropriating returns.

Much of the action in designing governance structures can involve containing or channeling spillovers. But this involves tradeoffs between enhancing appropriability and enabling legalistic gamesmanship. Characterizing tradeoffs and the ways in which contracts are relational enables us to achieve two goals: (1) to identify the discrete contractual mechanisms firms use to mobilize and govern collaborative R&D; and (2) to inform the literatures on incomplete contracting and appropriability by identifying the alignment of mechanisms across industries and venture objectives. The results suggest that while the allocation of property rights over innovations at the time of contracting between R&D partners is an important aspect of contract design, weak property rights admit scope for other dimensions of contract. These other dimensions may enable R&D partners to draw each other into relationships that extend beyond the term of formally contracted collaboration. The results accord with other recent work suggesting that property rights theory alone cannot completely characterize firm interactions in environments featuring weak intellectual property rights. Governance structures that complement property rights allocations are important.

2. INCOMPLETE CONTRACTING REMAINS POORLY UNDERSTOOD

Here is a puzzle. Some contracting literatures pose the incompleteness of contracts and find that it can motivate salient roles for ex post governance – that is, the institutional framework parties to complex exchange set up to manage their exchange relations subsequent to contracting. Other literatures have focused on formalizing how contracts can be incomplete. An interesting result is that in formalizing incomplete contracting, formal theory has been unable to motivate a role for governance. Rather, most of the action in formal models pertains to property rights allocations. This seems to be a contradiction. So, a question remains: What is it about incomplete contracting that formal theories miss (if anything) and that motivates a role for governance? The suggestion here is that appealing to weak property rights provides a way of motivating both how contracts can be incomplete and how governance can matter, the ultimate result being that there are important interactions between property rights allocations and the design of other dimensions of governance.

The idea that other dimensions of governance might complement how property rights are allocated – the idea that governance can matter all – might seem reasonable enough. One just need look out the window and observe that firms, joint ventures, contracts and other governance structures cannot be characterized by property rights allocations alone as formal theory might suggest. Governance structures are more than just lists of who owns what. But it turns out that the suggestion that governance can matter leads to fundamental questions about how and why contracts can be incomplete and about how incomplete contracting can enable us to understand economic organization.

These are persistent questions that seem central to understanding the design of the structures – firms, joint ventures, long-term contracts, "strategic alliances," and so on – that parties to complex exchange use to govern their relationships. Indeed, these are classic questions that have been explicitly explored in literatures on relational contracting (Llewellyn, 1931; Macaulay, 1963; Macneil, 1974, 1978; Williamson, 2000a), transaction cost economics (Williamson, 1971, 1975, 1985, 1996), property rights theory (Aghion & Tirole, 1994; Grossman & Hart, 1986; Hart & Moore, 1990; Moore, 1992), and related approaches to incomplete contracting (Hart & Moore, 1999; Maskin & Tirole, 1999a, b; Segal, 1999; Tirole, 1999).

These literatures set up ideas about how the design of governance structures does (and does not) matter in enabling complex exchange, and, indeed, there are some interesting tensions within and between these literatures about how governance matters. These tensions largely turn on how these literatures have conceptualized incomplete contracting. Some research has focused on crafting formal foundations about what we mean by incomplete contracting, and this gets into difficult matters about how to formalize other concepts such as "unforeseen contingencies" and "complexity" (see, for example, Maskin & Tirole, 1999a; Segal, 1999; Tirole, 1999). Other research poses ways, both exogenous and endogenous, in which contracts can be incomplete and works out implications for the design of governance structures from there (see, for example, Bajari & Tadelis, 2001; Crocker & Masten, 1988, 1991; Crocker & Reynolds, 1993; Goldberg & Erickson, 1987; Joskow, 1987, 1988; Masten & Crocker, 1985; Saussier, 2000; Tadelis, 2002; Williamson, 1971, 2000a). The principle thrust is sorting out tradeoffs that parties to complex exchange encounter when enabling adaptation to unforeseen contingencies. Adaptability has obvious advantages, but it comes at a cost. It can involve enabling contracting parties to impose renegotiation. Renegotiation itself can dissipate rents that would otherwise obtain to the parties. It can also disrupt the credibility of commitments parties make to enable exchange in the first place.

The various bodies of theory provide different lenses through which to examine and interpret the design and operation of governance structures. The perspectives they provide are largely complementary.[2] More importantly, these bodies of research are dynamic in that they continue to make progress in delineating concepts and expanding the range of phenomena to which they can be applied.

3. KNOWLEDGE SPILLOVERS AND STRATEGIC BLOCKING ARE SALIENT ASPECTS OF INNOVATION

Weak property rights can be characterized with respect to at least two factors – knowledge spillovers and strategic blocking – that disrupt the ability of firms to appropriate returns from their R&D. A very large literature has examined knowledge spillovers and their roll in productivity (see Griliches, 1994 for a review of a the literature), but less attention has been paid to how know-how can spillover. Knowledge spillovers can occur through many channels such as: (1) the theft of explicit and codified knowledge such as blue prints; (2) labor turnover and migration (Zucker, Darby & Brewer, 1998); (3) tacit know-how that resides in the heads of engineers but is uncodifiable and defies spillover detection (Teece, 1986); and (4) organizational routines and norms (Nelson & Winter, 1982). In the case of research joint ventures, an open question is how can the design of joint venture R&D mechanisms facilitate or impede appropriation?

More recent contributions, including the research presented here, suggest that a problem of strategic blocking can also contribute to the appropriability problem (Cohen, Nelson & Walsh, 2000; Hall & Ham-Ziedonis, 2001; Lanjouw & Lerner,

2001; Lanjouw & Schankerman, 2001). The kind of situation we have in mind is one in which some party marches in and asserts patent claims against another. We distinguish strategic blocking as the ability of one party to impose negotiation or renegotiation on another party. We use this label to explicitly distinguish the phenomenon from the canonical hold-up problem, although many authors confuse strategic blocking with hold-up.[3] The costliness with which third parties (the courts) can delineate and enforce property rights (if at all), enables other parties to game legal processes by engaging other parties in costly litigation as part of larger efforts to make demands (to impose technology licenses, say), impose negotiation, and extract quasi-rents. If the delineation of property rights were costless and could be costlessly demonstrated to the courts, then no scope for strategically motivated litigation would remain. But, just as Karl von Clausewitz suggested in *On War* (1838) that "War constitutes politics by other means," so, too, litigation may constitute negotiation by other means.

4. KNOWLEDGE SPILLOVERS AND STRATEGIC BLOCKING IMPOSE TRADEOFFS

The appropriability problem is an economic problem that implicates contract design in a way that can be understood by appealing jointly to the ex ante allocation of property rights and ex post governance. The appropriability problem induces tradeoffs between efforts to contain and channel spillovers and efforts to remedy strategic blocking. Knowledge spillovers can motivate efforts to contain or selectively channel them. Parties to collaborative R&D have various instruments available to them, including restrictions on the disclosure of intellectual property to third parties. Such restrictions can mitigate spillovers, thus inducing prospective R&D partners to contribute intellectual property as well as to participate in R&D. A difficulty is that disclosure restrictions enable parties to strategically block their R&D partners by asserting improper disclosures even after a joint venture has terminated. Such restrictions, for example, can enable parties to block the commercialization of technologies or R&D other parties pursue both during or after the term of R&D collaboration. At the same time, the prospect of strategic blocking can affect how parties allocate property rights over innovations, and it can affect how parties modify property rights over innovations as well as over intellectual property R&D partners contribute to collaborative efforts. Property rights might be modified, for example, by conditions that can expand or restrict the licensing of intellectual property.

The exercise around which this chapter is organized involves demonstrating interactions between property rights and certain dimensions of contract that

R&D partners use to manage their relationships subsequent to contracting. We have developed a dataset of 96 R&D joint venture contracts from filings made to the U.S. Department of Justice under the terms of the National Cooperative Research Act of 1984 (NCRA) and National Cooperative Research and Production Act of 1992 (NCRPA). These contracts establish the structures under which partners to a joint venture collectively manage R&D and govern both intellectual property R&D partners contribute to a joint venture ("Background Intellectual Property") and intellectual property generated by a joint venture ("Foreground Intellectual Property").

We enable analysis by crafting three types of data. First, we craft a taxonomy of contracts by distinguishing the ways in which contracts allocate and modify property rights over Foreground IP and modify property rights over Background IP. These allocations with modifications reflect the underlying objectives a joint venture might pursue. (Innovation per se does not fully characterize what is going on. For example, some joint ventures organized to encourage the adoption of particular technology standards. Others principally involve technology transfers from contract researchers to joint venture participants. Still others may involve collaboration on later-stage development.) Second, we take a page from the appropriability literature and distinguish differences in the appropriability of the underlying technologies within which the joint ventures operate (hereafter "appropriability regimes"). Technologies involving chemicals, for example, have long been understood to be more appropriable than technologies in electronics (Cohen, Nelson & Walsh, 2000; Levin, Klevorick, Nelson & Winter, 1987). Third, we distinguish dimensions of contract that pertain to strategic blocking. Specifically, we examine the duration of restrictions on the disclosure of Background IP and Foreground IP. Restrictions enable parties to impose negotiation subsequent to the termination of a joint venture. One party, for example, can assert misappropriation on the part of former R&D partners. If mitigating spillover were the only concern, then restrictions might be made to last indefinitely, which they sometimes do. Even so, strategic blocking suggests that an economizing balance between spillover and strategic blocking can be achieved by placing definite limits on disclosure restrictions, and, indeed, there is much variation in the duration of disclosure restrictions.

5. TRADEOFFS MOTIVATE A SIMPLE EXERCISE

These three types of data enable a simple exercise that demonstrates how the problem of balancing spillover and strategic blocking can be understood in simple ways that relate both to governance structures and appropriability regimes. We indicate a mapping from governance structures (indicated by contract type) and appropriability regimes (indicated by SIC codes) to the duration of disclosure restrictions. By appealing to the appropriability literature and to interpretations of the strategic objectives of joint ventures, we can characterize the mapping as a mapping from the strategic objectives of joint ventures (indicated, again, by contract type) and appropriability regimes to a dimension of ex post governance (indicated by the duration of disclosure restrictions).

What should one get out of this exercise? We suggest that four discrete contributions proceed from the analysis. First, the research provides clues to the answers of persistent question: What is the nature of incomplete contracting? How does ex post governance matter? Second, the analysis suggests that the governance of R&D joint ventures can be understood as treaties between joint venture partners to make commitments to contribute intellectual property, to develop technology, and to maintain the integrity of property rights. Treaties include two parts: a property rights scheme that establishes allocations of property rights (with modifications) over Background IP and Foreground IP and mechanisms that enable R&D partners to manage their relations both during the course of active collaboration and after the termination of collaboration. Third, the analysis demonstrates that property rights regimes matter by illuminating interactions between property rights schemes and dimensions of ex post governance. Finally, the research suggests that these interactions can be understood by appealing to phenomena that formal property rights theories are not yet equipped to accommodate: weak property rights and the legal gamesmanship (strategic blocking) that weak property rights enable.

6. WHERE DO LITERATURES STAND ON INCOMPLETE CONTRACTING?

Contracting literatures can be partitioned into two sets according to their orientations to incomplete contracting. First, as is well known, the literatures on relational contracting and on transaction costs economics (TCE) pose ways, both exogenous and endogenous, in which contracts can be incomplete and work out implications for the design of governance structures from there. A contrast is explicitly indicated in these literatures between complete contingent-claims contracting and a contract-as-framework or ex post governance perspective. Complete contingent claims contracting applies to environments in which adaptation of exchange relationships to unforeseen contingencies is not a concern. But, insofar as the problem of adapting exchange relations to unforeseen contingencies constitutes an important feature of complex exchange, then contracts can be understood as commitments to set up: (1) the framework parties use for enabling adaptation; and (2) the threat points around which parties craft adaptations. (See Williamson, 1971, p. 116 on processes that enable "adaptive sequential decision-making" is apposite.)

The second set of literature is composed of bodies of theory that have gone some way toward formalizing what we mean by "unforeseen contingencies" and "incomplete contracting." Indeed, the literatures on property rights theory (PRT) and the mechanism design approach are often recognized specifically as theories of incomplete contracting. These bodies of theory work out of a verifiability paradigm (see, for example, Laffont & Martimort, 2002, p. 32; Hart, 1995, p. 37), three features of which pertain directly to property rights: parties to exchange can costlessly delineate property rights, third parties (the court) can costlessly delineate property rights, and court ordering is frictionless. Finessing weak property rights is not a problem per se, because it enables analysis of other features of the economic environment which might go most of the way toward motivating how and why contracts are incomplete. Recent contributions, for example, have added to the literature a "describability" paradigm which enables a way of modeling "complexity" (Maskin & Tirole, 1999a; Segal, 1999; Tirole, 1999).

A thrust of the describability program was to establish the irrelevance of ex post governance by demonstrating that, even in environments characterized by complexity, contract design could be collapsed into two dimensions: the allocation of property rights between parties to exchange at the time of contracting and the alignment of incentives through sharing rules and auxiliary terms of contract (Maskin & Tirole, 1999a). Even so, the contract-as-framework perspective would seem to survive the irrelevance critique. The irrelevance theorems extend to stylized environments and do not obviously constitute a general critique (Hart & Moore, 1999; Kreps, 2000). Yet, while we await more general critiques or, at least, more general characterizations of the ways in which environments can be "complex," other researchers have appealed to more flexible conceptualizations of complexity to motivate phenomena that indicate a role for ex post governance. Usual suspects include Masten and Crocker (1985), Crocker and Masten (1985, 1991), Joskow (1987, 1988), Goldberg and Erickson (1987), and Crocker and Reynolds (1993). More recent contribution include Saussier (2000), Bajari and Tadelis (2000) and Tadelis (2002). This body of research highlights themes that have been established in earlier work in TCE: the role of renegotiation and other adjustment mechanisms in enabling adaptation over the course of a long-term exchange relationship.

A complementary way of organizing thinking about theories of contracting is to sort out the ways they relate to three dimensions of contract design: ex ante incentive alignment, ex ante allocation of property rights, and ex post governance.

Mechanism design literatures and other game theoretic literatures point up the role of ex ante incentive alignment. The Property Rights Theory inspired by Grossman and Hart (1986) points up the role of the ex ante allocation of property rights. The theory is sometimes understood as a theory of incomplete contracting in that it works out of a "verifiability" paradigm. Contracts are incomplete insofar as there are actions or other prospective dimensions of contract that cannot be verified by a third party (the court). The idea is that the unavailability of third-party verification renders such dimensions unenforceable, and only enforceable dimensions are deemed contractible.

The appeal to the verifiability paradigm has enabled progress in understanding of the role of property rights in aligning incentives and enabling exchange. Indeed, in the environments explored within the PRT literature, all of the action in contract design obtains to ex ante allocations of property rights. As Kreps (2000, p. 123) puts it, "ex post decision rights determine (implicitly) how the contract will be completed, with the ex post decision rights determined by aspects of the contract (most prevalently, the residual rights of control conferred by ownership)." Not only may no role for ex post governance obtain, but, surprisingly, no role for ex ante incentive alignment through the design of sharing rules may obtain. Aghion and Tirole (1994), for example, explore an environment in which the nature of output is uncertain but to which property rights can be costlessly assigned and enforced. Again, remarkably, all of the action obtains to property rights.

Empirical research suggests that property rights allocation do serve a role for aligning incentives even in environments featuring innovation (Lerner & Merges, 1998), but that still leaves open questions about the conditions under which roles for ex post governance, much less for other dimensions of ex ante incentive alignment, might obtain. The property rights theory provides clues, two of which we indicate here. First, the verifiability paradigm depends on a narrow concept of enforceability that itself depends entirely on frictionless court ordering. Yet contractual relations often depend on non-verifiable information and on costly, informal private ordering as well as on costly court ordering (Williamson, 1996, 2000c). Admitting alternatives to court ordering as well as a role for non-verifiable information raises difficult questions about how to understand enforcement but suggests directions in which the formal theory might be extended (Dixit, 2003). Second, the formal theory depends on property rights being absolutely strong in that they can be costlessly delineated and verified. Admitting roles for the ways in which property rights can be weak can in turn admit roles for ex post governance or other dimensions of ex ante incentive alignment. Anand and Galetovic (2000), for example, admit a role for tacit knowledge in a model featuring hold-up in an innovation environment. Tacit knowledge admits a role for other dimensions of contract in mitigating hold-up in addition to property rights allocations.

Finally, transaction costs economics emphasizes that contract mechanisms governing exchange play an ongoing role long after they are established. Much of the action here pertains to how parties adapt exchange relations to unforeseen contingencies. Applications to innovation environments include Oxley (1997) and Ryall and Sampson (2003). This line of research is not occupied with formalizing contractual incompleteness but is occupied with sorting out the implications of incompleteness for the organization of collaborative R&D. The research does not admit a role for ex post governance by assumption but rather looks for evidence that ex post governance does (or does not) matter in enabling collaboration. The research suggests that while the mechanisms of ex post governance might be costly to implement, they constitute an economizing match for collaborative efforts organized in more complex environments. Complexity induces the greater prospect of maladaptation subsequent to contracting, and the prospect of maladaptation invites remedies such as committee mechanisms and voting schemes that enable contracting parties to impose renegotiation and adjust their activities going forward.

7. RENEGOTIATION AND COLLABORATIVE R&D

One way to understand the relationship of the research presented here to contracting literatures is to indicate the different ways the literatures and this paper deal with renegotiation. An interesting feature of transaction cost analyses of contracting practices is that renegotiation can serve a purpose: enabling adaptation. Accordingly, contracts might feature mechanisms that enable one party to impose renegotiation on another in certain circumstances, and renegotiation provides a way for parties to contract to adapt their exchange relations in the event they find themselves painted into a corner. Renegotiation poses an interesting economic problem in that it entails tradeoffs. While it affords flexibility, renegotiation itself can be costly. At the same time, enabling flexibility can diminish the credibility of commitments parties must make to enable gainful exchange in the first place. "The problem," as (Crocker & Masten, 1991, p. 72) express it, "is to devise a [governance] structure that encourages rent-increasing adjustments (flexibility) but discourages rent-dissipating efforts to redistribute existing surpluses (opportunism)."

In game-theoretic analyses renegotiation is commonly something to be finessed in that it can pose an obstacle to equilibrium analysis (Kreps, 2000). Renegotiation entails bargaining over the division of surplus, and bargaining can introduce multiple equilibria. Game-theoretic analyses generally assume that the bargaining is frictionless; parties costlessly recover all available surplus, no surplus is dissipated through haggling, and so, no tradeoffs obtain.

This chapter features a complementary interpretation of the purposes that renegotiation can serve. Specifically, renegotiation may constitute an integral feature of mechanisms that are designed to promote the appropriability of intellectual property. The focus here is on nondisclosure requirements. They can be interpreted as mechanisms that enable parties to establish threat points. They enable parties to impose renegotiation on R&D partners, and renegotiation itself can serve as a means of blocking a partner's efforts to commercialize products or conduct follow-on R&D. Anticipating this, a partner might be more willing to reveal planned activities and to otherwise avoid initiatives that might invite the kinds of examination that disclosure restrictions enable.

The analysis boils down to sorting out interactions between dimensions of ex post governance (disclosure restrictions) and property rights regimes while controlling for differences in appropriability regimes. To enable analysis, we developed a dataset of R&D joint venture contracts.

8. JOINT VENTURES FILED UNDER THE NATIONAL COOPERATIVE RESEARCH ACT

We have collected the contracts of 96 research joint ventures from the files of the National Cooperative Research Act (NCRA). Congress passed the Act in 1984 to encourage firms to participate in research joint ventures by limiting their liability to antitrust damages.[4] Joint venture participants can choose to file their ventures with the Department of Justice and Federal Trade Commissions to qualify for reduced antitrust liability under the program.[5] Some ventures voluntarily include contracts with their submissions in the hope of increasing the likelihood of a favorable review. Ventures with contracts represent 16% of all NCRA approved joint ventures, and follow roughly the same time distribution of all NCRA joint ventures (Fig. 1).[6]

NCRA joint ventures vary in size, industry, term, and value (see Table 1). For example, some NCRA ventures have been formed by a very large coalition of firms, typically in a standards setting situation, while other NCRA joint ventures feature very few participants. The median venture in the sample was formed in 1995, had five participants, was anticipated to last two years, and had budgeted roughly one million dollars.

Ventures engage in a diverse array of research and production technologies. But making comparisons across NCRA ventures is made more difficult by the fact that for many ventures the subject of research (the inputs to the R&D process) is not the same as the field of ultimate use for that research. Examples include ventures involving computer software designed to control robotics ultimately used in the

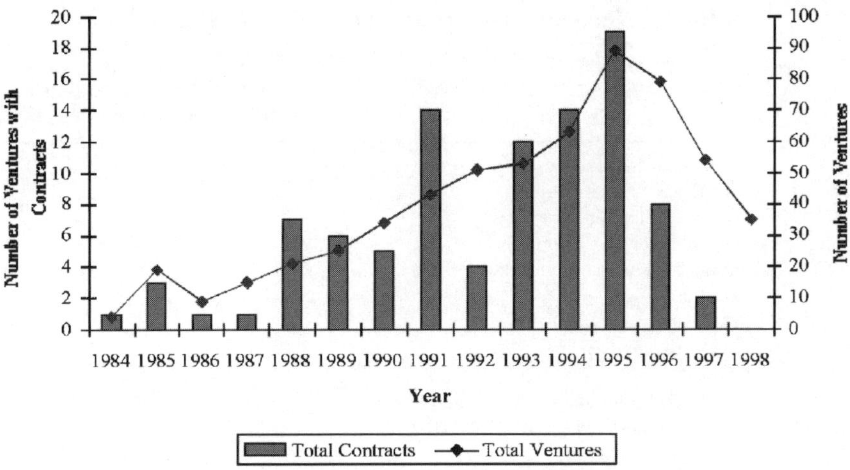

Fig. 1. NCRA Ventures and Contracts by Year.

automotive industry, ventures that develop plastic films ultimately used in aircraft industry, ventures that design bioengineered bacteria for remedying petroleum spills, and the design of advanced television technology implicating both home electronics and broadcast equipment. Table 2 contains a breakdown of ventures by technology of the R&D (product of the venture), and technology of the final goods market (uses of venture R&D).

Not only are the ventures heterogeneous in size, scope, and technology, they are also heterogeneous in the way they organize collaborative R&D. Two key factors that distinguish different joint venture governance mechanisms are the allocation of R&D responsibilities and the allocation of rights to Background IP and Foreground IP.

Table 1. Selected Summary Statistics for Joint Venture Observations.[a]

Number of observations	96
Range in number of firm-members	2–143
Median number of firm-members	5
Average nominal value of venture	\$6.7 M
Range in nominal values of venture	\$50K–\$45.8 M
Average anticipated length of venture	2.9 years
Range in anticipated length of venture	4 months to 10.5 years

[a] Not all ventures indicated their anticipated length or dollar values. Dollar values were indicated in only 49 ventures, the ex-ante term of the venture was indicated in 66 cases.

Table 2. Number of Observations by Industry (SIC Code).

SIC Code	Industry Description	JV Observation by Industry Describing Field of R&D	JV Observation by Industry Describing Downstream Field of Use
131	Crude petroleum and natural gas		1
138	Oil and gas field services	9	12
203	Preserved fruits and vegetables		3
281	Industrial inorganic chemicals	9	2
282	Plastics materials and synthetics	7	3
283	Drugs	5	4
286	Industrial organic chemicals	11	2
291	Petroleum refining	3	15
321	Flat glass		1
325	Structural clay products	2	
331	Blast furnace and basic steel products	1	1
332	Iron and steel foundries	1	
346	Metal forgings and stampings	3	
344	Fabricated structural metal products		2
348	Ordnance and accessories		1
351	Engines and turbines	6	6
353	Construction and related machinery		
354	Metalworking machinery	1	
355	Special industry machinery	1	
356	General industrial machinery	1	
357	Computer and office equipment	7	4
359	Industrial machinery		2
365	Household audio and video equipment		1
366	Communications equipment	1	5
367	Electronic components and accessories	8	7
371	Motor vehicles and equipment	1	5
372	Aircraft and parts	2	6
376	Guided missiles, space vehicles, parts		1
382	Measuring and controlling devices	1	
461	Pipelines, except natural gas		1
483	Radio and television broadcasting		1
596	Non-store retailers[a]		1
602	Commercial banks		3
632	Medical service and health insurance		4
679	Miscellaneous investing		1
737	Computer and data processing services	16	1
Total		96	96

Note: In many cases, joint ventures may conduct research on a certain technology (ex. plastics) to be used in a separate industry application (ex. aircraft components). This table states the number of joint venture observations per SIC code separately by the technological field in which joint ventures conducted R&D, and the technological field of the final products that used the results of that R&D.

[a]Includes internet retailing.

The degree to which joint venture members participate in venture R&D varies across different ventures. For example, some ventures entail member firms contributing R&D while the members of other ventures contract out research to various types of third parties. The way in which R&D is contracted out to third parties varies across ventures as well. In some ventures managerial control of R&D is assigned to a neutral third party that conducts the R&D itself. In others, the third party does not do R&D but rather coordinates the R&D activities of other (generally member) firms.

The role that third party "coordinators" play seems to be related to concepts of strategic blocking. In many cases the third party coordinator not only organizes R&D activities, but also assumes rights to subsequent IP. Coordinators are special in that they do not compete in the production, sales, or marketing of a downstream product that would benefit from venture R&D. In some cases, the coordinators are even non-profit entities. As a result, coordinators should be less motivated to engage in some sort of strategic blocking than other member firms (that compete in a downstream market) would be. Some ventures are even structured with member firms conducting R&D but then assigning title to the intellectual property back to the coordinator, suggesting the role of the coordinator extends beyond R&D capabilities and more toward issues of governance.

Given this richness in different organizational forms, a question for researchers is how to classify different venture structures into canonical "types." We attempt to do so by appealing to the concepts of strategic blocking and appropriability. One way to distinguish which parties can strategically block others is to examine the allocation of title to innovations. Any party retaining rights to portions of a venture's Foreground IP retains control over decisions to license third parties. When one venture participant strategically refuses to license a third party, their behavior impacts the value of partners' complementary IP. It also might impact the ability of venture partners to collaborate with third parties on future projects implicating venture technology. The mechanism that firms might use to neutralize strategic blocking of this sort is to assign relevant IP rights to third parties such as coordinators. We examine the allocation of rights to Foreground IP in our analysis of contracting technologies, described below.

Venture partners can attempt to minimize the ability of their partners to engage in strategic blocking by moving the locus of R&D to a neutral party or assigning rights to a neutral party. However, a tension arises when the skills or knowledge assets to conduct the required R&D are rare or unique and could be absorbed and exploited without detection or protection under the law. In these cases firms might prefer to conduct all R&D in-house and to individually retain the rights to their own Foreground IP (by implication retaining no rights to the Foreground IP of partners). By keeping strong control over portions of venture technology, member firms can minimize the hazard that their knowledge will be misappropriated, but they might also subject themselves to a greater risk of strategic blocking by other member firms. Another variable we examine in classifying canonical joint venture structures is the allocation of R&D projects across venture participants.

The tension between spillover and strategic blocking is not uniform across ventures, and this is evident in that different joint ventures use different structures. For example, some ventures employ scientists from member firms working side-by-side in the same facilities. Other ventures restrict scientist movement by allocating work in separate facilities and contractually establishing employee exchange (visitation). Not only are different structures used, but different structures tend to be used more or less often in different industries, reflecting industry differences in spillover potential and appropriability.

9. A TAXONOMY OF CONTRACTS

We construct a taxonomy of canonical contract types and use them as proxies for the underlying objectives of the joint ventures that we use in later empirical analysis. The dimensions of contract we use in the taxonomy relate to issues of appropriability and hold up: control over R&D programs, control over IP rights, and access rights to partner facilities.

To construct the taxonomy, we employ a decision tree using four binary indicators as displayed in Fig. 2. We identify five types of contracts according to: (1) whether member firms conduct R&D; (2) whether a coordinator is involved; (3) whether IP is assigned back to the creator of that IP or whether it is instead transferred to a third-party; and (4) whether scientists from member firms visit each others' facilities or work in isolation. These four questions examine how the ventures treat threats of spillovers and strategic blocking by partners and lend to interpretations that the objectives of the ventures pertain to know-how sharing, cost sharing, or standards setting.

At the first node, labeled "Do Members Conduct R&D for this JV?", we distinguish those contracts in which some subset of members (usually all members) perform R&D from those contracts in which no members conduct R&D. Ventures where no members conduct R&D have the benefit of cost or risk sharing but not the benefit (and risks) of knowledge sharing or synergies associated with multiple parties engaging in complementary research.

We next segment the cases where no venture members conduct R&D into those where the R&D generating entity is a coordinator ("Coordinator-led R&D"), and those that are not (which we label "Contract R&D"). The distinguishing features of coordinators are that they are entities that engage in multiple R&D projects

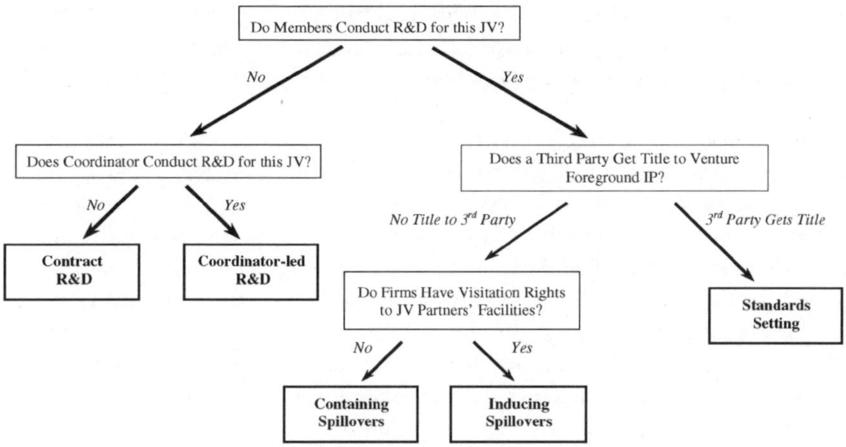

Fig. 2. Methodology for Determining Contract Taxonomy.
Note: Boxes in bold indicate five types of contract.

but engage in no commercialization activities. As a result, they function as R&D houses where R&D know-how could spillover across projects, but the coordinator is unlikely to act opportunistically on gained know-how since it does not commercialize any product. An example of a contract coordinator is the National Center for Manufacturing Sciences, a non-profit group with a variety of research projects supported by 150 corporations.[7] Ventures featuring contractor-led R&D also benefit from cost sharing, but because contractors typically are for-profit firms, they could possibly benefit from externalities associated with R&D production by using that know-how commercially in some related market, even if they are not explicitly granted title to venture IP. In many cases featuring contractor-led R&D, the contractor retains rights to subsequent IP, and members to the venture simply gain a license. Consequently, ventures that use contractors to conduct R&D benefit from cost sharing but may not fully appropriate the results of their R&D investment.

On the right side of the partition in Fig. 2, we partition contracts in which at least some members conduct R&D. These ventures are formed for reasons beyond shared costs. They allow some scope for R&D synergies as well as some risk of R&D spillovers. At the node labeled "Does a Third Party Get Title to Venture Foreground IP?" we distinguish those contracts that permit a non-member (a third party) to gain title to Foreground IP. These seem to be cases where venture members contribute R&D, but neutralize the prospect that a partner can engage in strategic hold-up by giving technology title to a more neutral third-party. In some cases, the neutral third party is the joint venture itself, where member firms have voting rights, often including veto rights, regarding actions the joint venture takes. We label these contracts "Standards Setting," because they often indicate they arise in a standards setting context where fear of hold-up is a greater concern than fear of spillover.

Finally, at the node "Do Firms Have Visitation Rights to JV Partners' Facilities?" we segment the ventures where members contribute R&D and retain title to Foreground IP into those involving or forbidding personnel exchange across partner facilities. Ventures that feature members contributing R&D, maintaining rights to their own R&D, and explicitly exchanging R&D personnel are probably crafted for the purpose of capitalizing on know-how sharing and R&D synergy despite creating the prospect for later-stage strategic blocking. We label this subset of contracts "Inducing Spillovers." Ventures that do not feature visitation of R&D personnel are labeled "Containing Spillovers."

Each of these five modes of contracting lends itself to a simple interpretation:

(1) *Contract R&D*: The joint venture is more likely to feature horizontally-related firms funding an R&D project. Members pool resources and avoid duplicative effort. They contract the services of a third party to conduct the R&D, and gain royalty-free licenses to the Foreground IP.
(2) *Coordinator-Led R&D*: Again, the joint venture is more likely to feature horizontally-related firms funding an R&D project. Members pool resources and avoid duplicative effort by contracting the services of a contract coordinator that will not commercialize any resultant or derivative technology. Coordinators exploit specialized know-how and competencies because they act as research houses that build competencies in particular technologies. All members gain royalty-free licenses to the Foreground IP.
(3) *Containing Spillovers*: Members conduct some or all of the R&D and individually retain all rights to their own research products. Members may contribute Background IP to enable the venture. The joint venture contains spillovers by denying scientific personnel exchange across member firms.
(4) *Inducing Spillovers*: As with Containing Spillovers, Members of ventures labeled "Inducing Spillovers" conduct some or all of the R&D, individually retain all rights to their own research products, and may contribute Background IP. However in this case, the ventures selectively induce spillovers by providing rights to members to visit each others' R&D facilities.
(5) *Standards-Setting*: The joint venture is vertical in inputs to the R&D process. Member firms innovate complementary technologies that collectively comprise a system. Members credibly commit to enabling the licensing of the system to non-member firms by assigning to a third party, e.g. the joint venture, title to all Foreground IP and assigning to the third party rights to freely license

the Foreground IP to any firms seeking to commercialize the system. Assigning title to innovations to a third party neutralizes the prospect of hold-up.

10. RESULTS

We exploit a feature of contracts, clauses that commit member firms to restrain from disclosing venture intellectual property to outsiders, to examine how the weakness of property rights – both from a spillover potential and from the threat of strategic blocking – correlates with how firms organize cooperative R&D. Clauses that describe restrictions on the disclosure of Background IP or Foreground IP are featured in all contracts, but contracts vary in the length of the term of nondisclosure across different joint ventures. The length of disclosure restrictions is a uni-dimensional measure of the relative importance of the hazard of spillovers vs. the hazard of strategic blocking. Longer restrictions are designed to prevent unwanted spillover, but err on the side of increasing the ability of partners to engage in strategic blocking by alleging breech of contract. In contrast, shorter restrictions reduce the time horizon in which partners can engage in strategic blocking, but increase the chances of spillover. Allegations of nondisclosure enable blocking because court ordering is lengthy, costly, and possibly incorrect (Williamson, 2000a). Firms can exploit the litigation process to block partner-competitors by leveraging the high costs of legal defense for more favorable settlement terms.

Table 3 reveals interesting patterns in the distribution of disclosure restrictions on Foreground IP and Background IP across modes of contracting.[8] Ventures using long-lived Foreground IP disclosure restrictions are most often ones that the taxonomy assigns as "Containing Spillovers." Stated differently, firms in ventures where members do R&D, retain the rights to their own Foreground IP, and restrict employee visitation also tend to write contracts that contain long nondisclosure agreements on Foreground IP. In contrast, all of the seven contracts corresponding to "Standards-Setting" feature short-lived or no restrictions on Foreground IP disclosure.

The table also shows that disclosure restrictions on Background IP are used differently than those on Foreground IP. Restrictions on Foreground IP are more common than those on Background IP, with half of all contracts (48) featuring no Background IP restrictions at all.

Most of the contracts featuring indefinite restrictions on Background IP disclosure correspond to "Containing Spillovers," but some (three of the seven) are identified as "Standards-Setting" contracts. The overall pattern is that firms identified in ventures designed to contain spillovers use severe disclosure restrictions on both foreground and Background IP, while those in ventures

Table 3. Number of Joint Venture Observations by Joint Venture Classification Determined by the Taxonomy, and by the Length of Disclosure Restrictions on Foreground IP and Background IP.

Term of Disclosure Restrictions (In Years)	Contract R&D	Coordinator-Led R&D	Containing Spillovers	Inducing Spillovers	Standards Setting	Total
Restriction on disclosure of foreground intellectual property						
0	13	5	9	4	4	35
1	1	1		1	1	4
2	3	7	3			13
3	1	2	1	4	1	9
4						
5	1	4	6	2	1	14
6						
7	1	1				2
8	2					2
9						
10	4		1	2		7
Indefinite	2		6	1		9
Total	28	20	26	14	7	95
Restrictions on disclosure of background intellectual property						
0	23	9	8	6	2	48
1				1	1	2
2		1				1
3	1			1	1	3
4						
5		10	8	1		19
6						
7			1			1
8						
9						
10	2		1	3		6
Indefinite	2		8	2	3	15
Total	28	20	26	14	7	95

Note: Disclosure restrictions on Foreground IP prohibit disclosure of IP that is a result of the JV research program to third parties. Disclosure restrictions on Background IP limit disclosure of IP that pre-dated the JV and was contributed to the joint venture by JV participants.

designed around standards-setting use mild Foreground IP restrictions often with tough Background IP restrictions.

How might these patterns be understood? Disclosure restrictions constitute one type of instrument for containing knowledge spillovers. Indeed, if disclosure restrictions could enable member firms to contain spillovers without inducing negative externalities then all joint ventures would choose to set infinitely-lived disclosure restrictions.

How firms set limits on disclosure restriction depends on two factors we observe in the data. First, we suggest that the ability of member firms to behave opportunistically relates systematically to the relative strength or weakness of intellectual property. In some industries, the ability of firms to appropriate returns from their technology is relatively weaker than in other industries (Cohen, Nelson & Walsh, 2000; Levin, Klevorick, Nelson & Winter, 1987). Difficulties in detecting infringement due to the nature of the underlying technology, the rapid pace of technological change in certain industries, and lower royalties from hold-out by other licensors when products infringe a thicket of patents all weaken the ability of firms to appropriate returns from their IP. In other industries infringement is easy to detect, the technologies are long-lived, and IP rights are well-defined. In balancing the tension between preventing a partner's ability to spillover core technologies and ceding excessive hold-up power to the partner, firms must be sensitive to the ability to appropriate returns from the underlying technology.

In areas where appropriability is weak venture members will desire to shorten restrictions on the disclosure of IP to reduce the ability of their partners to engage in later-stage strategic blocking. Indefinite restrictions could enable a partner to impose litigation over a legal technicality of disclosure long after the useful life of that IP is over. Accordingly, we expect restrictions to be shorter where appropriability is weak, spillovers are difficult to detect, and property rights are ambiguous and difficult to enforce, and longer where appropriability is strong, property rights are well-defined, and enforcement is easy.

Second, venture members will impose disclosure restrictions on IP that vary according to the underlying objectives of the venture. For example, ventures engaged in Standards-Setting will want to impose short disclosure restrictions on venture Foreground IP to promote the dissemination and adoption of the technology.

We are able to test these hypotheses by using industry as a proxy for the strength of intellectual property rights, and modes of organization as proxies for joint venture objectives.[9] Table 4 shows the results of a duration model regressing the duration of disclosure restrictions on industries and modes of contracting. In each case we apply a Weibull specifications to the duration data.[10] We treat instances of indefinite restrictions as censored observations.

The regression results indicate that the taxonomy's proxies for venture objectives and the industry code proxies for strength of appropriability are systematically related to the length of nondisclosure agreements in the contracts.

Table 4. Regression of the Duration of Disclosure Restrictions for Foreground IP and Background IP on the Mode of Contract and Industry.[a]

Dependent Variable – Length of Disclosure Restriction on Foreground IP and Background IP	Foreground IP	Background IP
Constant	0.282*** (0.078)	0.371*** (0.088)
Coordinator-led R&D	−0.165 (0.103)	−0.280** (0.122)
Containing spillovers	0.204** (0.088)	0.300*** (0.104)
Inducing spillovers	0.104 (0.106)	0.160 (0.126)
Standards-setting	−0.339*** (0.125)	0.296* (0.170)
SIC 138	−0.250** (0.120)	−0.179 (0.140)
SIC 281	0.283** (0.125)	−0.063 (0.148)
SIC 282	−0.147 (0.122)	0.042 (0.168)
SIC 283	0.026 (0.159)	−0.025 (0.173)
SIC 286	0.317*** (0.115)	−0.201 (0.122)
SIC 351	−0.151 (0.148)	−0.135 (0.170)
SIC 357	0.017 (0.134)	−0.068 (0.159)
SIC367	−0.437*** (0.116)	−0.594*** (0.135)
SIC 737	0.096 (0.099)	−0.001 (0.117)
ln p	1.356*** (0.084)	1.247*** (0.088)

Note: Regressions employ a hazard model using the Weibull distribution, with parameter p. Estimates of $p > 1$ indicate an increasing hazard.

***, **, and * indicate statistical significance at the 1, 5, and 10% levels, respectively.

[a] The dummy for contract R&D is the omitted mode, and dummies for several SIC codes with only 1–2 observations were the omitted SICs.

10.1. Ventures that Contain Spillovers Employ Longer Restrictions

Ventures the taxonomy labels as Containing Spillovers (those where member firms do R&D and no visitation across partner facilities occurs) employ relatively longer restrictions on IP disclosure of both Foreground and Background IP. This is consistent with the notion that containing spillovers is important, and that, for these ventures, containing spillovers is more important than the increased risk of strategic blocking.

10.2. Ventures that Use Coordinators Employ Shorter Restrictions

Joint ventures using coordinators employ relatively shorter restrictions on the disclosure of Background IP, and the coefficient on the restrictions on Foreground IP disclosure suggests shorter restrictions, although it barely misses significance at the 10% level (it is only significant at the 11%). One interpretation is that firms using coordinators may be engaging in cost-sharing on relatively less strategic or core technologies, and so their spillover is not strategically significant. On the other hand, because coordinators and venture members do not compete against each other in down-stream markets they have less incentives to engage in opportunistic strategic blocking of each other, and so one might think that imposing longer nondisclosure restrictions would be relatively costless. The fact that coordinators are associated with relatively shorter nondisclosure durations suggests that spillovers of these technologies may be desirable or insignificant.

10.3. Ventures Designed to Create Standards Employ Shorter Restrictions on Foreground IP

Interestingly, joint ventures that the taxonomy labels as Standards-Setting – those where members conduct R&D but assign title to a third party – employ shorter nondisclosure restrictions on Foreground IP, consistent with a standards motivation, but they also employ longer restrictions on Background IP.

10.4. Characteristics of the Underlying Technology Matter

The signs and coefficients on the industry variables for chemicals, SIC 28, and electronics, SIC 36, are broadly consistent with the existing empirical literature on appropriability (Cohen, Nelson & Walsh, 2000; Levin, Klevoric, Nelson & Winter, 1987). Joint ventures in chemicals, where appropriability is considered to be strong, employ relatively longer nondisclosure restrictions, and joint venture members in electronics, where appropriability is relatively weak, employ shorter nondisclosure restrictions. Where appropriability is weak, shorter restrictions reduce the prospect that partners could engage in strategic blocking, as has been observed in electronics in the area of patent litigation (Hall & Ham-Ziedonis, 2001).

Collectively, we interpret these results as evidence that joint venture participants design contracts to weigh and balance objectives with spillover potential and the potential for strategic blocking. They also provide evidence that appropriability conditions are not uniform across all technologies, and further that while property rights matter, the weakness of property rights matters as well.

11. CONCLUSION

The focus of this paper is on how and why property rights can be weak and what weak property rights implies for the design of R&D joint venture contracts.

Weak property rights are interesting in themselves, but they are also interesting because they are not something that all contracting theories are prepared to accommodate. Property rights theory suggests how judicious ex ante allocations of property rights alone can enable collaboration. It admits no scope for ex post governance, partly because it has focused on environments in which property rights can be costlessly delineated and enforced. In environments, such as the ones we study, property rights can be weak, and they can be weak in different ways. Weakness, it turns out, invites dimensions of governance that complement property rights allocations.

This paper is an empirical exercise that relates the design of governance mechanisms to objectives of joint venture participants and to the strength of property rights, including characteristics of the underlying technology. Not only does the allocation of property rights matter, but the strength of property rights matters as well. The challenge for scholars of contract design is to extend models beyond assumptions of crisp, clear, well-defined rights so as to more accurately reflect the diversity of joint venture structures we observe in the world and to more carefully explore the relative importance of complementary mechanisms to contract.

NOTES

1. There have been a handful of legal cases and much debate in the legal literature centered on the issue of leveraging intellectual property through patent litigation beyond the intended scope. See, for example, the FTC's case against Intel: "In the Matter of Intel Corporation," Docket No. 9288, http://www.ftc.gov/os/1998/06/intelfin.cmp.htm. See also testimony submitted in Joint Department of Justice – Federal Trade Commission Hearings on Competition and Intellectual Property Law and Policy in the Knowledge-Based Economy. http://www.ftc.gov/opp/intellect/020501xscript.pdf.

2. Some authors recognize transaction cost economics, the property rights theory, and mechanism design approaches to incomplete contracting collectively as "the literature on incomplete contracting" and might suggest that the latter two bodies of theory formalize and subsume transaction cost economics. (See, for example, Maskin & Tirole, 1999b, p. 141; Salanié, 1997, p. 176.) The claim would derive from identifying incomplete contracting with the canonical "hold-up" problem first articulated by Klein, Crawford and Alchian (1978). Other authors suggest that identifying the scope of transaction cost economics with the hold-up problem misses the most important action, much of which pertains to processes that enable contracting parties to adapt their relationships to unforeseen contingencies. (See, for example, Hart & Moore, 1999; Tadelis, 2002, p. 433; Williamson, 1971, pp. 115–117; 1996, pp. 31–32, 100–103, p. 605, 2000b, p. 34, 2000c.) More on this below.

3. The canonical hold-up problem involves contexts in which at least one party to prospective exchange must sink relationship-specific investments to enable that exchange.

(Again, see Klein, Crawford & Alchian, 1978.) The counter-party can "hold-up" the investing party subsequent to contracting by imposing renegotiation and extracting quasi-rents. Anticipating this, the investing party underinvests or foregoes exchange entirely unless the counterparty can credibly commit to not expropriating quasi-rents. Follow-on literature indicates the role of long-term contracting in mitigating hold-up (early examples include Joskow, 1987, 1988; Masten & Crocker, 1985) and the property rights theory inspired by Grossman and Hart (1986) indicates the role of ex ante allocations of property rights in mitigating hold-up.

4. The National Cooperative Research Act of 1984 was amended to extend limited antitrust immunity to firms involved in production joint ventures under the National Cooperative Research and Production Act (NCRPA) of 1993. While our data is derived from both programs, all the ventures we analyze include some R&D.

5. The NCRA and NCRPA extend single rather than treble damages to firms later found guilty of violating antitrust laws if they filed their ventures with and were given approval from the Department of Justice and Federal Trade Commission.

6. Both the time distribution and industry distribution of NCRA ventures submitting contracts are roughly comparable to the time and industry distributions of international research joint ventures in the MERIT/CATI data collected by John Hagedoorn. (See, for example, Hagedoorn, John (2002). Inter-firm R&D partnerships: an overview of major trends and patterns since 1960. *Research Policy*, *31*(4), 477–492.)

7. NCMS indicates the following information on its web site (http://www.ncms.org/) "The National Center for Manufacturing Sciences (NCMS) is a non-profit, membership organization dedicated to advancing the global competitiveness of North American industry through collaboration. Formed in 1986 and now with 150 corporate members and a diverse array of public partners, NCMS is widely regarded as the nation's premier provider of collaborative manufacturing research.

Mission: To increase the global competitiveness, economic growth, and operating efficiency of North American manufacturing industry by accelerating the creation and commercial use of advanced manufacturing intellectual capital – knowledge, know-how, processes, practices, tools, and technologies.

Why collaborate?

By collaborating, companies can carry out programs that they could not accomplish on their own.

The National Cooperative Research and Production Act of 1984 (later amended by the Production Amendments in 1993) provided significant legal protection for U.S. companies to collaborate in pre-competitive research. NCMS was formed to develop the collaborative model and to make it a reality for a wide variety of industry sectors."

8. We are missing disclosure data for one RJV. The other 95 contracts are represented.

9. The analysis exploits SIC codes pertaining to the technological area of R&D conducted by the ventures, rather than the industries to which the research results are ultimately used. We do this because appropriability theory relates to R&D inputs.

10. The model here aggregates durations into segments: 0–5 years, 6–10 years, indefinite. An alternative model on a subsample that drops observations with zero duration (which cannot be handled in logs) but uses the discrete duration values results in all the coefficients having the same signs and significance as these in the table with few exceptions: the coefficients on Containing Spillovers and on Standards-Setting loose their significance in the Background IP regression, the coefficients on Containing Spillovers

and on SIC 286 drop to the 10% significance level in the Foreground IP regression, while the coefficient on SIC 13 gains in significance on the Foreground IP regression.

ACKNOWLEDGMENTS

The views expressed in this paper do not necessarily reflect those of the U.S. Department of Justice. Our research has benefited from comments by David Mowery at the annual conference of the Academy of Management, by Nicholas Argyres at the annual conference of the International Society of the New Institutional Economics, and by participants of the annual conference of the Association for Public Policy Analysis and Management. We have also benefited by comments from participants of the Workshop on Intellectual Property Rights sponsored by the National Science Foundation and organized by Nicholas Vonortas. Authors appear in alphabetical order. All errors remain our own.

REFERENCES

Aghion, P., & Tirole, J. (1994). The management of innovation. *Quarterly Journal of Economics*, 1185–1209.

Anand, B., & Galevtovic, A. (2000). Weak property rights and holdup in R&D. *Journal of Economics and Management Strategy*, *9*, 615–642.

Bajari, P., & Tadelis, S. (2001). Incentives vs. transaction costs: A theory of procurement contracts. *RAND Journal of Economics*, *32*, 387–407.

Cohen, W. M., Nelson, R. R., & Walsh, J. P. (2000, February). Protecting their intellectual assets: Appropriability conditions and why U.S. manufacturing firms patent (or not). NBER Working Paper No. 7552.

Crocker, K. J., & Masten, S. E. (1988). Mitigating contractual hazards: Options and contract length. *RAND Journal of Economics*, *19*, 327–343.

Crocker, K. J., & Masten, S. E. (1991). Pretia ex machina? Prices and process in long-term contracts. *Journal of Law and Economics*, *34*, 64–69.

Crocker, K. J., & Reynolds, K. J. (1993). The efficiency of incomplete contracts: An empirical analysis of air force engine procurement. *RAND Journal of Economics*, *24*, 126–146.

Dixit, A. (2003). On modes of economic governance. *Econometrica*, *71*, 449–481.

Goldberg, V. P., & Erickson, J. R. (1987). Quantity and price adjustment in long-term contracts: A case study of petroleum coke. *Journal of Law and Economics*, *30*, 369–398.

Griliches, Z. (1994). Productivity, R&D, and the data constraint. *American Economic Review*, *84*, 1–23.

Grossman, S., & Hart, O. (1986). The costs and benefits of ownership: A theory of vertical and lateral integration. *Journal of Political Economy*, *94*, 691–719.

Hall, B., & Ham-Ziedonis, R. (2001). The patent paradox revisited: An empirical study of patenting in the U.S. semiconductor industry, 1979–1995. *RAND Journal of Economics*, *32*, 101–128.

Hart, O. (1995). *Firms, contracts, and financial structure*. Oxford University Press.

Hart, O., & Moore, J. (1990). Property rights and the nature of the firm. *Journal of Political Economy, 98*, 1119–1158.

Hart, O., & Moore, J. (1999). Foundations of incomplete contracts. *Review of Economic Studies, 66*, 115–138.

Joskow, P. L. (1987). Contract duration and relationship-specific investments: Empirical evidence from coal markets. *American Economic Review, 77*, 168–185.

Joskow, P. L. (1988). Price adjustment in long-term contracts: The case of coal. *Journal of Law and Economics, 31*, 47–83.

Klein, B., Crawford, R., & Alchian, A. (1978). Vertical integration, appropriable rents, and the competitive contracting process. *Journal of Law and Economics, 21*, 297–326.

Kreps, D. M. (2000). Markets and hierachies and (mathematical) economic theory. In: D. J. Teece & G. R. Carroll (Eds), *Firms, Markets, and Hierarchies: The Transaction Cost Economics Perspective* (pp. 121–155). Oxford University Press.

Laffont, J. J., & Martimort, D. (2002). *The theory of incentives: The principal-agent model*. Princeton University Press.

Lanjouw, J. O., & Lerner, J. (2001). Tilting the table? The use of preliminary injunctions. *Journal of Law and Economics, 44*, 573–603.

Lanjouw, J. O., & Schankerman, M. (2001). Characteristics of patent litigation: A window on competition. *RAND Journal of Economics, 32*, 129–151.

Lerner, J., & Merges, R. P. (1998). The control of technology alliances: An empirical analysis of the biotechnology industry. *Journal of Industrial Economics, 46*, 125–156.

Levin, R. C., Klevorick, A. K., Nelson, R. N., & Winter, S. G. (1987). Appropriating the returns from industrial research and development. *Brookings Papers on Economic Activity, 3*, 783–831.

Llewellyn, K. N. (1931). What price contract? – An essay in perspective. *Yale Law Journal, 40*, 704–751.

Macaulay, S. (1963). Non-contractual relations in business: A preliminary study. *American Sociological Review, 28*, 55–67.

Macneil, I. R. (1974). The many futures of contracts. *Southern California Law Review, 47*, 691–816.

Macneil, I. R. (1978). Contracts: Adjustment of long-term economic relations under classical, neoclassical, and relational contract law. *Northwestern University Law Review, 72*, 854–905.

Maskin, E., & Tirole, J. (1999a). Unforeseen contingencies and incomplete contracts. *Review of Economic Studies, 66*, 83–114.

Maskin, E., & Tirole, J. (1999b). Two remarks on the property-rights literature. *Review of Economic Studies, 66*, 139–149.

Masten, S. E., & Crocker, K. J. (1985). Efficient adaptation in long-term contracts: Take-or-pay provisions for natural gas. *American Economic Review, 75*, 1083–1093.

Moore, J. (1992). The firm as a collection of assets. *European Economic Review, 36*, 493–507.

Nelson, R. R., & Winter, S. G. (1982). *An evolutionary theory of economic change*. Harvard University Press.

Oxley, J. E. (1997). Appropriability hazards and governance in strategic alliances: A transaction cost approach. *Journal of Law Economics and Organization, 13*, 387–409.

Ryall, M. D., & Sampson, R. C. (2003). Do prior alliances influence contract structure? Evidence from technology alliance contracts. Working Paper.

Salanié, B. (1997). *The economics of contracts*. MIT Press.

Saussier, S. (2000). Transaction costs and contractual incompleteness: The case of electricité de France. *Journal of Economic Behavior and Organization, 42*, 189–206.

Segal, I. (1999). Complexity and renegotiation: A foundation for incomplete contracts. *Review of Economic Studies, 66*, 57–82.

Spence, M. (1984). Cost reduction, competition, and industry performance. *Econometrica*, *52*, 101–121.
Tadelis, S. (2002). Complexity, flexibility, and the make-or-buy decision. *American Economic Review*, *92*, 433–437.
Teece, D. J. (1986). Profiting from technological innovation: Implications for integration, collaboration, licensing and public policy. *Research Policy*, *15*, 285–305.
Tirole, J. (1999). Incomplete contracts: Where do we stand? *Econometrica*, *67*, 741–781.
Williamson, O. E. (1971). The vertical integration of production: Market failure considerations. *American Economic Review*, *61*, 112–123.
Williamson, O. E. (1975). *Markets and hierarchies*. Free Press.
Williamson, O. E. (1985). *The economic institutions of capitalism*. Free Press.
Williamson, O. E. (1996). *The mechanisms of governance*. Oxford University Press.
Williamson, O. E. (2000a). Revisiting legal realism: The law, economics, and organization perspective. In: D. J. Teece & G. R. Carroll (Eds), *Firms, Markets, and Hierarchies: The Transaction Cost Economics Perspective* (pp. 197–234). Oxford University Press.
Williamson, O. E. (2000b). Some reflections. In: D. J. Teece & G. R. Carroll (Eds), *Firms, Markets, and Hierarchies: The Transaction Cost Economics Perspective* (pp. 32–36). Oxford University Press.
Williamson, O. E. (2000c). The new institutional economics: Taking stock, looking ahead. *Journal of Economic Literature*, *38*, 595–613.
Zucker, L. G., Darby, M. R., & Brewer, M. B. (1998). Intellectual human capital and the birth of U.S. biotechnology enterprises. *American Economic Review*, *88*, 290–306.

WILL MP3 DOWNLOADS ANNIHILATE THE RECORD INDUSTRY? THE EVIDENCE SO FAR

Stan J. Liebowitz

ABSTRACT

This paper investigates the impact of peer-to-peer networks that promote the unauthorized downloading of MP3 files exert on the recording industry. Although Napster was the most famous of these systems, its progeny have continued to allow millions of music listeners to download music (and other) files without remuneration to the copyright owners. Economic theory does not allow us to determine what the impact of such downloading will be. Using data on the historical sales of prerecorded music I examine in detail the recent decline in record sales and attempt to gauge the importance of various factors that have been put forward to explain this decline. My conclusion is that the evidence is most consistent with a claim that MP3 downloads decrease sales. Nevertheless, it is unclear whether the damage to the industry will be fatal.

1. INTRODUCTION

After Napster exploded into the public's consciousness in 1999, concerns with MP3 downloads, their impact on the recording industry, and the industry's attempt to thwart such downloads, have been in the news almost continuously. The

Intellectual Property and Entrepreneurship
Advances in the Study of Entrepreneurship, Innovation and Economic Growth, Volume 15, 229–260

ISSN: 1048-4736/doi:10.1016/S1048-4736(04)01507-3

Napster case was only the most public front in a wide ranging battle between the entertainment industries (movies and records) and the millions of individuals who were using the Internet to download, without authorization, copyrighted works, but who otherwise were thought to be the putative customers of these industries.

This fight has spilled over into the political arena. The manifestations were seen in a bill proposed by Representative Howard Berman from California in 2002 that would have allowed copyright owners to engage with impunity in otherwise illegal activities "disabling, interfering with, blocking, diverting, or otherwise impairing" the "unauthorized" distribution of copyrighted items.[1] The bill did attempt to restrict these otherwise illegal activities to those that do not "alter, delete, or otherwise impair the integrity of any computer file or data residing on the computer of a file trader" but it still amounted to authorizing "hacking" behavior that is disallowed in other venues.

This is only a small part in a flurry of activities taken by the copyright-based industries to protect themselves against what they see as online pilfering of their products. The most recent strategy announced by the recording industry is to prosecute those making their files available to others over peer-to-peer systems. As this article is being written, the RIAA is supposedly collecting the names of hundreds or thousands of users in order to bring suits against them in the next few months. How this will play out is yet to be seen.

At the center of the current controversy is the fear by the recording industry that great damage, perhaps even mortal damage, will be done to them if they do not stop online "trading" of music files, a fear I refer to as the "annihilation hypothesis." The movie industry is taking a backseat to the record industry for the moment because the size of the files needed to view movies make online movie trading prohibitively time consuming at current connection speeds and because most users do not yet have the equipment to write DVDs.

As this is all coming to a boil, various possibilities are emerging as potential outcomes. One outcome would be for the recording industry to beat back the threat of peer-to-peer systems by engaging sufficiently in successful public prosecutions of users that the fear of punishment ends this short term experiment in massive copyright infringement. A second possibility is that peer-to-peer downloading proves impossible to prevent by threats of punishment, and an alternative to market based copyright systems arises. Such a system might include a tax on ancillary products, such as blank CDs, CD burners, or stereo equipment, with the proceeds of the tax to be distributed based on the number of downloads of individual titles.[2] A third alternative would be the rise of digital rights management systems, which would be an advanced form of copy protection which allows varying degrees of use of the product depending on the price paid by the user. In order for this to work, however, the copy protection would need to be sufficiently difficult to break that peer-to-peer files sharing systems would lose their appeal. Finally, another alternative would be for things to continue down the track that they are on, with peer-to-peer file sharing proving to be an unpreventable act which reduces the revenues of the recording industry but not sufficiently so to destroy or alter the industry in any meaningful way.

Central to any discussion of these hypotheses should be empirical examinations of the industry. Yet there have been few studies attempting to examine the impact of MP3 downloads. In the Napster case the recording industry presented evidence purporting to demonstrate that MP3 downloads were harmful to record sales and Napster presented evidence purporting to demonstrate that online file-sharing was beneficial to the record companies.[3]

As I have explained elsewhere, the evidence provided by the recording industry in the Napster case, in my opinion, failed to support any claim of harm, to say nothing of supporting the annihilation hypothesis, although the judge nevertheless ruled against Napster.[4] I also argued that although there was not yet good evidence to support such harm, there were good economic reasons to believe that online file sharing would be harmful to the industry.[5] At that time there was little evidence of any decline in record sales and I provided rationales as to why the evidence to support this claim had not yet surfaced.[6]

Peer-to-peer file sharing, although anathema to the recording industry, is providing a wonderful natural experiment to economists or anyone else who might be interested in the impacts of unauthorized copying. The longer it goes on, the more evidence that can be accumulated on the impact of the copying on the market for authorized works. The fact that there appear to be actual measurements of the number of files downloaded provides a metric on the size of the copying activity that had never existed with any prior copying technologies.[7] The natural experiment is still ongoing and results are still coming in. With all the energy and heat being generated on these topics, however, this seems like a good time to take stock and examine what we currently know about the impact of MP3 downloading on record sales.

My conclusions, in a nutshell, are that MP3 downloading does appear to be causing harm. No other explanations that have been put forward seem to be able to explain the decline in sales that have occurred since 1999. Still, it is not clear that the harm will be fatal.

2. BACKGROUND

Records are just a particular category of consumer goods, and not a terribly large one in terms of revenues. Even so, listening to music is one of the major daily activities for many Americans. Estimates indicate that the average American spends 45 minutes a day listening to recorded music and almost three hours listening to the radio – where music is the major programming ingredient.[8] Thus even if its share of GDP is small, its share of the collective consciousness and behavior of consumers is very large. It is also possible that its value to consumers might be much larger than that indicated by market revenues, just as air or water provide enormous values but very little, if any, market revenues.[9]

As is the case with most economic goods, we expect the demand in the market to be impacted by economic factors such as changes in taste, income, prices of substitutes and complements, demographics, and changes in the nature of the product. Many of these factors are discussed in Section 3.1. The factor that is of particular relevance in this market is the impact of copying.

The analysis will consist, therefore, of simply examining the sales of albums and trying to determine the factors responsible for changes in sales. We can then focus on sales in the last few years, since that is the period in which MP3 downloading began, and see whether sales decline in an otherwise unexplained manner, or whether they fail to increase as we might have expected them to.

3. THE CURRENT NUMBERS

Data on quantities of records sold in the United States on a yearly basis are available from the Recording Industry Association of America (RIAA), as are data on revenues. The data on revenues are estimates based on the list price of recordings – actual transaction prices are not available.[10] Although these numbers are for just the United States market, we can presume that any negative impacts of MP3 downloading would be at least as likely to show up in the U.S. market as any other, due to the large number of computers, Internet connections, and CD burners and also to the fact that most of the population can afford to purchase CDs legitimately.

The data reported by the RIAA include the number of units of recorded music sold in various recording mediums (e.g. CDs, audio cassettes and vinyl records) as well as the revenues generated by each of these mediums. It also contains a breakdown of sales by playing time – singles or albums.

Naturally, the RIAA tend to use these numbers in a way that is most consistent with the picture they wish to portray about the conditions of the industry. At the present time the condition they wish to portray is one where the industry is reeling from the impact of MP3 downloads.[11]

Before a more complete analysis of the data is undertaken, it is useful to examine the evidence at its most basic level. This, however, requires some decisions about how the "output" of the industry is measured. Various measures of

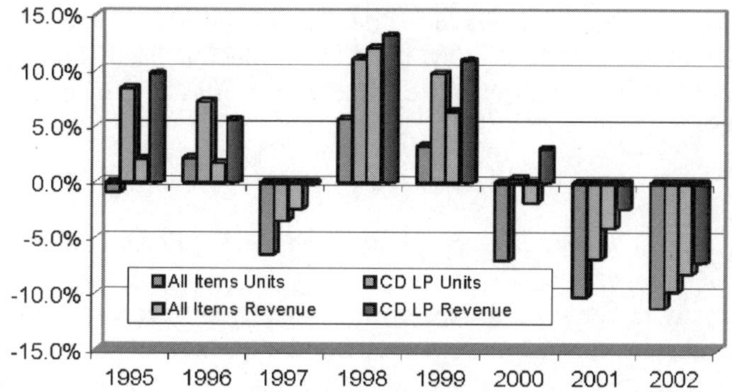

Fig. 1. Four Views of Yearly Changes.

"sales" can lead to somewhat different views of the industry environment as is seen in Fig. 1.

Figure 1 presents yearly changes in sales from 1995 to 2002, measured both as changes in revenues and as changes in units. A further classification breakdown compares just CD-based albums (excludes CD singles) to all items, which would include singles and albums together for all recording medias (CDs, cassettes and vinyl).

Because of a very lengthy and severe decline of singles (documented below), as well as a very pronounced decline in cassettes, the "all items" categories show weaker sales growth numbers than the categories based only on CDs. Further, unit sales performed worse than revenues since prices rose somewhat during this period.

In 2001, for example, sales fell by 2, 4, 7, or 10%, depending on which classification one chooses as the basis for comparisons. CD revenues fell the least, total units fell the most. It is not surprising that the RIAA chose total units as the measurement trumpeted in its headlines, experiencing about a 10% drop in sales that year, since that is a far more impressive a drop than the 2% fall in CD revenues. The fact that the RIAA choose this particular number, however, does not necessarily make it wrong or even misleading.

Determining the proper unit of measure is not straightforward. Although the total industry might seem to be best represented by the "all items" category, that overall focus can be misleading when there is a change taking place in the individual sectors. For example, sales should be artificially inflated during a period when a new media is rolling out and consumers are replenishing their music collections in the new media. Alternatively, when singles are being eliminated from the market the "all items" category will show a smaller growth rate than the non-single categories, as was the case above.

The decline in singles appears in the data beginning in the 1970s although the decline was not monotonic. Clearly, most of this decline has nothing to do with MP3 downloads; thus I will treat the decline in singles as an exogenous event and remove it from the analysis.[12]

3.1. Economic Factors Possibly Influencing Record Sales

There are many possible influences on the quantity of records sold. The prime textbook candidates for factors likely to alter the quantity of records sold would normally be changes in tastes, record prices, incomes, population, the prices of substitutes and complements, and any other factors that might be deemed important. There are several major factors in this market that could be examined:

(i) The price and quantity of records: Actual transaction prices for record sales are not, to my knowledge, available. Instead, one must rely on RIAA estimates of list prices and also for estimates on the number of units sold.[13]
(ii) I am not aware of any good data on yearly production costs.
(iii) For most products, income changes are expected to be related to consumption since as income goes up consumption is likely to rise.[14]
(iv) Changes in recording formats – RIAA provides data the number of units in different formats (i.e. tapes, vinyl disks, CDs) sold in each year and one can try to incorporate this into the analysis.
(v) Changes in the "quality" of music or in musical "taste" would be difficult to ascertain but clearly would have an impact on the market. It is often stated, although usually in a form that is not testable, that some years are better than others in terms of musical innovation.[15] I try to control for this variable by including information on concert sales as a measure of interest in music. This is found in Section 5.5 below.
(vi) Changes in the size and distribution of the population potential users – estimates in the distribution of album purchasers are available from the RIAA and these will play a role in the analysis, in Section 5.6.
(vii) A change in media portability can change the nature of the demand, opening up or closing markets. When LP records became the dominant form of recorded music there was no portable form of recorded music. LP records were the domain of the home stereo since they were too bulky to be played in automobiles or other mobile venues.[16] With the advent of 8-track tape, a new market opened up to record producers who were then able to sell tapes for the automobile as well as vinyl records for the home. Prerecorded audio cassettes were another new market, one that allowed listening to music

while exercising or just walking around, as well as expanding the automobile market. This important factor is discussed in Section 5.3.

(viii) Changes in the price of complements/substitutes, such as television, movies, radio, videogames and so forth.[17] I am not aware of any "major" change in these categories over the period of analysis. The less substantial changes that did occur are examined in Section 5.4.

(ix) Finally, sales might be influenced by the nature of the distribution channels. As authorized digital downloads play an increasing share of this market, the sales of "albums" would be likely to change, particularly if pricing is no longer "per unit" but instead more of the "per month" variety. On a more prosaic level, if distribution shifts from record stores to major discount chains, such as Wal-Mart and Best Buy, as seems to have been the case in the last decade, unit sales might increase because retail margins lower the effective price to consumers. If these large retail chains are able to negotiate lower wholesale prices than record stores were able to, the revenue growth using list prices would overstate the actual health of the recording industry.[18]

4. SOME UNDERLYING TRENDS

The record industry, as is the case with most industries, is fairly complex, with many factors changing at once. Choices need to be made to make the analysis tractable. At the most basic level we have seen that sales in the recording industry have dropped during 2001 and 2002, although they had risen during 1999, Napster's first year of existence, and gave mixed results for 2000, depending on how record sales were measured. I try now to perform a somewhat more exacting analysis. To do this I will limit the market to full-length albums, regardless of the media on which an album is housed. First, however, we will examine the market for singles to see what might be left out of an analysis that focuses on albums.

4.1. The Market for Singles

One of the causes of the decline in record sales is the decline of singles that has been ongoing for quite some time. This can be seen in Fig. 2.

There are three lines and two vertical axes in Fig. 2. On the left hand axis are the dollar revenues in the singles market. The line with triangle markers represents the nominal dollar revenues for singles each year. With no adjustment for inflation, revenues for singles increased from 1973 to 1997, at which time they experienced a precipitous drop. A very different story is told by the line with

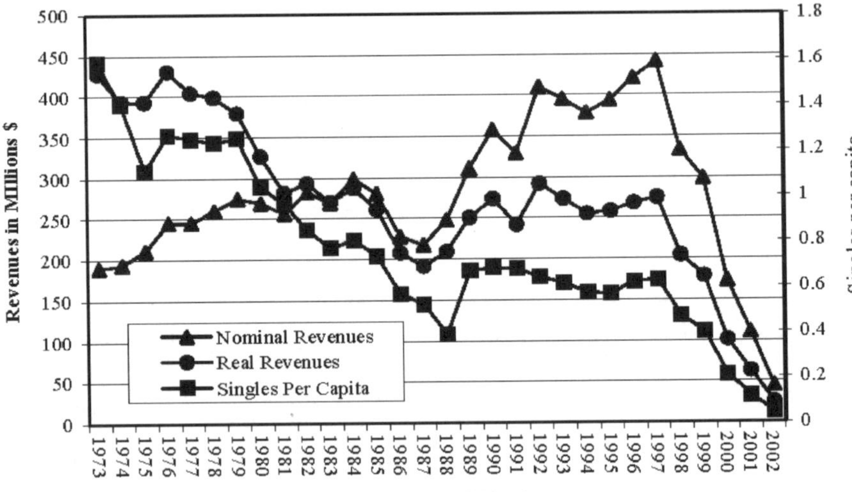

Fig. 2. Sale of Singles.

square markers representing the number of singles per capita (read off the right hand axis).[19] In the early 1970s the typical consumer purchased approximately 1.5 singles per year. By the mid-1990s that had dropped to 0.5 singles and by 2001 that number had further dropped to 0.1 singles per person per year.

The line with circles represents inflation adjusted (real) revenues in the singles market, and it tells a story quite similar to the singles-per-capita.[20] Real revenue can be seen to have fallen almost continuously from 1973 to the mid-1980s, whereupon it held at a relatively constant but much lower level until 1997, at which point it began a steep descent. This, of course, fits with our understanding of this market. Vinyl singles were popular in the 1960s, but when vinyl records lost popularity, so did singles. With the advent of CD singles the market was able to stop its decline, although it did not recover its previous heights. In 1998 the bottom fell out.

Although the decline in singles has been a long time in the making, there is little doubt that 1998 marked a major acceleration of that trend. The bottom began to fall out of the single's market a year before Napster came into existence. And from Fig. 1 we know that 1998 was a good year for the record industry as a whole. One possible explanation is that MP3s were becoming popular prior to their availability online. Since MP3s are individual songs, they would be closer substitutes to singles than to entire CDs. Further, if the purchase of singles was mainly to try out individual songs with the expectation of later purchasing a CD, then MP3 might achieve that same function. This argument was the essence of the Napster defense – that MP3s were being used to "sample" songs and that

CD sales would not be hurt by such sampling. Another possibility, argued in a magazine article, is that record companies decided that singles did not pay.[21]

For all intents and purposes, singles are now practically non-existent.[22] It seems possible that MP3s are responsible for killing off the singles market, even if what was left of the market was very small (in 1997 singles accounted for merely 3.6% of record revenues). Nevertheless, the downloading of MP3s was clearly not the primary factor in the fall of singles, although it might have been the final nail in the coffin.

From this point on I will focus attention on "albums" which include long playing vinyl records, long playing cassettes, eight track tapes, and full length CDs. Since the market for singles has been undergoing a lengthy decline, its exclusion is likely to provide results less severe than would be found with statistics for all forms of records as typically reported by the RIAA.

4.2. Some History of Album Sales in this Market

An album is generally a collection of songs with a playing time in the vicinity of fifty minutes. I will use the term "album" to describe either long playing vinyl records, full length CDs or full length cassettes. The thirty year decline in singles stands in stark contrast to the generally increasing sales of full-length albums which can be seen in Fig. 3. Inflation adjusted revenues, either overall or per capita, rose significantly during the period. Album revenues rose from $4 billion to over $7 billion after adjusting for inflation, and per capita sales rose from $28 to almost $40.[23]

Another way of restating these findings is to look at the impressive increase in unit sales over this time period as illustrated in Fig. 4. Total units sold increased

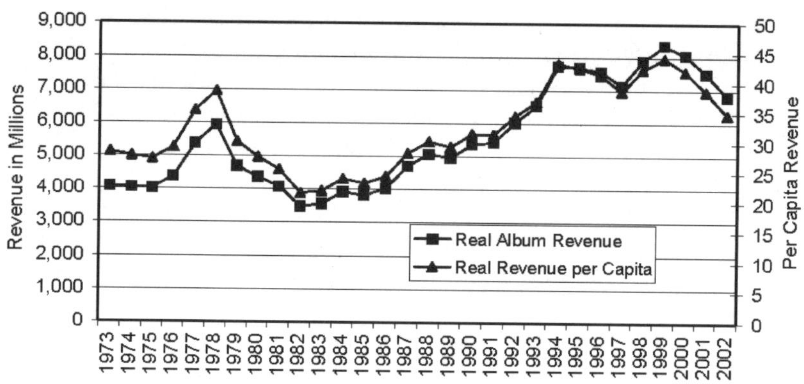

Fig. 3. Album Revenues.

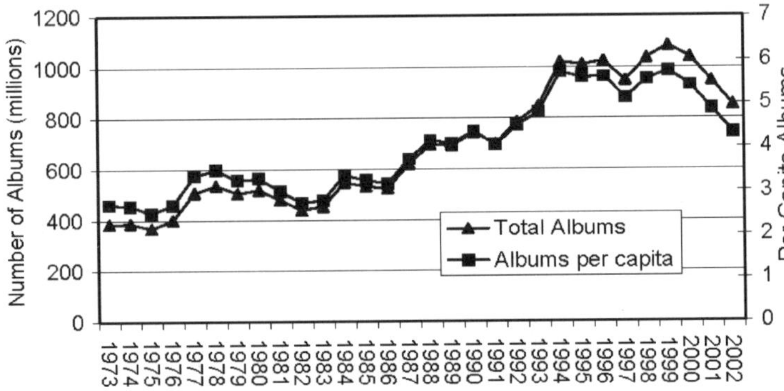

Fig. 4. Album Unit Sales.

by a factor of 2.5, from 400 million to approximately one billion units. Yearly sales of albums per capita (defined over the population between ages 10 and 60) increased almost as much, from 2.7 albums in 1973 to over 5 during much of the 1990s. Since the population (as defined) only increased by 37% over this period, total and per capita measures cannot greatly differ.

The second feature of interest is the non-linearity of the plot. There are at least four dips in sales prior to the current dip which is underway. These dips occurred in 1978–1982, 1984–1986, 1991, and 1994–1997 with the current dip having begun after 1999. The fact that sales are currently falling, by itself, would not appear to be a cause for alarm, since this has happened several times in the past, only to be followed by further sales increases.

The final important feature of this plot is the realization that the great majority of the increase in per capita sales took place within a narrow interval, from 1982 to 1993. Per capita sales were largely flat before and after this period. One question that we need to answer is whether the growth during 1982–1993 was unusual, or whether the stagnation during 1973–1982 and 1994–2001 was unusual.

There were three technological changes that took place prior to MP3 downloading that might influence the sales of albums. First, the recording medium changed – from vinyl to cassettes to CDs. Second, the advent of cassette recorders allowed the taping of pre-recorded material, giving consumers the first potentially serious ability to replace the purchase of albums with copies. Third, cassettes players (and shortly after CD players) allowed the first portable use of prerecorded media since there were no truly portable LP devices.

An examination of the shifts in recording media will complete this section on the background of the industry.

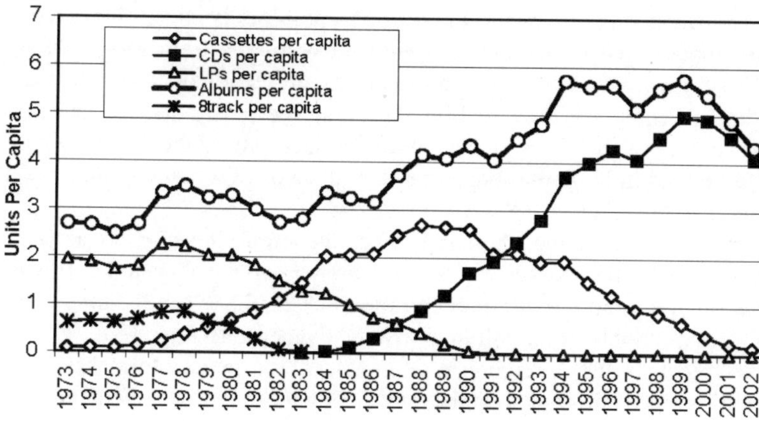

Fig. 5. Format Changes and Album Sales.

4.3. The Shift in Recorded Media

I present yearly sales for albums for different media types in Fig. 5.

From the early 1970s until the mid-1990s it was common for consumers to have at least two major media formats available at one time, and for a few years in the mid/late 1980s there were three formats with non-trivial market shares. What this means is that for most of the period since 1978, when cassettes began to replace vinyl records, part of consumer demand should have been devoted to replicating libraries in a new media format. This follows a period of relative tranquility in media dominated by the 331/3 rpm vinyl album.[24]

Figure 5 also provides some evidence of "home" demand vs. "automobile" demand for albums. LP records sales were for the home since they were too large to fit in mobile players, whereas 8-track sales were primarily, if not exclusively, for the automobile. In the mid-1970s this worked out, on a per capita basis, to approximately 2 albums for home use and one for the automobile. I do not have evidence on the penetration of 8-track units in automobiles, although I presume they were not as ubiquitous in automobiles as cassettes and CD players are now. Thus the full demand for music in automobiles may be understated by these values.

Changes in media might be thought to lead to changes in demand due to a "librarying" motive that would consist of individuals wishing to update some of their favorite recordings to the new medium, particularly if the new medium has either sonic or other utilitarian advantages. The strength of this librarying motive is an empirical question, however.

The empirical support for the librarying hypothesis is mixed, however. When cassettes first started to appear in significant numbers, from 1977 onward, LP (and 8-track) sales fell by about as much as prerecorded cassette sales grew, precluding a temporary upward blip. This may be partly due to the impact of copying, discussed below. It wasn't until the last half of the 1980s, when CDs began to be sold in large numbers, that overall sales of albums began their fairly dramatic rise, from 3.2 in 1986 to 5.6 in 1994.

But even in this decade, the support for the librarying hypothesis is mixed. According to Ebrain market research, a branch of the Consumer Electronics Association, the rapid growth in penetration of CD players ended after 1995, yet record sales did not begin to fall in a serious manner until after 1999.[25]

If the typical consumer replenishing his library for the new format took 5 years or more to achieve this goal, then the current downturn in sales might be attributed to the librarying motive. One might test this by examining whether a higher percentage of CD purchases were of relatively old recordings after 2000 than was the case before. This would be a particularly sharp test since it is claimed that a larger number of "old" songs are available for download than can be found in the typical record store, indicating that if MP3 downloads were responsible that there would be fewer old records being purchased after 2000.

Although I have not performed such a test, I suspect that librarying is not the key factor at work because there is little evidence that the switch from LPs to cassettes caused much of an increase in sales. Still, the impact of unauthorized copying possible with cassettes might have clouded that finding as well.

Although the peak of cassette sales was only slightly higher than the peak of LP sales, the peak of CD sales was much higher than either of the older formats. Why did CD sales peak at a level so much higher than the previous recording formats? I will return to this question in Section 5.3.

5. OTHER FACTORS AFFECTING SALES

5.1. The Role of Price and Income

As reported, the data provided by the RIAA do not list actual prices but only a derived list price. It is unclear how serious a problem the use of list, as opposed to transaction prices, would be in this case.[26] Examination of these prices indicates that the inflation adjusted list price has remained largely unchanged for the last twenty years, although there were some significant price changes during the 1970s, as seen in Fig. 6. Although there was a fairly significant fall in the inflation adjusted list prices from 1978 until 1982 (a time of highly variable

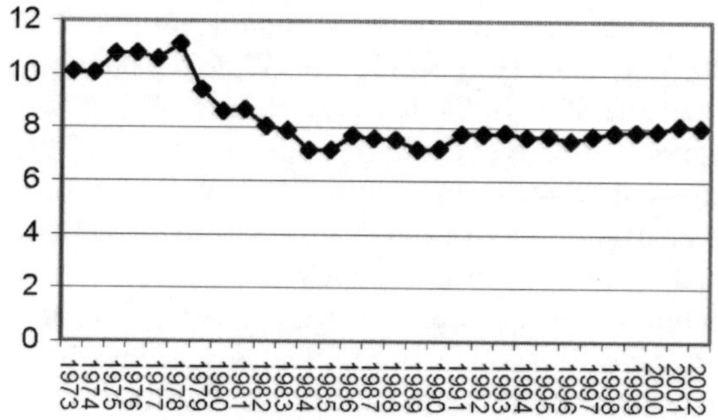

Fig. 6. Real List Price Per Album.

vinyl prices, unusually high inflation, and a changeover from vinyl records to the less expensive prerecorded cassette), inflation adjusted list prices have remained remarkably constant since 1981. The consistency of list prices is almost as if the record industry had decided to keep prices in line with the inflation rate.

With prices remaining almost constant during the last 20 years, changes in record sales would seem to be, of statistical necessity, due to other factors. One of those factors would be income.

Unfortunately, yearly variations in income are also not particularly large, although they are not constant the way that price is. More importantly, there has been a fairly steady upward trend in income over this period, at the same time that there has been a rather large increase in the purchase of records. This means that the common trend in the two series would make it misleading to run a regression comparing income with record purchases and price.

Instead, I ran a regression with first differences of per capita album sales as the dependent variable and first differences of per capita real income and first differences in price as the independent variable (with a constant term). The results are reported in Table 1.

Table 1.

	Coefficients	Std. Error	*t*	Sig.
Constant	−0.06123	0.078	−0.788	0.438
Change in real price of album	−0.152	0.139	−1.098	0.282
Per capita change in income	0.0007286	0	2.455	0.021
R square	Adjusted *R* square	Dependent variable: change in albums per capita		
0.201	0.14	$N = 29$		

The coefficient on change in income was significant ($t = 2.46$). Price has the correct sign, but is not statistically significant. What are the implications of these coefficients as an explanation of the recent decline in albums?

The largest decline in real per capita income in the last 30 years (1978–1980) was $300, which would have been expected to lower per capita record sales by only 0.2 units. Per capita record sales have declined by about a full unit since 1999. Income has actually gone up slightly from 1999 to 2002.[27] Even if the record buying public was skewed toward a demographic component particularly hard hit by income declines, it is inconceivable that income could have been responsible for any but a trivial component of the recent fall in records. For price, similarly, the rise in real list record prices since 1999 of 30 cents per unit might be expected to lead to a decrease in per capita unit sales of 0.05, and we can not have any statistical confidence even in this small result.

There is another element involved in listening to music, and that is the constraint of time. Listening to music requires time, and higher incomes do not necessarily lead to greater amounts of free time. If an income constraint is not binding for many consumers of albums, then income changes will not have much impact on album purchases.[28]

5.2. Copying Using Cassettes

The issue of home copying using cassette recorders was a precursor to the current controversy about MP3 downloading. Both would have been expected to lower sales if they were substitutes for the purchase of additional prerecorded material. It might be argued that home copying of audiotapes was a better substitute for an original purchase than is an MP3 download since cassettes could be played on the main home stereo system, whereas MP3s are typically not playable on the main stereo system unless converted back to the original format found on CDs.[29] At the time, dire predictions were made about the health of the recording industry by no less than Alan Greenspan.[30]

What does hindsight tell us about the impact of audiotaping?

Figure 7 repeats earlier information on albums and prerecorded cassettes, but adds in the number of blank cassettes, all per capita.[31] I have only been able to get sales numbers for blank cassettes starting in 1978, although sales obviously occurred prior to that. Cassette recorders allowed consumers to use taped copies to replace the purchase of originals. During the period of growth of cassettes, however, there was also a large increase in the sales of prerecorded albums. Clearly, the existence of cassette recorder/players seems to have had a positive overall impact on record producers, although we can not say that the growth might not have been more robust had the recording (copying) potential of cassettes been removed.

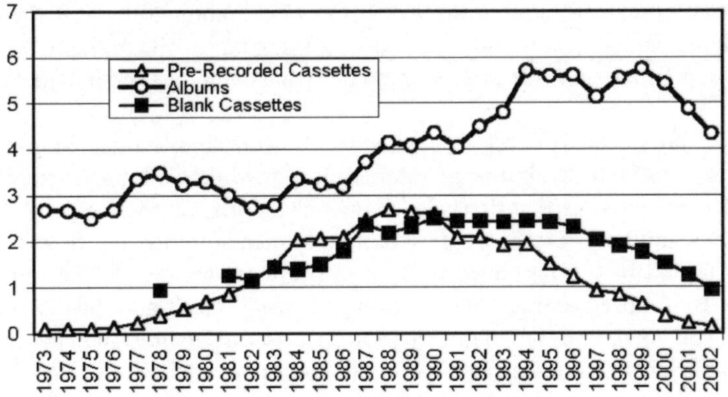

Fig. 7. Albums and Cassettes, Per Capita.

Of course, copying using cassettes requires having an original handy. That means either borrowing one or purchasing a legitimate copy. The difficulty of borrowing might have been sufficiently great that most copies would have been made from originals purchased by that individual. One of the important differences between this episode and the current MP3 controversy is the ease with which one can "borrow" music from others.

Why should the existence of cassettes have increased the sales of albums that had been relatively flat before cassettes came on the scene? The ability to penetrate new markets is the key. Copying may have diminished the need to purchase a second original for the new location, but the extra music-listening time from having new music listening locations available appears to have translated into the purchase of more originals, as explained in the next section.

5.3. The Impact of Portability on Purchase of Albums

We have seen that almost all of the increase in sales of albums during the 1973–2001 period came about during the years 1981–1993. What might be responsible for this increase?

Sales will increase when consumers wish to spend more time (and money) listening to recorded music. The industry would either need to improve the musical experience relative to other forms of entertainment, or they would need to create new consumption opportunities. I believe that it is the latter path that they have successfully taken in the last few decades. The opening of new venues for music listening due to media changes would show up as an increase in sales, everything else constant. This would be particularly true if the recorded mediums used in these new venues (automobiles or portable devices) were incompatible with the devices used in home stereos, as was the case with 8-track tapes, since then it would not possible to purchase a single pre-recorded item that could be used in both the home and car.[32]

In particular, music listeners, during the 1980s, were able, in increasing numbers, to listen to their favorite albums in automobiles, or while walking or exercising. It was the 8-track cassette that first opened up the auto market to prerecorded music.

Cassettes, unlike 8-tracks, were used in more than automobiles, however. They were also used in the home as well as in portable devices. Some increase in sales due to a more complete penetration of automobiles undoubtedly occurred, although I do not have detailed information on cassette penetration in automobiles.

We do have detailed information about the portable market, however. Although it was the cassette that first allowed listening to portable prerecorded music, the portable cassette player was soon followed and eclipsed by the portable CD player.

Figure 8 illustrates the penetration rates of portable cassette players and portable CD players. Cassette players began their penetration in 1983, about seven years after non-portable cassette players began to have an impact, whereas portable CD players began their market penetration in 1988.

Figure 9 calculates yearly changes in the summed penetration rates for portable cassettes and CD players. This may be a somewhat more informative method of determining the movement toward this new portable market.

The peak in the increase in penetration rate of portable players came during 1990–1995. The per capita sales of albums also increased from 4 to 5.5 during this period, to reach its all time high. This is consistent with the view that the increased sale of albums was due to the opening of new markets that could use prerecorded music in situations where portability was important.

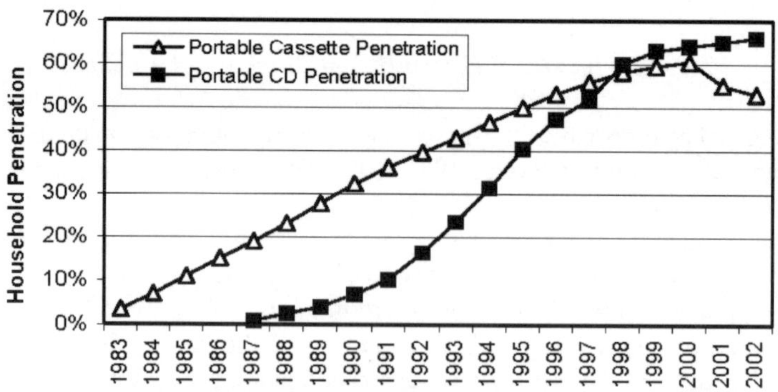

Fig. 8. The Portable Market.

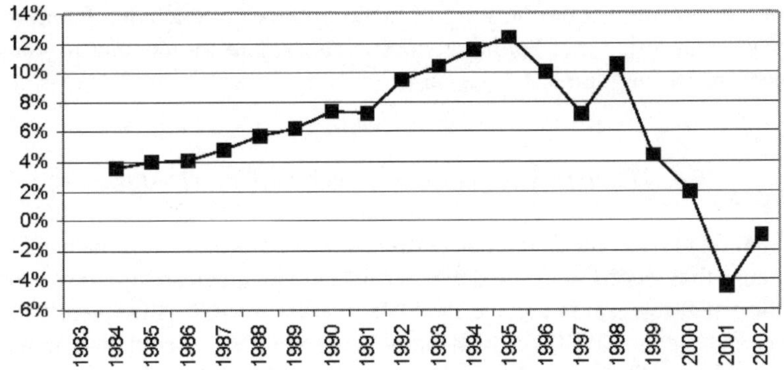

Fig. 9. Yearly Penetration Changes in the Portable Market.

Running a regression that includes a variable for the change in portable penetration (both cassette and CD) yields the results found in Table 2.

Although we should not make too much of a data set that consists of 29 observations (in first differences), the results indicate that portable penetration is related to sales (as we knew from visual inspection).

The impact of portable equipment penetration would appear to be fairly large since the penetration went from zero to over one hundred (the summed shares of cassettes and CDs), which is capable of explaining an increase of 3.2 albums per capita, slightly larger than the increase that actually occurred during that 15 year period.[33] The impact of income seems to be less important. Although per capita income rose from approximately 10.4 thousand to 14.5 thousand over the entire period, from 1982 until 1995 it increased from 11 to 13.3. The coefficient on income would predict an increase of 1.6 units during this period.

Although the data indicates a recent decline in the penetration of portable devices (due to fewer portable cassette players), the size is too small to have

Table 2.

	Coefficients	Std. Error	*t*	Sig.
Constant	−0.177	0.087	−2.033	0.053
Change in portable penetration	0.031	0.013	2.339	0.028
Change in real price of album	−0.195	0.129	−1.51	0.144
Per capita change in income	0.00064	0.00028	2.322	0.029
R square	Adjusted *R* square	Dependent variable: change in albums per capita		
0.345	0.266		$N = 29$	

any but a minor impact on the sales of albums. We can conclude, therefore, that neither income, price, nor format changes can explain the decline in sales that have occurred during the last few years.

5.4. The Impact of Substitutes, such as the DVD Market

One claim that is encountered if one peruses discussions of the decline in CD sales, is the claim that record sales are down because videogames are up, movie viewership (theatrical release) is up, and the DVD market is up. It is frequently pointed out, for example, that DVDs cost about the same as CDs, yet contain both pictures and sound, giving them an advantage that explains the shift away from CD sales.

Keep in mind that there are two ways in which these products could substitute for CDs. First is the financial side – money spent on DVDs cannot be spent on anything else, including CDs. Of course, both are only small portions of budgets and they might not have any impact on one another. If consumers tend to budget a certain amount for entertainment, however, each might then become an important constraint upon the other. The second way in which they might impinge upon one another is if they both use the same time budgeted for entertainment. Either the financial or time constraints might play a role.

First, we can examine changes in box office revenues with changes in album revenues. I find a positive correlation (0.24) over the period 1972–1999, indicating a tendency for the two to move together.[34] This result is incompatible with the claim that movies are substitutes for CDs.

A similar result holds for changes in videogame software sales and changes in the sale of albums. Unfortunately, data for videogames only go back to 1990. If we limit ourselves to the period before 2000, before any impacts of MP3 downloads, there is a positive correlation of 0.16, which indicates that the two move together and not separately. This relationship also fails to support the claim that videogames are particularly close substitutes for CDs.[35]

Finally, the same story holds for prerecorded movies (VHS). From 1990 to 1999 there is a slight positive correlation between yearly changes in per capita unit purchases in the two markets (0.12).[36] Again, there is little evidence that VHS tapes and CDs were substitutes, and thus there is little reason to believe that DVDs and CDs are substitutes.

Admittedly, there is presumably some common factor at work causing sales of CDs, prerecorded movies, box office receipts, and videogames to all tend to change in the same direction.[37] But it does not appear to be overly strong.

The second constraint in engaging in any of these activities is one of time. Published estimates indicate that listening to recorded music took approximately

45 minutes of a person's time per day, whereas going to movies took 2 minutes, watching prerecorded movies took 9 minutes, and playing videogames took 7 minutes.[38] I do not believe it would be reasonable to argue, at the low time-levels of usage for these non-music activities, that changes in movie attendance, DVD usage, or videogames usage, for the population as a whole, could be responsible for more than a small portion of the changes in album sales discussed below.

Although there are reasons to suspect the veracity of these time-use numbers, I am not aware of any other estimates that could be used in their place.[39]

5.5. Has the Music Changed?

When the record industry has a decline, critics of the industry will often claim that sales fell because the music was not interesting. Thus, it is not surprising that this has emerged as an answer to the question: "if MP3s are not hurting the industry, what is?"

Clearly, this hypothesis could be true. It would not be surprising if there were peaks and valleys in the creative output of the record business. The problem with the hypothesis, however, is determining a way to test it.

The clearest market measure of the vibrancy of the industry is sales of records. But since that statistic is now tarnished by the possibility of MP3 downloading, and we wish to test the hypothesis that MP3 downloads hurt sales, some other measure market valuation of the music is needed.

One source of data that I have been able to access is the financial success of concerts. Pollstar Magazine provides data on concert tour revenues. Concert tours suffer from one important defect – many of the acts are well past their recording prime although they still draw very well in concert venues (e.g. the Rolling Stones). Nevertheless, there is a positive correlation between concert revenues changes and changes in record sales (0.37 from 1991 to 1999) although it is not as high as would be preferred.

Interestingly, the years 2000 and 2001 had the largest real increases in concert revenues – the same time the record industry was experiencing unusually large decreases in revenues. Although this is not sufficiently strong to say that interest in music increased after 2000, it is certainly inconsistent with the claim that interest in music fell after 1999.

A second potential source of data would be time spent listening to radio. Of course, radio consists of non-music activities and listening to radio is a function of other activities such as time spent in automobiles. Nevertheless, data available from Arbitron allow us to focus on the use of radio in the home, where competition from alternative sources of entertainment would be most pronounced.

Table 3. Fall 1998 to Fall 2002.

Group	Decline in Radio Listening (%)
Persons 12+	7.98
Persons 12–17	11.50
Persons 18–34	9.41
Persons 25–54	7.47
Persons 35–64	6.98
Persons 65+	6.49

Data from the last 5 years reveals a fairly serious decline in listening – 8%. Do we finally have some support here for the claim that music has lost its inherent interest? Digging somewhat deeper, we find little support for this claim in Table 3.

First, this decline in listening is across the board – it exists for all age groups. It is unlikely that those over the age of 65 would find their attraction to listening radio diminished because contemporary musical compositions were not to their liking. The age groups most likely to be affected by a decline in music quality should be those who listen to new compositions. Those would most likely be heavily skewed toward lower age groups. Although there is some evidence that youthful listeners have tuned out somewhat more than other age groups, the difference is not very substantial (approximately 4%).

Further casting doubt on the claim that radio reveals a declining interest in current music, can be found by examining the listening trends for different types (formats) of music. If contemporary music is in the doldrums, than the listening trends that should show the largest declines are for the contemporary formats. Yet, the largest decline in listening was in "adult standards" (40%) which hardly seems to be a format that reflects current musical quality. Further, of the two formats that represent contemporary music, "Adult Contemporary" had a mere 4% decline and "Contemporary Hits" had an 11% increase. Urban music, another category which contains contemporary music, had a 12% increase. It is true that "country" music had a 13% decline, but when current critics suggest a current state of musical malaise, they do not usually seem to be talking about country music.

Another possible measure of interest in music might be the audience size for the television show where awards are given out for best record (Grammies). Unfortunately, I have been unable to put together a time series of audience size for the Grammies. Even if such data were readily available, however, the audience size for the Grammies would likely be impacted by other television-specific factors, such as the increased competition from relatively new networks and cable services, weakening its value as a measure of the raw appeal of the music.

Table 4. Distribution of Consumers.

Age Group	1987	1995	2001
10–14	7.3	8	8.9
15–19	24.2	17.1	13.3
20–24	19.1	15.3	11.5
25–34	25.3	24.4	20.2
35+	23.7	34.4	45.2

This evidence as I read it, although admittedly weak, fails to support the suggestion that the current decline in record sales is due to some dissatisfaction with the state of music itself.

5.6. Changes in the Audience

There has been a flattening of the age distribution of purchasers during the last few decades. The broadening comes from the wider age distribution of record purchasers, as reported in surveys.[40] This broadening of the consumer market might be the result of a generation that grew up with LP records (the baby boomers) and which continued to purchase albums as they aged, whereas previous middle-aged generations did not purchase albums. This can be seen in Table 4 which lists the age distribution of *record* purchasers (note that this is more than albums).[41]

In 1987, 43.3% of purchases were made by consumers between the ages of 15 and 24. In 2001 this particular age group was responsible for a far smaller share of overall record sales, 24.8%. Further, consumers at the extremities of the distribution, those aged 10–14 or those older than 45, increased their share purchases.[42] Much of this change is due merely to the change in size of these cohorts, however, which I address in Table 5.

Table 5. Purchase Intensity of Age Group.

	1987 (%)	1991 (%)	1995 (%)	2001 (%)
10–14	108.00	119.54	111.52	115.93
15–19	313.58	251.69	246.88	182.68
20–24	229.19	232.66	223.61	176.54
25–29	160.74	169.09	170.99	164.02
30–34	124.70	142.40	145.70	141.84
35–39	106.71	122.82	127.30	130.60
40–44	74.51	94.74	97.29	128.80
45+	34.81	38.00	49.28	64.77

The differences in purchase behavior by age group, and the flattening that has occurred, can be more clearly seen in Table 5. Here I compute the ratio of a group's share of purchases with that group's share of the overall population, a ratio I define as "purchase intensity." A group with a purchase-intensity of two would mean that the typical member purchased twice as many records as a typical member of the population as a whole.

What Table 5 tells us is that in 1991 those between ages 10 and 14 purchased records at a rate that was 19.54% higher than the typical purchaser from the overall population. Those aged 15–19 purchased records at a rate that was 152% higher than that of the overall population. Those aged 45 and up purchased records at a rate that was only 38% of the overall population rate.

If we assume that core listeners, defined as those aged from 15 to 24, continue to purchase records with a constant intensity over time, then the fact that other segments of the population have increased their purchases would imply that per capita sales should have been increasing.

Of course, there is a more sinister interpretation to this flattening of the age distribution of purchasers. It could be, and in fact it probably is the case, that use of MP3s is most prevalent among the younger age groups. The RIAA evidence against Napster (known as the Fine report) focused on college students with the expectation that these users would be most likely to use MP3s.[43] Is it possible to use these data to test the hypothesis that MP3 downloading is responsible for this flattening? The answer is yes, but I caution the reader to avoid placing too much weight on this analysis since these data are based on surveys, which are often quite unreliable.

Table 6 presents data for this purpose. Table 6 takes the intensity of purchase estimates, as found in Table 5, and calculates the change in these intensities over time.[44] Examining changes in purchase intensities for the entire 1995 to 2001 period (the second column) provides evidence that would seem to support

Table 6. Change in Purchase Intensity by Age Group.

Age Group	1995–2001 (%)	1995–1998 (%)	1998–2001 (%)
10–14	3.81	12.75	10.25
15–19	−35.14	−12.99	−19.61
20–24	−26.66	−19.89	−5.64
25–29	−4.25	−3.09	−1.12
30–34	−2.72	4.62	−7.70
35–39	2.53	15.45	−15.29
40–44	24.46	5.08	20.42
45+	23.90	7.68	17.57

the RIAA charge that MP3 downloading is injurious to its health. Consumers in two age categories have disproportionately large declines – consumers in age groups 15–19 and 20–24. Their intensity of purchase drops 35 and 27% respectively. These are clearly the two groups that would have been thought most likely to use MP3 downloading. It is somewhat surprising, however, under this reading of the data, that those under the age of 14 have not also experienced a decline in purchase intensity since they too would have been thought to have downloaded MP3s.

Yet Napster and other Internet file sharing systems did not come online until the second half of 1999, so their impact could not have been felt until at least later in 1999. The next two columns in the table allow us to address the impact of MP3 downloads more directly. In those two columns I break the change into a pre-Napster 1995–1998 period and a Napster impacted 1998–2001 period. The results are not consistent with the MP3 damage hypothesis.

In the pre-Napster 1995–1998 period sales fell among the 15–19 and 19–24 groups by 13 and 20% respectively.[45] The post-Napster period has a smaller decline. Sales intensity fell by 20% for the 15–19 group, but were down only 6% in the 20–24 group. MP3 downloading, if it had an impact, should have made the decline larger in the second time period, yet it did not. Nor do individuals in their late 20s appear to have been impacted the way one might have suspected if MP3s were impacting their habits.

Note also that some part of this result may be driven by the decline in singles, as opposed to sales of albums. If the consumers of singles, which we know fell during this period, consisted mainly of individuals in the 15–24 age groups, then that could account for at least some of these results.

In sum, the evidence from purchase intensities seems consistent with a claim that MP3s hurt record sales, but the evidence is much weaker for the claim that the ability to download MP3s over the Internet harms sales. Yet it is difficult to believe that Internet downloading would not dramatically increase the number of MP3s relative to personal exchanges. Thus these results are not consistent enough to warrant much confidence.

6. THE CURRENT SITUATION

This brings us up to the present, which is the time of the writing of the article, the Spring of 2003. Without a doubt sales have fallen from 1999 to 2002, as we have seen. Yet the 2002 value of 4.33 albums is still relatively high by historical terms. The questions that I think we would need to be able to answer, if we are to provide useful guidance to public policy, include the following:

(a) Is the current decline in record sales due to MP3 downloads, MP3 use, or neither?
(b) If the current decline is due to MP3s, should we measure the impact as the decline from 1999 levels or from higher levels that might have been expected to occur if not for MP3 downloads?
(c) Is the current decline just the tip of the iceberg?
(d) If the current decline were due to MP3s, could we generate a metric indicating the number of "lost sales" for each MP3 album equivalent?

What I believe the lengthy examination of the industry that I have just imposed upon the reader can accomplish, however, is to allow us to make some educated guesses about how the recording industry would have evolved without MP3s, and thus to better gauge the impact of MP3 downloads from that information.

What would have been the prognostications for albums sales if MP3s didn't exist? I believe that the movement in the market to but a single recording format, and the maturation of the portable and automobile markets imply that sales would at least have leveled off, with or without MP3s. Whether they would have fallen is less clear, but since purchases of dual recording formats would no longer have been necessary, there is some possibility of a decline in sales.

What, then, is the impact of MP3 downloading? Given the enormity of the whole MP3 download enterprise it should be easy to recognize its impact on album sales if its impact is large. What do I mean by "large?" If each MP3 song substituted for a purchased song, as has often been claimed by the record industry in the case of blank tapes, we would clearly have a large impact.[46] If each two MP3s substituted for one purchased song, that would be large. Even if each four MP3s substituted for one purchased song, that too should probably be considered a impact large. If, on the other hand, each hundred MP3s substituted for one purchased song, that would not, in my opinion, qualify as a "large" impact and it would be most difficult to measure, given all the factors influencing the sales of albums. There is, of course, a vague "in-between" range that is neither large nor small.

It is useful to recall that the emergence of Napster occurred in the fall of 1999. The decline in albums sales beginning in 2000 would seem to fit the timing of Napster's birth quite well. The growth in sales from 1997 to 1999 would seem to argue against a claim that MP3 use, independent of Internet downloads, damaged record sales. This is contrary to the findings in the material in and around Table 6 on the alteration of the distribution of users, where an independent impact of MP3 downloading as opposed to MP3 use, was not discernable. Still, the evidence, in my opinion, currently supports a claim of damage, although the case is far from airtight.

Table 7.

	Albums	CDs	Cassettes
1999	5.749	4.974	0.655
2000	5.418	4.915	0.396
2001	4.874	4.545	0.232
2002	4.334	4.092	0.158
Decline	1.416	0.883	0.496

The decline in albums from 1999 to 2002 is the largest decline that has occurred over the last 30 years. But we need to expose these results to somewhat greater scrutiny.

The decline in prerecorded cassette sales would seem to have little to do with MP3 downloads. The decline in *album* sales per capita since 1999, as illustrated in Table 4, is 1.41 units, but the decline in the album subcategory of "cassettes" is 0.50, leading to a decline in albums, net of cassettes, of 0.91 units.[47] A three year decline totaling 0.91 albums per person excluding cassettes, would still be the largest and steepest on record, but not extraordinarily so. Since there are no other potential causes of this decline that held up under examination, we must conclude that MP3 downloads are harming the industry (Table 7).

Harm is not the same as fatal harm, however. What can we say about the conversion rate of MP3 downloads as replacements for purchase? Certainly, it does not appear to be zero. But neither does it appear to be "large." Given the enormous number of MP3 downloads (if the numbers can be taken seriously), which are themselves an incomplete portion of the entire MP3 phenomenon, it seems safe to say that the CD equivalent of MP3 downloads is at least equal to the entire sales market for CDs.[48] If MP3s downloads replaced sales at a 1 to 1 ratio, there would be no CD market to speak of. In that case we would be talking about a drop in CDs of 5 units, not 0.89 units. If MP3 downloads converted at 4:1 the album market would have dropped by 25%. Removing the impact of the decline in cassettes, which seems appropriate, it would appear that the conversion rate at the moment is on the order of 5:1 or 6:1. Not large but not small.

The more important question is whether the decline is sales will continue.

One possibility is that MP3 downloads will continue to increase. If this is the case, even if the conversion rate remains constant, at 6:1 say, the damage to the industry will continue to increase. Obviously, the degree to which individuals engage in MP3 downloading depends, among other things, on potential deterrent effects of legal penalties and perhaps more importantly, on any social opprobrium associated with this activity. It is difficult to judge whether or how these factors will change in the next few years, although these are the dimensions where the

RIAA has focused its proscriptive activities.[49] A second reason MP3 downloading might grow more is the possibility of greater broadband access, making the downloading of large files less burdensome.

Even if the activity of MP3 downloading remained constant, however, there is an additional reason that the impact of MP3 downloads might have become increasingly severe over the last few years. This reason is the spread in CD burners. Without burning a CD, MP3s can not substitute for the purchase of CDs in the many devices that were made to use CDs, such as automobiles, home stereos, and portable devices.[50] If MP3s are only used on computers they are not likely to replace the sale of CDs. To the extent that the steady state share of CD burners has not yet been reached, however, the measured conversion rate will be understated.

A survey by the RIAA in August of 2000 suggested that 30% of music consumers under the age of 54 had CD burners.[51] As I am writing this paper, in the late spring of 2003, it seems reasonable that the penetration of CD burners must be quite large since they have been quite inexpensive and standard on most new computers for several years now. Although the number of households with computers is usually put at about 60%, this number is presumably higher for music users under the age of 54, meaning that between 30 and 50% of music users with computers possessed CD burners in 2000. As a guess, by 2002, with approximately two years having passed since this survey, it would seem reasonable to believe that the share of music users with CD burning capability could easily have been at 75% or more of its short-term potential maximum.[52] Further, since we might expect that this option would have been used first by those most interested in burning CDs, the additional impact from providing CD burning capability to that computer owners who did not yet have it would be smaller than the share of such owners.

The decline in sales due to the increased penetration of CD burners would be largely complete, if this reasoning were correct. If the decline in albums continues unabated in 2003, meaning another decline in the vicinity of 0.4 albums per capita, this hypothesis will be largely discredited. On the other hand, if the decline drops to below 0.25 albums, we can presume the decline in sales is largely over.

7. CONCLUSIONS

If the analysis in this paper is correct, MP3 downloads are causing significant harm to the record industry. It is not clear, however, whether such downloading in our current legal environment will cause a mortal blow to the industry. I suspect that the worst damage to the industry is behind us, but we will know soon enough as new data are made available. It is important to understand the impact of MP3 downloads if we, as a society, are to determine appropriate responses to allow the record companies to take in defending their economic position from the encroachments brought about by MP3 downloads.

It is also understandable that the record companies may not be willing participants in this laboratory for economic analysis. Why should they risk their revenues in an experiment that may end badly for them? How many of us would volunteer to undergo surgery without anesthesia to test a theory that claimed it wasn't really necessary? I think it is important to understand that a decline of 20–25% is a considerable number.[53]

Nevertheless, how far should we be willing to go to protect the record industry from a 20–25% decline in business? Are recent proposals, such as the Berman bill going too far? My own view is that such proposals are going too far. Should we switch to a non-market alternative as suggested by Lessig, Natenal, and Romer?[54] I believe that switching to a non-market alternative should be an absolutely last-resort policy, for reasons I enunciate in my 2003 paper.[55] Allowing record companies to protect their wares using digital right management technology seems to be a far more reasonable alternative.[56] If, on the other hand, future evidence indicates that in fact the record industry would not survive MP3s, we might need to rethink the answers to these questions.

This paper has not taken a particular position on these issues. The goal of this paper was to provide some empirical analysis to replace what has often been little more than emotional wrangling and haranguing on both sides. Answers to difficult policy questions are not likely to be well informed without such an understanding of the empirical backdrop to the issues.

NOTES

1. The text of the bill along with its justification can be found at: http://www.house.gov/berman/pr072502.htm.

2. Such a system has been proposed in detail by Neil Weinstock Netanel "Impose a Noncommercial Use Levy to Allow Free P2P File-Swapping and Remixing" draft, 2002, and mentioned by Lessig (2001) in his book *The Future of Ideas*, p. 201, and in two recent papers: Raymond Ku, "The Creative Destruction of Copyright: Napster and the New Economics of Digital Technology," 69 U. Chi. L. Rev. 263 and a working paper by Mark S. Nadel "Questioning The Economic Justification For (And Thus Constitutionality Of) Copyright Law's Prohibition Against Unauthorized Copying: §106" draft, Nov 11, 2002. I am writing a paper entitled "Alternative Copyright Systems: The Problems with a Compulsory License" suggesting that these papers are far too optimistic about the regulator's ability to emulate the workings of a market.

3. Napster and its experts claimed that file-sharing merely allowed the customers of record companies to "sample" or "try out" songs in advance of purchase. With the better information provided by such sampling, customers would be more likely, according to

Napster's experts, to purchase the songs that they knew they liked. The recording industry and its experts, on the other hand, claimed that file sharing substituted for the purchase of a record and that the downloading of MP3s would and had decreased the purchase of records.

4. See Chapter 7 in my book "Rethinking the Network Economy" Amacom, 2002 or my article "Back to the Future: Can Copyright Owners Appropriate Revenues in the Face of New Copying Technologies?" in The Economics of Copyright: Recent Developments and Analysis, edited by Wendy Gordon and Richard Watt, Edward Elgar Publishing, 2003. Expert reports were put forward in the Napster case by David Teece and Robert Hall, working for opposite sides and coming to opposite conclusions from one another. Neither seemed to conduct a serious empirical analysis, although the Teece paper was not made public.

5. There are certain theoretical conditions (e.g. indirect appropriability) under which unauthorized copying might in fact enhance, or at least leave undiminished, the revenues to the copyright owner. See S. J. Liebowitz "The Impact of Reprography on the Copyright System," Copyright Revision Studies, Bureau of Corporate Affairs, Ottawa, 1981 but the term "indirect appropriability" was coined in my 1985 paper "Copying and Indirect Appropriability: Photocopying of Journals," Journal of Political Economy, October 1985, pp. 945–957. Indirect appropriability can only work under certain non-general conditions, although Michele Boldrin and David K. Levine ("The Case Against Intellectual Property," *American Economic Review*, 209–212, May 2002) have recently put forward a similar idea without, in my view, taking note of these real-world limitations. (See the critique by Benjamin Klein, Andres V. Lerner, Kevin M. Murphy in "Intellectual Property: Do We Need It? The Economics of Copyright 'Fair Use' in a Networked World," *American Economic Review*, 206–208, May 2002.)

6. At that time, I was quoted in various interviews as suggesting that our theories might be missing some important elements since I thought that we should have begun to see harm and yet we were not seeing any. Since then the statistics from the recording industry, as I discuss below, have provided the evidence to support the supposition that MP3 downloading would be harmful to the industry.

7. Although it is not clear just how accurate these estimates are. See note 48 for details on some of these post-Napster estimates. These estimates indicated that at Napster's peak downloads were the equivalent of 250 million CDs per month (worldwide). According to RIAA numbers, U.S. national sales ran approximately 80 million CDs per month. So the potential impact of MP3 downloads on sales should have been large. See "Music Downloads Soar," Reuters, September 6, 2001, available at: http://news.cnet.com/news/0-1005-200-7080479.html.

8. See Table No. 909 "Media Usage and Consumer Spending: 1993 to 2003" in the 2000 U.S. Statistical Abstract.

9. I felt the need to remind readers of the diamond water paradox because of a statement by Paul Romer who, in an otherwise thoughtful paper, ignores the distinction between revenues and value. He states: "The stakes in the battle over the music business are small enough to get lost in rounding error for world GDP of about $30 trillion. However, this battle creates a "teachable moment" that could help frame policy in more important areas." See "When Should We Use Intellectual Property Rights?" *American Economic Review*, May, 2002, 213–216.

10. Data on quantities were reported beginning in 1973. For prior years only the industry revenues were reported.

11. Although not likely an issue with the historical data, some concern about the legitimacy of the current and future numbers arises since the industry is trying to make a case of MP3 damage based upon these numbers. The RIAA informs us that the numbers are compiled by PricewaterhouseCoopers, but somewhat greater concern must be in evidence when there is so much depending on these numbers.

12. Even if MP3 file trading exacerbated the decline in the single, we should still be able to get an unbiased impact of MP3 downloads by limiting our attention to full length recordings.

13. There is an "alternative" market for music that operates outside of normal distribution channels and often doesn't use barcodes on CDs. If this alternative market were to grow to become a significant source of CDs, then the official statistics on CD sales would be understated. Unfortunately, I am not aware of any estimates on the size of this market.

14. In other words, "normal" goods, in the rhapsodic lingo of economists.

15. There is an article in Slate by Mark Jenkins that makes this claim. See "Hit Charade: The music industry's self-inflicted wounds." Tuesday, August 20, 2002 available at http://slate.msn.com/?id=2069732. He also claims, however, that CD prices have outpaced inflation, a statement inconsistent with my analysis.

16. Some small number of automobiles actually did have record players (so I learned in doing this research) but they never achieved any substantial level of penetration.

17. In this regard it is interesting to note that a large decline in record sales occurred as radio gained popularity in the early 1920s. Record sales fell by about half from 1920 to 1925. This tends to refute the common misconception that payola – record companies paying to get airplay – reveals the positive impact of radio to record sales. Instead payola should be thought of as indicative of competition between companies within the record industry. Imagine, for example, that there was no radio. The only way to listen to music in automobiles would be to listen to prerecorded music, which would certainly increase record sales along the lines suggested later in the paper. See "Off the Record" by David Morton, Rutgers University Press, 2000, for additional evidence on the impact of radio. For a history of Payola See Ronald Coase "Payola in Radio and Television Broadcasting," Journal of Law and Economics, October 1979, 269–328.

18. I thank Craig Newmark for making this point to me.

19. The measure of population used here is not the entire population since it excludes individuals younger than 10 as well as individuals older than 60.

20. The consumer price index is the measure of inflation used to adjust revenues.

21. See Eric Boehlert in Salon.com "Why the record industry is killing the single" December 19, 2001 at http://dir.salon.com/ent/music/feature/2001/12/19/music_industry_sidebar/index.html.

22. They are in the vicinity of twenty million per year in the US. Music videos can be thought of as a substitute for singles since they do not contain the music from an entire album. The RIAA provides data on music videos but their sales numbers are even lower than those of pure music singles.

23. These values are all expressed in 1983 dollars.

24. I am relegating the 8-track tape phenomenon to a lesser status since its sales were never anywhere as large as the major media formats and 8-tracks were popular for little more than the decade of the 1970s.

25. The growth in market penetration, beginning with 1984 is: 0.30, 1.10, 2.70, 3.90, 4.40, 5.20, 6.00, 6.60, 8.80, 9.00, 8.00, 8.00, 3.10, 1.90, 2.00, 1.00, 2.00%.

26. If discounts were relatively constant from year to year, list prices would be good proxies for actual prices. On the other hand, if discounts varied considerably from year to year (the growth of Wal-Mart might be just one reason to suspect that discounts have not stayed constant) then the use of list prices could cause misleading results.

27. In 2002, per capita income fell by $40, but it was $245 higher than the 1999 value.

28. I say this fully aware that during the depression sales dropped from 74.8 million in 1929 to 5.5 million in 1932. There are good reasons that the sales downturn was so strongly related with the fall in income at that time, however: it was a far more severe economic downturn, radio had hurt record sales considerably, and the penetration rate of record players was likely to have been far lower. Therefore, the income constraint would likely have been more important than the time constraint during the depression.

29. Some CD and DVD players have recently allowed the playing of MP3 files on CDs, but I do not know the penetration of these machines and suspect it is not yet that large. Similarly, portable MP3 players are reported to have extremely small market shares, in the single digits.

30. "At present... severe economic damage [is being done] to the property rights of owners of copyrights in sound recordings and musical compositions... under present and emerging conditions, the industry simply has no out... Unless something meaningful is done to respond to the... problem, the industry itself is at risk." From Greenspan's testimony in 1983 on the Home Recording Act. Hearings before the Subcommittee on Patents, Copyrights and Trademarks, October 25, 1983.

31. Information on the number of blank cassettes (beginning with 1981), as well as information below on the penetration rates of portable technologies all come from Ebrain Market Research, which is a service of the Consumer Electronics Association. The blank cassette value for 1978 comes from the text of the 1990 Edition of "Entertainment Industry Economics" [Harold Vogel, Cambridge University Press] states that in 1978 approximately 150 million blank tapes were sold.

32. A similar impact on sales would occur if cassettes were not considered of equal quality to LPs or CDs, thus relegating them to non-living room status.

33. The portability coefficient might be too large due to a questionable measured decline in portability penetration that coincided with the decline in album sales. To control for this possibly coincidental relationship, I ran this regression excluding data beyond year 2000. The portability coefficient dropped by one third, and it lost statistical significance ($t = 1.4$).

34. This correlation drops to 0.07 if the period is extended to 2002. The drop in the correlation occurs, however, because record revenues have done badly since 2000 but movie box office revenues have done well. This sudden divergence would be better explained by MP3 downloads than by a sudden alteration of the relative complementarities of the two products.

35. Since 2000, videogame sales have continued to increase whereas CD revenues have fallen. The correlation for the entire period is negative (−0.16).

36. When the years 2000–2002 are included, the correlation becomes negative (−0.38). These last 3 years recorded particularly large increases in DVD growth and also negative changes in CD sales. But the recent large increases in prerecorded movies were not so large as to be entirely unprecedented.

37. It wouldn't appear to be income. The correlation of income changes and other variables is 0.20 for box office changes, 0.24 for videogame revenue changes, and −0.20 for prerecorded movies.

38. Table 909 in the U.S. statistical abstract. Original source is Veronis, Suhler & Associates.

39. While examining the numbers for time spent listening to recorded music, I was struck by the very close association between these estimates and the sales of albums. In particular, the time spent listening to music fell in the last few years at the same time that album sales have fallen. If this had been true, then the annihilation hypothesis would clearly be false since MP3 downloading is an act that should tend to lead to increased listening of recorded music at the same time that album sales were falling. If listening to music was falling, then the decline in albums could be attributed to this decline in listening and MP3 could be exonerated. Alas, when I investigated I found the two series tracked one another so closely that it appeared that record sales might have been virtually the only ingredient used in estimating time spent listening to music. When I contacted Veronis, Suhler, they confirmed that these estimates of time spent were not taken from surveys of use, but instead were based on a formula using other statistics that they agreed probably included record sales to an important extent.

40. Although surveys are often not reliable, and I often argue against anyone taking them too seriously, they are the only mechanism that can generate information on the age distribution of consumers. As long as the questions being asked are very simple and as long as respondents have no reason not to tell the truth, surveys might give useful results.

41. This information comes from the RIAA and is based on polls taken by Peter Hart Research. More information is available at the RIAA website www.riaa.com. I have not seen the surveys used to generate this information so I can not form an opinion about their likely veracity. For our purposes, the implications of the results in the table will hold as long as any imperfections in the survey remain consistent over time.

42. Although Table 1 has 35+ as the oldest age group, because the 1983 data topped out at 35+, more recent data provides more detailed breakdowns and has 45+ as the oldest category. The share of purchases made by consumers in this age group has increased significantly in the last decade, from 11.8 to 25.7%.

43. The Fine report compared sales at record stores near college campuses with sales at more typical record stores. It found that sales fell in stores near college campuses, but this was true for the year prior to Napster's existences as well as for the year following its birth. I discuss this report (and others) in detail in Chap. 7 of Rethinking the Network Economy.

44. It simply takes the ratio of the early data intensity to the later date intensity and subtracts this ratio from 1.

45. The pre-Napster fall in purchase intensity is consistent with the findings found in the Napster case's Fine Report that retailers near colleges had reduced sales relative to other retailers. One explanation might be that individuals in these groups were sharing MP3 files with one another before the advent of anonymous online file sharing. This result would, however, seem to discredit a hypothesis put forward to explain the Fine report, that is, that college students were purchasing CDs online instead of purchasing them at their local brick-and-mortar retailer. Online or offline, the statistics on purchase intensity would be equally impacted, as long as the sales occurred.

46. When the recording industry made estimates of the harm from copying, each blank audio cassette sold tended to be counted as a lost sale of an album, clearly leading to an overestimate of harm.

47. The decline in CDs is somewhat less than the net decline in albums because of vinyl long-playing records and music videos.

48. At its peak, Napster downloads were estimated to be in the vicinity of 2.8 billion files per month. It was also claimed that the four largest Napster replacements, FastTrack, Audiogalaxy, iMesh and Gnutella, were responsible for 3.05 billion downloaded files a month in August of 2001, although not all of these were songs. The data come from Webnoize, a company that tracks Internet usage as reported by Scarlet Pruitt, IDG News Service\Boston Bureau, September 07, 2001 in "Study: Look out labels: file swapping growing." See http://www.idg.net/idgns/2001/09/07/StudyLookOutLabelsFileSwapping.shtml. Assuming ten songs per CD, this would imply 300 million CDs per month worth of songs. According to RIAA numbers, U.S. national sales ran approximately 80 million CDs per month, less than one third the CD equivalent number of downloads. Thus the potential impact of downloading should have been large. According to the IFPI, total sales of CDs worldwide are less than two and a half billion, which is less than the CD equivalent of MP3 downloads.

49. The RIAA has sent out large numbers of letters threatening legal action if, for example, schools and businesses do not try to restrict the downloading of songs through their networks. They have threatened individuals who have made available large numbers of songs to peer-to-peer networks. They have run advertisements equating peer-to-peer downloading with theft.

50. Note that treating MP3s as a new format is possible, but that it has, at least through 2001, a trivial impact in the portable market, with fewer than three million units in use.

51. It was this or a similar statistic, reported in USA Today in 2002 but not dated or attributed to the RIAA, that made me suspect that perhaps MP3 downloading might not cause great harm to the record industry. After all, the RIAA numbers for 2001 were not yet out and the decline in 2000 was quite small given the well publicized extent of MP3 usage. I had thought that the lack of CD burners was a logical explanation for why harm had not yet shown up at a scale similar to the MP3 phenomenon, but the report of 30% CD ownership seemed to weaken that explanation. Shortly after this, when the 2001 numbers came out, the decline in CD sales became far more apparent. See my interview in Salon Magazine at http://www.salon.com/tech/feature/2002/06/13/liebowitz/index.html?x.

52. Not every music user has a computer, so the maximum penetration is less than 100%.

53. The recording industry is likely to complain that if revenues fall by 20%, that profits will have to fall far more. While there is some truth to this in the "short run," the recording industry appears to be quite scalable, with no minimum efficient scale that would require giant producers in order to be able to sustain a market. Therefore, a new equilibrium should eventually be achieved with normal profits but with revenues 20% lower.

54. Lawrence Lessig, "Just Compensation: Congress should help artists get paid without delivering the Internet into the hands of the big labels." The Standard, Apr 06 2001; Neil Weinstock Netanel "Impose a Non-commercial Use Levy to Allow Free P2P File-Swapping and Remixing" Working Paper; Romer in note 9.

55. Stan J. Liebowitz, "The Production of Music in the face of peer-to-peer downloading: Does there exist something better than the market?", working paper.

56. I discuss this at length in my Cato policy analysis and in my book "Rethinking the Network Economy."

STRATEGIES THAT WORK WHEN PROPERTY RIGHTS DON'T

Bharat Anand and Alexander Galetovic

ABSTRACT

In many sectors property rights over knowledge and information are weak as they are embodied in employees, competitors can copy or customers can pirate. Yet comprehensive studies show that firms systematically invest in these assets. We offer a simple taxonomy of strategies that firms use to cope with weak property rights.

We classify these strategies in three groups: (i) Some firms threaten offenders with strong competition. (ii) Other firms exploit complementarities and offer potential offenders a better deal than they can get elsewhere. (iii) And yet other firms exploit weak property rights and make profits on complementary assets or products that they can own.

We go beyond taxonomy by showing when a particular strategy works. It depends systematically on the characteristics of both the asset and the investing firm.

1. INTRODUCTION

In 1999, the music industry was first "Napsterized." Thanks to the proliferation of peer-to-peer file-sharing protocols, consumers began to freely copy and distribute digitized files of songs on a massive scale. As a result, in fiscal year 2002, the

Intellectual Property and Entrepreneurship
Advances in the Study of Entrepreneurship, Innovation and Economic Growth, Volume 15, 261–304

ISSN: 1048-4736/doi:10.1016/S1048-4736(04)01508-5

major recording studios saw their sales drop by almost 10%. To combat piracy, the studios attempted to take copyright offenders to court and looked for increasingly sophisticated technologies that might make it difficult to pirate digitized content. But the legal and technological approaches had limited success, and the industry increasingly searched for "new business models"[1] that might allow them to capture the value created by their investments into the financing and marketing of music. For several years, however, firms remained mystified about where such business models would come from, what form they might take, and what logic should underpin the strategic response to piracy. Indeed, in May 2002, Disney CEO Michael Eisner declared: "[T]he only thing I can say is it's very hard to create a business model against something that is a perfect copy for free. It would even challenge the greatest business man that ever was."[2]

As it turns out, the music industry is not alone in its dilemma. Indeed, firms in many industries are finding it increasingly difficult to establish property rights over the information-intensive assets that they invest in – knowledge, information, ideas, and human capital – or the products that they develop. Nevertheless, they continue to invest in assets or products that they cannot own. So, how is it that these firms can cope with the challenge of insecure ownership rights over these assets? And, how can they recoup their investments in these assets?

We examine the strategies that firms use to capture value from such assets, and offer a simple taxonomy that can assist managers in tackling similar problems in different settings. Implicit in this approach is the claim that the problem facing the music industry (and, increasingly, the movie studios) is not unique. Imposing some structure on the problem and identifying common solutions may also assist firms in their search for an effective solution rather than relying purely on an Archimedean stroke of genius. Understanding how firms in other industries resolve similar dilemmas may not, of course, yield the magical solution for the recording industry. But, as we show later, it can be used to shed light both on the limits and promise of different classes of strategies, and on why it is that some firms have responded differently than others when facing a similar dilemma.

As we describe below, the strategies that firms use are both rich and varied, and can work by relying on market incentives rather than legal ones. In some cases, market-based strategies work by successfully *minimizing* the threat of third-party appropriation; in other cases, they work by *exploiting* the incentive of others to appropriate value. In some cases, firms adapt their internal organization and governance of projects in ways that allow them to cope with weak property rights. In other cases, firms alter their organizational boundaries and use their ownership of complementary assets to capture value created by assets they cannot own. Or, the resulting market structure may endogenously adapt to grant firms enough market power to capture value. But we go beyond taxonomy in this paper by showing that whether or not a particular strategy works depends systematically on the characteristics of both the asset and the financier.

In Section 2, we study a simple model that examines how firms can capture value by investing in ideas despite weak property rights. We use the model's central results to extract four basic principles that organize the taxonomy, in Section 3, of how firms tackle similar problems in a variety of settings. Section 4 applies these principles to examine how firms in the music industry might confront piracy. Section 5 concludes with some decision traps that managers should avoid when investing in assets that the firm cannot own.

2. THE FINANCING OF IDEAS

Property rights over "ideas" are virtually impossible to establish. Consequently, the financing of projects in which the key input is an "idea" has long been considered a problem for financiers and, by implication, entrepreneurs looking for finance. Hart and Moore (1994) examined the difficulties in structuring financing arrangements for "an entrepreneur who needs to raise funds . . . but cannot commit not to withdraw his human capital from the project." But this threat of repudiation or *holdup* is not specific to small entrepreneurial ventures alone. Indeed, many large, established companies have found it difficult to prevent employees from leaving and commercializing ideas. Perhaps the most cited example of a firm that consistently failed to capture value from the ideas it financed is Xerox and its research arm, Xerox PARC. As *The Economist* noted, "Xerox PARC is famous for having pioneered ideas (including a superior personal computer, the facsimile machine, the Ethernet, and the laser printer) that made fortunes for many of its Silicon Valley neighbors but little for itself." Elsewhere, "people at Bell Labs have still not forgotten how, half a century ago, William Shockley took the transistor idea, which he and his colleagues had invented at Murray Hill, to Palo Alto in California and started a company (Fairchild Semiconductor) that eventually became Intel." Indeed, the notion that ideas walk with individuals rather than residing in the black box of firms was strikingly illustrated in a survey of 100 founders of the 1989 *Inc* 500 fastest-growing private companies in the United States. Bhide (1994) found that 71% of these founders "replicated or modified an idea encountered through previous employment."

In Anand and Galetovic (2000a), we studied the financing of ideas when property rights are weak. The premises of the model are that ideas are embodied in individuals (we call them "researchers") and that it is difficult to establish property rights over these ideas. The reason is that the asset in question (the idea) cannot be well defined in advance, *and* the embodied asset (the researcher's human capital) cannot be owned. Weak property rights make ex-post hold-up by the researcher possible because, after receiving financing, the researcher may walk away from the firm to develop the idea elsewhere. Financiers are aware ex-ante, and so the question is why would they ever finance in the first place. As we will show, different financiers are differently endowed to deal with weak property rights. These endowments determine whether they can deal with weak property rights and, if they can, how.

The simple model described in the next section is unlikely to capture all the particular features of varied settings. For example, the challenge of insecure property rights faced by the recording industry primarily stems not from the difficulty in establishing secure contractual relations with human capital (artistes) but from the difficulty in preventing consumers from easily copying and disseminating the final product. But, as we show later, the lessons extracted from this simple model still extend to many of these other settings.

2.1. A Simple Model[3]

The basic structure. Consider a model with only two periods: a research phase and a development phase; and three dates: date 0, before research; date 1, in-between research and development; and date 2, after development. The output of the research phase is an idea. This idea is necessary to develop a marketable innovation whose value is S. To obtain the idea it is necessary to invest R. The development cost is D^{r} if the researcher partners with the venture capitalist (VC) (call this "independent development") and D^{c} if the researcher partners with the corporation (call this "joint development").

In both the research and the development phases, the researcher can obtain funding from either a VC or a corporation, and it can switch financiers at date 1. Switching in-between research and development opens the door to hold-up – the idea becomes embodied in the researcher, who can leave to develop it elsewhere. We model the strength of property rights as follows. If the researcher switches at date 1, after the research phase, to pursue development elsewhere, the financier can establish its rights over the innovation's cash flow as specified in the contract signed at date 0 with probability v; with probability $1 - v$, the financier is unable to establish its rights. Thus, if v is close to 1, then property rights are "strong"; if v is small, then property rights are "weak."

Complementarities, account opaqueness and spillovers. Corporations and VCs are different in fundamental ways. Of course, these differences do not stem from the capital that each provides, but from the way they organize the projects they finance. Specifically, a corporation encompasses multiple projects that are joint or non-separable, whereas VCs finance projects that are legally and functionally separate.

Jointness or non-separability of activities may lead to "synergies" or complementarities across projects that the corporation can exploit when projects are more closely integrated. Examples include shared lab facilities, common purchasing facilities, and a brand name. We will see that, from the perspective of the holdup problem wrought by weak property rights, this is important because corporations may have more resources with which to cope and bribe the researcher to stay. We can model situations in which corporations can exploit complementarities assuming that $D^c < D^r$.

But in other situations, jointness may be a curse rather than a boon. Integration typically dilutes incentives by decreasing the autonomy of managers, which implies that stand-alone financing may give stronger incentives.[4] Similarly, joint activities make proper accounting of the costs and revenues specific to a given project more difficult.[5] Account opaqueness limits the ability of the corporation to share surplus with the researcher. We can model situations in which jointness imposes net costs by assuming that $D^c > D^r$.

There is a third implication of jointness: it may lead to internal spillovers from the researcher to the corporation so that part of the knowledge accumulated by the researcher may leak to the corporation. If the researcher leaves, spillovers give rise to a form of "reverse holdup," as the corporation can threaten to compete with the researcher.[6] We model internal spillovers as follows. If the researcher leaves the corporation, she is able to develop the innovation with probability b. In that case, the corporation earns surplus S^c and the researcher surplus S^r, with $S^c + S^r \leq S$. By contrast, the VC cannot exploit local spillovers.

Thus, we have determined that financiers differ in three relevant dimensions: complementarities or diseconomies of scope; ability to commit to share surplus; and spillovers. We now show that these characteristics interact with the strength of property rights and determine: (a) whether a corporation, a VC or neither is the most appropriate source of finance; and (b) what strategy the financier should follow to cope with weak property rights.

2.2. *Key Results*

Figure 1 presents the key results. The x-axis denotes the strength of property rights, v, over the idea, which ranges from 0 (weak property rights) to 1 (strong property rights). The vertical axis denotes the magnitude of complementarities

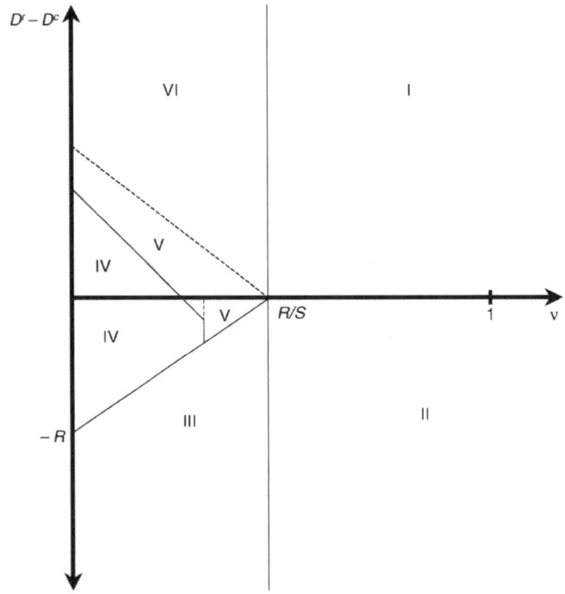

Fig. 1. The Financing of Ideas.

(the positive quadrant) or diseconomies of scope (the negative quadrant) within the corporation. Within regions I and II, where property rights are strong, the VC always finances research. When property rights are weak, by contrast, all depends on complementarities and spillovers. When these are significant, the corporation is willing to finance research (regions V and VI). Similarly, when diseconomies of scope are strong (region III), the VC is willing to finance. But in region IV, where both financiers are alike, competition and weak property rights kill financing. We discuss each case in turn.

Strong property rights. First consider regions I and II, where $v \geq R/S$. Then the VC always finances research, because

$$vS \geq R;$$

that is, when property rights are sufficiently strong, the financier can recoup its costs in expected value. How strong is "strong enough" depends on the characteristics of the project being financed. To see why, note that the financier can never be sure of getting back any surplus if $v < 1$. Instead, the maximum payment it can expect to receive is vS, which is sufficient to induce ex ante finance as long as it exceeds the upfront financing cost R, that is $vS \geq R$. This implies that as v falls and the strength of property rights declines, the tendency to finance projects that generate large surpluses S – i.e. the "sure hits" – increases.

We may note that in regions I and II, corporate financing of ideas is never observed in equilibrium. This may seem strange. Surely corporations are willing to finance an idea that they can own! But it turns out that with strong property rights the researcher prefers independent financing by the VC. Why? Recall that the corporation cannot commit to share any surplus with the researcher at the time of financing. This implies that the researcher's bargaining power stems only from her outside option of possible breakaway. Precisely because property rights are strong, this outside option would be of little value to the corporation. By contrast, the independent financier can commit ex-ante to share surplus with the researcher and, thus, is the preferred financing alternative.

Note that these results suggest that the apparent dominance of VC financing of research does not stem from any intrinsic advantage to venture capital finance. Nor does it depend on any superior ability of VCs to search, screen, or identify promising projects. Rather, it derives from the fact that any attempt by corporations to more closely integrate projects to extract complementarities also has a dark side – namely, the increased *incentive* to manipulate accounts and reduce surplus sharing with employees or managers. Numerous examples of conflicts between corporations and employees (or suppliers more generally) can be traced to this characteristic.

For example, consider the following account of a lawsuit against News Corporation and Fox Television Network by David Duchovny, the star of one of the most popular programs on the network, *X-Files*:

> Duchovny's suit . . . accuses Fox of limiting the potential back-end gross of the show – and thereby his take through profit participation points – by cutting sweetheart deals with Fox-owned entities at below-market prices.[7]

Profit participation in Hollywood has always been a messy business. Showbiz lawyers have long charged that studios often resort to creative accounting practices and downright fraud to hide the true grosses of hit movies and TV shows. But the suspicion has never been greater on the part of many artists than in the new era of vertical integration: "Intra-company transactions are inherently fraught with the potential for conflict."[8] Similar disputes between employees and firms over account manipulations and surplus sharing have contributed to strikes in professional sports. For example, foreshadowing the lockout in professional basketball in 1999, owners claimed that "we need to slow the growth" in salaries for high-end players, whereas players pointed out that "[O]wners never have volunteered to open their books and let the bottom lines for all 29 teams be made a public record."[9]

It is important to note that strong property rights stimulate efficient switching to corporate financing *after* research. As we show in Anand and Galetovic (2000a, Proposition 3.1), in region I it pays both the researcher and the VC to allow a breakaway to the corporation to take advantage of complementarities after the idea has been developed. Essentially, the VC is willing to allow efficient breakaway because strong property rights allow her to capture complementarities realized by the corporation through an appropriate breakaway fee. Because property rights are strong, the VC is willing to finance, anticipating that the researcher will switch to the corporation in the development stage. Of course, in region II, the VC also finances the development stage – it would be inefficient to switch to a corporation and suffer diseconomies of scope.

Weak property rights and the importance of complementarities. We now turn to weak property rights, that is $\upsilon S < R$. First note that, even though the corporation and the VC compete after research, the project may be able to get financing. But for this to occur, one financier must offer a sufficiently large cost advantage in development. To see this, assume that complementarities are strong and that the corporation has a large cost advantage in development (i.e. $D^{r} - D^{c} > 0$ and large). Then, if the researcher gets first-stage financing from a corporation, the corporation will be able to prevent holdup if the surplus from the project, $S - R - D^{c}$, exceeds the researcher's outside option in the second stage, $(1 - \upsilon)S - D^{r}$. The reason is that the corporation's competitive advantage creates a surplus large enough to entice the researcher to stay. Intuitively, weak property rights improve the researcher's outside option and make projects more expensive to finance. Complementarities create a surplus that the corporation can use to sweeten the researcher's deal and make him stay – carrots to prevent holdup. *Mutatis mutandis*, exactly the same argument explains when a VC may be willing to finance with weak property rights.

But why might financiers have any advantage over others? It's not deeper pockets, but differences in the nature of related assets that financiers typically can offer. In the case of a corporate financier (region VI), a researcher may be able to leverage existing operations to offer superior access to customers, channels, or supplies. It is not surprising, for example, that Kids R Us was founded by Toys R Us so as to leverage the latter's industry-specific skills to select styles and manufacturers, buy, display, and sell products, process information, control inventories, and even benefit from a related brand name.

Conversely, if access to relationships and knowledge, rather than operations and physical assets, are key, then independent financiers may have advantages (region III). For example, *Fortune* magazine noted that some venture capitalists offer a combination of broad-based expertise from being successful entrepreneurs themselves and a unique network of personal relationships, and that "Kleiner Perkins' pals ... include the founders of Netscape, Cisco, AOL, Intuit, Intel, TCI, Sun, Lotus, and Comcast."

Region V: when internal spillovers help financing. Can financing ever occur when property rights are weak and no type of financier has an appreciable cost advantage in development? The model suggests a third avenue by which the corporation can mitigate the threat of holdup – the ability to exploit the internal spillovers of research. When spillovers are sufficiently large, a corporation may be able to compete with the researcher that it financed, even after he leaves.

To see the effect, recall that the corporation will be able to compete in the product market with probability b, so that the expected surplus to the researcher (gross of development costs) is

$$(1 - v)[(1 - b)S + bS^{\mathrm{r}}],$$

where $S^{\mathrm{r}} < S$ is the researcher's surplus when competing with the corporation. If this surplus does not cover the development cost D^{r}, no financier will be willing to finance independent development. In other words, some trivial algebra shows that holdup does not occur if

$$(1 - v)[(1 - b^{*})S] < D^{\mathrm{r}},$$

where $b^{*} = b(1 - S^{\mathrm{r}}/S)$.

As Fig. 1 (region V) shows, local spillovers enlarge the area of corporation finance because the threat of product market competition reduces the value of the researcher's outside option and makes holdup less attractive. This helps him to get financing for research. In other words, ex post competition is used as a stick to prevent holdup. Notice that product market competition does not actually occur in equilibrium; the mere ability to do so is enough to forestall breakaway by the researcher. But the key point is that the threat of a "reverse holdup" by the corporation can neutralize the threat of holdup by the researcher.

Spillovers are often thought to prevent researchers from appropriating the value created by their ideas. Yet, this result suggests that spillovers can help researchers by enlarging the scope for financing them. The intuition is that both researchers and financiers confront problems of appropriability – one due to weak property rights, the other due to spillovers – which cancel out instead of adding up! More generally, the example suggests that strategies to overcome certain types of market imperfections may rely on creating other imperfections or frictions, an application of the well-known second-best principle.

2.3. Five Basic Principles

This simple model yields five basic principles for the broader analysis of strategies to tackle the challenge of weak property rights:

(1) *The limits of the law*. Legal remedies will not always work; nor are they always needed. Regions I and II, which correspond roughly to the case in which laws are "strong enough" to protect the financier, do not encompass the entire region of financing.

(2) *The use of carrots to prevent holdup*. Financiers need not have monopoly power to ensure that they can recover their costs. Indeed, regions V and VI suggest that market power per se is not sufficient; rather, market power in the relevant complementary asset is key to preventing holdup. The reason is that complementarities enable firms to increase the pie. Offering the researcher the carrot of sharing in this larger pie provides her with incentives not to leave. Below, we examine particular strategies in more detail to shed light on what these complementary assets might be in practice.

(3) *The use of sticks to prevent holdup*. Appropriability problems can cancel out rather than add up. As said before, this is a special case of a more general second-best principle in economics: when you add a second distortion, welfare may increase. In this case, internal spillovers undo the deleterious effects of the initial distortion – weak property rights. Region V suggests that this principle may be useful in devising strategies to cope with weak property rights over assets. Common examples include strategies that increase the threat of retaliation to the researcher or increase "reverse-holdup."

(4) *Strategies that exploit weak property rights*. Financiers may recoup their costs even if there is breakaway by the researcher in equilibrium (region I), and they may thus share in a larger surplus. Consequently, strategies need not be directed solely at preventing breakaway. Region I shows that financiers may recoup their research costs even if the researcher leaves after getting finance. Below, we build on this principle and examine a class of strategies that exploit, rather than mitigate, holdup.

(5) *Diversity: what works for one does not necessarily work for another*. A central lesson of the model is that diversity of strategies rules in the financing of ideas. But, as illustrated above, we can go beyond this observation, noting that whether or not a strategy works will depend systematically on the nature of the asset being financed – i.e. the strength of property rights over it – and the nature of the rest of the assets or activities in which the financier is involved.

In subsequent sections, we apply these simple principles to examine strategies that firms use in a variety of different markets to overcome the problem of weak property rights over assets.

3. STRATEGIES TO COPE WITH WEAK PROPERTY RIGHTS

The array of strategies discussed in the rest of the paper are summarized in Fig. 2.

3.1. Law and Disorder

The first, and often last, line of defense that many firms have used to deal with weak property rights has been to call for either stronger laws or stronger enforcement of existing laws. The fact that property rights are being violated, so the argument goes, is merely evidence that existing laws must be allowing it. Toughening penalties for infringement or endowing rights holders with broader protection would, therefore, prevent such infringements from occurring.

While legal remedies have obvious merits and can be effective in many circumstances, in many others, they are not. The difficulty in relying on the law to secure, or the courts to enforce, property rights over certain types of assets or resources has been documented repeatedly. In his summary on "Property Rights," Armen Alchian noted in 1967 that

> [t]he cost of establishing private property rights so that I could pay you a mutually agreeable price to pollute your air may be too expensive. Air, underground water, and electromagnetic radiation, for example, are expensive to monitor and control. Therefore, a person does not effectively have enforceable private property rights [over these assets]. The inability to cost-effectively

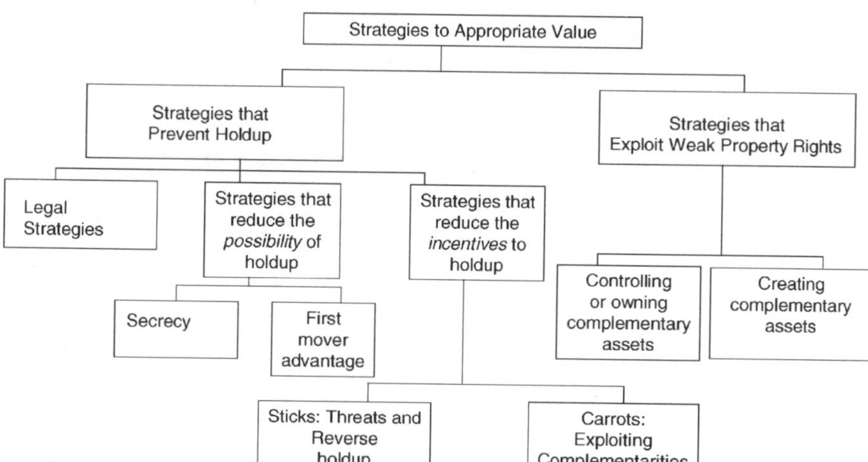

Fig. 2. Strategies to Capture Value in the Presence of Insecure or Non-Existent Property Rights.

> monitor and police uses of your resources means "your" property rights over "your" land are not as extensive and strong as they are over some other resources, like furniture, shoes, or automobiles.

Similar issues prevent legal protection of many types of intangible assets. Consider patents, for example. Key questions that have been addressed in the intellectual property literature are the optimal length and breadth of patent protection (that is, for how long should an inventor receive statutory protection for her invention, and how broad should the coverage of the patent be), and, as a consequence, what should count as an infringement? But this question – and, consequently, the extent of protection offered de facto by the law – are often very hard to answer in practice. As Merges and Nelson (1990, p. 841) point out, "the legal principles and objective evidence often leave considerable room for discretion," both in decisions involving the scope of patents by the Patent Office and in the litigation that accompanies allegations of patent infringement. For example, should a patent be awarded for new uses of a previously patented product? Should an alleged infringement be treated as such on the grounds that it is *equivalent* to an earlier patent, even if not *exactly identical*? Indeed, a more fundamental question is whether or not it is even possible to precisely circumscribe the boundaries of patent coverage.[10]

The limits to legal remedies were most strikingly documented in a pair of surveys by economists at Yale (Levin et al., 1987) and Carnegie-Mellon (Cohen et al., 2000). Interviews with over 600 managers across 130 lines of business in 45 industries revealed that

> [t]he picture is striking. For new processes, patents were generally rated the least effective of the mechanisms of appropriation. Generally, lead time, learning curves, and sales or service efforts were regarded as substantially more effective than patents in protecting products. (Moreover), in only one industry, drugs, were product patents regarded by a majority of respondents as strictly more effective than other means of appropriation (Levin et al., pp. 794–796).

In other words, managers in most industries observed that existing patent laws were rather ineffective to protect their investments in R&D. And these surveys concerned the effectiveness of patent law, not copyright or trade secret law! In these cases, or where firms attempt to recoup investments in assets that are as difficult to describe and define as "ideas" or "skills" or "relationships," many have argued that constraints to the law are significantly greater.[11] O'Donoghue et al. (1998) note similar findings from additional studies:

> There is at least some evidence that effective patent lives are short. Mansfield (1984) reports from survey evidence that in some industries 60% of patents are effectively terminated within 4 years, which is considerably less than the statutory life of 17 years .... Further evidence of short effective patent lives comes from patent renewal data: Schankerman and Pakes (1986) conclude that European patents lose on the order of 20% of their value per year, and Pakes

(1986) reports that only 7% of French patents and 11% of German patents are maintained until the patent expires (see also Pakes & Schankerman, 1984).

The story is similar in other industries. Bhattacharyya and Nanda (2000) note that despite the profusion of new financial products in the last decade, "financial products cannot be patented and all details of a new security become publicly available once the offering is filed with the SEC."[12] And, in the context of recent piracy in the music industry, the difficulties in relying solely on legal protection to solve the problem of infringement has been elegantly noted in a recent article by Shirky (2001). He argued that "the music industry is losing not the right to enforce copyright but the ability to do so."

What does this imply for firms and their ability to recoup investments in assets over which property rights are weak? As Alchian (1967) argued,

> [W]hen private property rights are unavailable or too costly to establish and enforce, substitute means of control are sought. Government authority, expressed by government agents, is one very common such means.

But there are other means. As Gallini and Scotchmer (2002) note, intellectual property is merely one of a variety of incentive schemes to promote R&D, and other protections can substitute for a formal reward system. Indeed, managers often argue that limits to legal remedies make it necessary to consider strategies that encompass so-called "creative business models" or market-based solutions. However, in many cases, it is not clear what these market solutions might be and, more importantly, whether these solutions can be designed systematically. "Creative business models" are typically given no more than lip service. The rest of this paper aims to fill this gap by providing a systematic approach to *market-based* solutions that firms can use to cope with weak property rights.

3.2. Market-Based Solutions: Secrecy and First-Mover Advantages

Secrecy. It is often claimed that firms may rely on secrecy (Horstmann et al., 1985). In the previously cited Levin et al. survey, for example, firms often claimed to rely on some form of secrecy to appropriate value from innovations. An additional attraction of trade secrets is that they achieve an indefinite period of exclusive use for the firm's invention or discovery. In practice, however, von Hippel (1988, p. 54) argues:

> [Secrecy] is an option only for innovations that can in fact be kept secret: the holder of a trade secret cannot exclude anyone who independently discovers it or who legally acquires the secret by such means as accidental disclosure or reverse engineering .... [T]rade secrets have proven to be effective only with regard to (1) product innovations that incorporate various technological

barriers to analysis (e.g. the formula for Coca-Cola) or (2) process innovations that can be hidden from public view (e.g. process equipment kept behind factory walls).

Indeed, secrecy presumes that firms can choose not to disclose ideas to the market but still somehow extract rents from their commercialization.[13] Another reason why secrecy might be of limited practical use is that it may be difficult, even impossible, to keep secrets from employees because these secrets typically reside with the individuals and are not "in the firm." As *The Economist* notes:

> [T]he term 'technology transfer' is something of an oxymoron. Real innovations do not move from laboratory to shopfloor as patents, research reports, or even working prototypes. To stand any chance of success, they have to be transferred as concepts embedded in people's heads.[14]

Consequently, the ability to keep a secret either relies on a "black box" conception of firms or presumes that firms somehow have access to employee contracts that are strong enough to establish property rights. This may work sometimes. But in many other cases it simply assumes away the problem caused by weak property rights.

A first-mover advantage. Exploiting a first-mover advantage in commercializing assets is another way for firms to appropriate value from assets over which property rights are weak. The logic is straightforward – being first to market allows a firm to capture monopolistic rents, albeit during short time windows, until reverse engineering, imitation, or piracy occurs. Palm and Netscape offer two examples. But companies like Intel have taken this a step farther and base their strategy in *consistently* being the "first to market."

Case Study 1: Intel's Repeatable First-to-Market Strategy

Innovations in Intel's core business – microprocessors – have been notoriously difficult to patent or protect via other legal mechanisms. Levin et al. (1982, p. 80) noted that while certain innovations could be patented in semiconductors, "what appears impossible to patent is the actual physical layout of an integrated circuit of even moderate complexity. The principal reason for this is the difficulty of rendering a full verbal description of the circuit layout." Similarly, Casadesus-Masanell et al. (2002) note that technology moves so rapidly that patents quickly become economically obsolete. Also, because most companies draw from common underlying technology bases, it is not always clear who might be infringing upon whom. To compound these problems, for a typical cohort of chips, prices fall by almost 30% within a few months of introduction and by up to 75% within two years. In spite of these difficulties, Intel's cumulative profits between 1987 and 1997 exceeded $20 billion, making it one of the most profitable firms in the world.

So, how did Intel sustain its first-mover advantage across product generations? It followed a multi-pronged approach: first, it made massive investments in product design and manufacturing and process technologies to ensure that it launched new products ahead of competitors in each new generation; then, to allocate chips that were typically in short supply in the initial months after launch, it employed a specific rationing rule (viz., tying current-period allocations to previous-period purchases) to ensure that its OEM customers would have little incentive to switch their purchases away from Intel; it coupled these moves with an aggressive branding campaign directed at end-users that was further meant to increase switching costs for OEMs; and, last, it made numerous investments in complementary software applications (such as videoconferencing systems) that relied on high-speed chips, so as to stimulate demand for new chips in each generation by making customers feel a continued need to upgrade their computers. As a result of these actions, Intel has consistently ensured that profits from the short time window between product launch and imitation by competitors were more than sufficient to recoup the billions of dollars of capital investment that it incurred in fabrication facilities for each new generation of chips. Indeed, in an industry where patent protection was limited, imitation was fast, and competitors could manufacture at lower cost, Intel continued to be the overwhelmingly dominant supplier of microprocessors, with a market share of almost 90%. Of course, Intel did not eschew vigorous legal defense of its intellectual property in keeping would-be infringers at bay. But that was not the sole, or even the primary, line of defense.

Like Intel, companies in other industries have designed ways to *repeatedly* exploit first-mover advantages across product generations, and thereby extract rents from innovation. Consider, for example, Capital One Financial Corporation, one of the most successful credit card issuers over the past decade. Between 1993 and 2000, Capital One introduced new products at an unprecedented rate in the industry, while still maintaining the lowest chargeoffs in the industry. "Of its $750 million marketing budget (in 1999), over 55% was spent on marketing some product that did not exist six months before, 80% on products that did not exist a year ago, and 95% on products that did not exist two years before" (Anand et al., 2000). Moreover, by 1999, the company was performing over 30,000 experiments annually that were aimed at generating new products and services – and all without any patents or formal property rights to protect any of these new ideas! Underlying Capital One's success was its "information-based strategy" (IBS), that entailed an intense use of information on customer behavior to develop detailed credit scoring models, generate new products, and match products with appropriate target customer groups. Imitation by rivals was made even more difficult when one considers the entire set of supporting activities that the firm engaged in to fully exploit IBS, and the manner in which it aligned its entire internal organization to do so. This included creating a novel organizational structure that involved business analysts working cross-functionally, combining authority for credit and pricing decisions into a single division, exploiting IBS not just in product development but in customer service and internal recruiting as well, and developing novel "horizontal accounting" methods for tracking the performance of its products (by customer cohort in addition to product features).

The examples of Intel and Capital One offer a useful lesson on how firms may cope with weak property rights. Each of these companies created, over periods spanning a decade or more, a sustainable means to appropriate value from what were essentially unprotectable innovations. The key was to ensure that its leadership position in one period could be leveraged into a leadership position in subsequent periods, thereby ensuring a first-mover advantage that was repeatable across product generations. Nevertheless, first-mover advantages rely on a long-enough time window before the second mover enters. For many products where property rights are weak, such windows are typically much shorter, maybe even negligible. Consider, for example, that test versions of Microsoft's new operating system, XP, were available on the web two months *before* Microsoft planned to make it available.[15] Or that, Harry Potter III was available on the same day that it appeared in theaters.

In other words, both secrecy and first-mover advantages work only if the appropriability problem that it is meant to solve is not too severe after all. In many cases that have received attention recently, the problem of appropriability appears to be more severe. Indeed, time windows before entry of competitors may be hours rather than months, and secrecy or disclosure may be endogenous to the actions of employees rather than of the firms themselves.

Next we consider market-based solutions for overcoming the extreme case of weak property rights – where an asset, once developed by a firm, can be used freely and immediately by others. The strategies considered below may be effective even when the aforementioned ones are not.

3.3. Strategies that Prevent Holdup By Relying on Incentives

In contrast to strategies that attempt to reduce the possibility of holdup, those discussed in this section work by *reducing the incentives* for holdup. In other words, there is no presumption that holdup can be prevented or restricted. Instead, holdup does not occur because incentives are set so that it does not pay. The simple model in Section 2 suggests a natural classification of incentive-based strategies: sticks and carrots.

Sticks: Threats and Reverse Holdup

The model in Section 2 suggests that the threat of reverse holdup may mitigate holdup. The relevant tradeoff is in whether or not reductions in future benefits offset the short-term gains from holdup. In practice, firms employ a variety of threats when they deal with suppliers, buyers, or competitors.

Threats to suppliers. The canonical case is holdup by employees. Employees may leave with ideas, knowledge, or relationships that were financed by a firm and join competitors or managing independent firms. Partners from consulting or law firms often leave to start their own boutiques. Successful editors from publishing houses leave and start their own imprints elsewhere or join independent publishing firms. Advertising agencies also have been vulnerable to the loss of client relationships by the departure of key executives. And Silicon Valley is littered with cases of research scientists, software programmers, and engineers leaving large firms to launch their own entrepreneurial ventures. Indeed, a recent McKinsey survey of over 6,000 managers at fifty of the largest U.S. firms revealed that "20% of managers said there is a strong chance they will leave their current company in the next two years, and another 28% said there is a moderate chance of leaving."

In settings like these, the relevant question is not why holdup occurs, but why it does not occur more often. One reason is that not all the key assets are lost when employee leaves. Projects are often conducted by teams of individuals; research is often done in shared facilities or even common labs; informal channels of communication and interaction between employees occur on a daily basis, giving rise to local spillovers in ideas or knowledge; and relationships with clients are often partly embodied in firms, not just in their employees. In each of these cases, organizations typically retain the ability to develop projects, commercialize ideas, and compete for client relationships even after key employees depart.[16] But the point is that the threat of competition by the employer can effectively deter holdup in many such cases.

Microsoft shows how companies might use the threat of competition to reduce the rents in the product market to departing employees. Rob Glaser, a ten-year Microsoft veteran, left in 1994 to set up Real Networks and develop software that would enable real-time streaming of audio and video content over the web. Within three years, Microsoft commercialized its own competing product, Windows Media Player. In response, Glaser alleged that Microsoft was competing unfairly by including programming code in its software to disable Real Networks' product, by intentionally ensuring that its own product was unable to play content based on the latest versions of the Real Networks technology, and by refusing to renew its license with Real Networks that would allow it to do so. But the point is that, even without anticompetitive practices, competition itself can work to reduce product market rents to departing employees.

Similarly, publishing houses sometimes find it difficult to retain relationships with well-known authors whom they have nurtured over many years. In a recent episode, an editor at one of the largest publishing houses recounted:

> We were in the midst of renegotiating a contract with one of our authors with whom I had established a relationship more than twenty years ago and published more than fifty of her books. It turned out that another publisher came in and offered her an incredible amount of money. The negotiations got so frustrating that during one of these conversations, I threatened that if she left, we would repackage and publish at least three of her previous books (that we owned the rights to) every single year, in both hardcover and paperback, and not have to put a hand in our wallet to do so. Would we have been able to free-ride on the marketing dollars aimed at supporting her new books? Absolutely. Would this have cannibalized her new books? Maybe, but only to a limited extent. Might it have caused some confusion amongst readers? Maybe in the near term. But the point is that it was enough to convince her to stay.

Threats to buyers. Firms often license or sell the right to commercialize key assets to downstream buyers. Anton and Yao (1994) have elegantly described the holdup problem in these settings:

> A buyer may be hesitant to buy or contract for an unknown commodity, but if the inventor first reveals information regarding the invention, a lack of property rights makes it possible for the newly informed party to steal or expropriate the invention.

But even then, the authors argue, the inventor may have a credible threat: reveal the invention to a competing buyer, manufacturer or licensee. This, in turn, would stimulate the original buyer to voluntarily negotiate a contract with the inventor, as the buyer would be concerned about losing its monopoly.

The threat of competition has been effective in extracting value in many different settings.[17] Indeed, even when buyers are consumers rather than firms, such threats can work. But to be effective, threats must be credible: the firm must be not only able to carry out the threat, but also willing to compete against the buyer. Of course, in equilibrium, threats should dissuade holdup.

Threats to competitors. Going back to Arrow (1962), competition has been considered anathema to firms that cannot secure property rights. The reason, of course, is that competitors could compete away rents after another firm has sunk its investments. Examples of competitor hold-up include the reverse-engineering of innovations and free-riding on key information.[18] As discussed earlier, firms that invest in intellectual property, in industries ranging from electronics to computer software, typically have been the focus of many studies. But a similar issue arises in many professional service industries, including advertising, law, and investment banking. In these markets, the key assets typically are embodied in mobile employees. Consequently, the threat that competitors might poach key employees is severe.

Can the threat to retaliate if a competitor exploits key assets or poaches key employees work in these settings? Yes, under certain conditions. First, notice that it consists in competing away the profits to the poaching firm in future periods. In other words, in deciding whether or not to poach, a firm will trade off the increase in profits from poaching or free-riding in the current period with the fall in future profits. This immediately implies that no firm in the industry can be too small, for, if it were, its maximum gains from poaching on others to increase its current-period market share will be large relative to the future loss of rents, making poaching attractive. Thus, the smallest firm in the market must be "big enough" to ensure that its future losses are large enough to make poaching unattractive. Interestingly, for a similar reason, no firm in the industry can be too big; otherwise, the other firms' shares will not be big enough to deter poaching. This simple logic suggests that retaliatory threats by firms will be effective only if there are a few firms of similar size. That is, these markets will tend to be *natural oligopolies*.

Case Study 2: Investment Banks and the Control of Relationships

The investment banking market offers a natural setting for examining how reverse holdup threats might work, as well as their consequences for the structure of markets. To begin, note the following characteristics of investment banks, their key assets, and investments in these assets. First, Wilhelm and Downing (2001) note, in their comprehensive recent account of the investment banking industry, that the services that investment banks offer their clients – including merger and acquisition advice, advice on restructuring deals, equity and debt underwriting, and other advisory services – are "relationship-based, human-capital intensive" (p. 32). Second, Eccles and Crane (1988), in their seminal account of investment banking deals, observe:

> [T]he most distinctive [economic characteristic of the investment bank business] is the loose linkage between value provided to customers and fees earned. Investment bankers provide information, advice, ideas, and studies, much of it unsolicited, to both investors and issuers without the firm receiving any direct compensation. Issuers compensate the firm only when a deal is done.

This implies that investments in establishing relationships with clients are sunk at the time of deals, when banks can hope to recoup these investments from deal fees.[19] Third, relationships between banks and firms need to be considered at both the institutional and individual levels since they reside in both banks and bankers. Indeed, some banks (most recently Deutsche Bank in the 1990s) have periodically attempted to build their client presence in particular areas by simply hiring away staff en masse from other banks. Conversely, the threat of massive poaching can be devastating to a bank's market share.[20]

How do investment banks overcome the problem of weak property rights over client relationships? In several ways, argue Wilhelm and Downing (2001). These include the reliance on cooperative arrangements between banks – for example, underwriting syndicates on equity deals, or alliances that maintain online debt platforms. These arrangements limit banks' incentives to free-ride or cheat on each other by holding their feet to the fire in various ways, including the loss of reputation. Another central mechanism that banks use to cope with the holdup problem is the threat of retaliation on competitors – if they were to poach key managers or undercut them on prices. In an insider account of investment banking, Rolfe and Troob (2001) note the prevalence of implicit threats and the lack of price competition:

> There has always been an unspoken agreement among the bankers that when it comes to underwritings they won't compete on price. The spreads are sacrosanct. He who cuts spreads will himself become an outcast. The community of investment banks has always been small enough so that if one bank were to break ranks on the pricing issue, the others would quickly join forces and squash the offender . . .. Every banker knows that the pricing issue is a slippery slope best avoided because once the price cutting begins, there's no telling where it will end.

Indeed, the consequences of implicit threats are documented in several recent studies of the industry. First, note that, in spite of apparently brutal competition on non-price dimensions, there is virtually no price competition in the industry. This ensures that firms can set prices high enough to recover their upfront investments and overcome otherwise "loose linkage" between fees and costs. Negotiated underwriting syndicates have always superceded competitive bidding for deals, and underwriting fees have, for decades, been virtually unchanged. Indeed, Ritter and Chen (2000) reports that, strikingly, 90% of initial public offering deals raising between $20 and $80 million have gross spreads of exactly 7.0%.

The market share consequences of such retaliatory threats are also strikingly similar to the simple predictions described earlier in this section. In Anand and Galetovic (2000b), we document market shares of the top banks over four decades and note that they are quite stable. Moreover, no bank is dominant: in every (five-year) period; the industry leader has a market share of less than 20%, and the top banks are of similar size. Third, numerous regulatory attempts over several decades to increase competition in the industry have had virtually no effect on aggregate market structure – which is what one would expect if the prevailing market structure were necessary for making key investments rather than merely a convenience to extract market power. And, last, the presence of a large number of small or boutique investment banks has had virtually no impact on the dominant oligopolistic banks – in contrast to the view that they serve as a "competitive fringe" that restrains large bank behavior. The reason is that these two sets of banks serve essentially different markets: large, "bulge bracket" banks serve predominantly large corporations, while small banks serve small firms. In other words, the market is characterized by a "funnel-shaped" structure, with a few very large banks and numerous very small ones.

The evidence suggests that reverse holdup threats can be quite effective in markets where firms deal with a small number of competitors. But, more strikingly, the logic of the argument suggests that, even without regulatory intervention ensuring "small numbers," the voluntary entry and exit of firms can itself ensure that such market structures obtain in equilibrium.

Some of these phenomena described in the context of the investment banking market also can be seen in other intermediation markets – such as law, consulting, and advertising – that display similar characteristics of the underlying production technologies. These are markets in which significant investments are made in assets such as client relationships (that are largely embodied in human capital) over which property rights are weak, and in which investments in these assets are largely sunk at the time of deal pricing. Indeed, what appears to be strikingly similar across each of these markets is that: (i) there is vigorous non-price competition in sales effort, advertising or marketing; (ii) there is limited price competition; (iii) the lack of price competition may be efficient rather than welfare-reducing because it neutralizes another pre-existing imperfection, weak property rights; and (iv) the two characteristics of investment banking market structures described above – specifically, the natural oligopoly comprising a few large banks combined with a "funnel-shaped" aggregate market structure – are observed in virtually every one of these other markets, as well.

We have argued here that understanding how firms solve the underlying problem of weak property rights over their key assets in these markets can help to shed light not only on the particular strategies that firms employ in these and related markets, but also on the industrial organization of these markets. A key point is that in each of these settings, either market-based strategies, or, as Gallini and Scotchmer (2002) have noted, natural market forces can protect inventors, making formal protections or other incentives unnecessary. Indeed, these arguments suggest that even when legal mechanisms for establishing property rights are ineffective, regulatory intervention to establish monopoly power in these markets may be unnecessary.

Carrots and Complementarities

In some cases, firms can make holdup relatively less attractive by offering carrots or rewards. Rewards work when the firm can offer a share in a pie that is larger than available elsewhere. Of course, this raises the question: what allows a firm to offer something that is better than the outside option? The answer lies in precisely what makes a "firm" distinctive, its ability to exploit *complementarities* that come from performing multiple related activities or from owning multiple assets. Thus, the key point is that sometimes complementarities allow firms to offer better carrots to employees, customers or suppliers who might have held them up.

In some cases, complementarities are purely technological. That is, a product or asset may simply not work without its complement. But as Shapiro and Varian (1998, p. 127) point out, in many settings complementarities are entirely a construct of firms' strategies. Gillette, for example, chooses to make its Mach-3 razors compatible with (and thereby a complement to) its own razor blades. Similarly, many firms offer loyalty programs to their customers. By purchasing multiple units from the same firm, the customer is offered the reward of a discount on the bundle, thereby making her purchase of the first unit a complement to her purchase of the second and third and so on. In the same spirit, a firm may offer rewards to key employees or suppliers to retain them. For example, companies often structure compensation contracts that mimic consumer loyalty programs – so that wages might increase well above productivity with tenure, or stock options may vest only after several years.

But more relevant here, complementarities can be intangible as well. Thus, employees may stay on with a firm, even when pecuniary benefits are greater elsewhere, because of personal relationships or the company's unique "culture." Goldman Sachs, for example, has long been considered the leading investment firm with, as Endlich (2000) puts it, "a solid-gold reputation and a first-class list of clients." Goldman, he argues, could retain talent for a number of reasons. Its partnership status focused the firm on long-term performance; the resulting family feeling and collegial atmosphere separated it from its peers; and the firm's close client contacts not only wrought superior information, but allowed it to "control egos and monitor arrogance ... [since] the client, not the salesman, banker, or trader, [was] the focus of any transaction." Industry observers agree with the conclusion that "the firm's culture ... more than anything else, sets it apart from its competitors, making it the organizational glue that induced employees to stick."

This example points to a key lesson about when complementarities are useful to prevent holdup: *they cannot be easily available or accessed either through the market or by other firms*. In other words, complementarities must be either embedded within an organization or difficult to contract over at arms-length. The reason, quite simply, is that if they were not, then the available surplus would be no different inside and outside the firm, making them useless in preventing holdup.

At first glance, this might seem to imply that firms possessing no particular unique complementary asset are doomed. But that need not be the case. The reason is that, in many settings (including the example of Goldman Sachs), complementarities stem from an entire collection of activities or assets. Hence a firm may have an advantage not only from owning a particular complementary asset or product, but also from coordinating these multiple activities, as the case of International Management Group (IMG) strikingly shows.

Case Study 3: How IMG Exploits Complementarities and Appropriates Value from Mobile Talent[21]

IMG has been in the business of managing talent – athletes, opera singers, supermodels, authors and speakers – for over forty years. It started as a single-agent-managed firm in 1962 and by 2003 had grown to more than 85 offices and 3,000 employees in 85 countries, establishing itself as the most successful talent management and representation company in the world. Two facts about IMG are striking. First, even though IMG's business exemplifies the hazards of a "free agent nation," relatively few of its customers switch to other firms. Second, IMG's size and scope are quite different from those of similar human-capital-intensive firms. For one, most firms in the business of talent representation in other industries (e.g. Hollywood) appear to be relatively small and specialized. For another, most large professional service firms are owned by partner-employees who appropriate most profits. In stark contrast, IMG is neither a partnership nor small, and its agents are employees. And, it has a strikingly broad vertical and horizontal presence in the value chain.

It is clear that IMG is not intrinsically immune to holdup from employees or competitors. Over the years, sports agents have left IMG, and clients have switched to other firms. For example, in 1992, IMG agents John Simpson and Julian Jacobi left to found Stellar Management because they "didn't have any equity in this business."[22] And in 1999, IMG agent Jeffrey Schwarz quit, taking with him three of IMG's tennis stars: Pete Sampras, Martina Hingis, and Marcelo Rios. Similarly, golf stars Nick Faldo, Greg Norman, and Nick Price left IMG at the peak of their success in the 1990s. Moreover, these departures were costly for the firm. One industry expert, following the poaching of classical music singers by different agencies in the 1990s, noted that "it could take ten years for an agent to build a singer's career and another five before star fees come cashing in.... You lose one artist and you're upset. You lose three stars and you're in trouble." But the question is not why some agents and clients have switched away from IMG over the years, but why so few do. So, how does IMG do it? To understand it, let's focus on IMG's strategy within one of its core businesses, golf.

IMG's prominence in golf began in 1958, when its founder, Mark McCormack, represented golf's biggest star, Arnold Palmer and, within a few years, its other two major stars, Jack Nicklaus and Gary Player. But IMG soon expanded into hosting and managing golf events (starting with the World Match Play Championship in England), then into the design, management, and marketing of golf courses, which was followed by the production and distribution of television programming of golf events, and, finally, by training academies for golfers of different ages. By 2002, IMG had hosted several major golf events, created more than 400 golf courses worldwide, broadcast live international coverage for some of golf's most prestigious events, developed the Sony world ranking system for professional golfers, and run the largest training academies for golf instruction.

At first glance, one might conclude that IMG's "diversifying" moves were intended to reduce its reliance on mobile talent. But, if anything, they made IMG *more* dependent on stars. To see why, consider, for example, that the most important determinant of a golf event's success is the presence of top stars. In 2000, television audiences for the 17 tournaments played by Tiger Woods (the top-ranked professional golfer) were more than twice as big than those of the 17 events that Woods skipped! In other words, expanding into golf events was likely to increase IMG's vulnerability to holdup. Similarly, production and distribution of television programs was already a relatively mature business at the time of IMG's entry, and IMG did not have any intrinsic skills or expertise at the time. Moreover, neither the logic of training academies as providing an early window on future talent, nor the argument that they might increase the loyalty of future stars (by establishing relationships with them early on), withstand scrutiny. Fewer than 500 golfers participate per year, so academies were unlikely to be an effective way to search for future talent. And, finally, early relationships did not preclude or deter athletes from leaving IMG after rising to stardom.[23] If anything, establishing relationships early on increased their investments in these relationships and would, therefore, appear to increase the cost from holdup.

It turns out that the main benefit of IMG's diversification was to enhance complementarities with its key asset – talent – thereby enlarging the pie to be shared with them. First, consider golf events. Entry into this activity increased competition for talent by increasing the prize money offered at such events. Moreover, because IMG's clients won approximately 50% of all prize money, and the firm received 10% of their winnings, IMG could afford to offer larger purses than competitors, which in turn offered them an advantage in bidding for the rights to host these events in the first place.[24] In addition, IMG created various "non-standard" events to showcase their stars (for example, the made-for-TV "Skins" event that featured four prominent golfers competing for exorbitant prize money). And IMG's presence in broadcasting ensured television coverage for these events. Then, IMG would typically grant its "host exemptions" for each event to its own lesser-known golf clients, thereby allowing them access to events that they might not otherwise have. And last, by choosing golf for its events, IMG might favor those courses that had been designed by its own clients (typically, aging golf stars).

IMG's presence in television also offered opportunities to showcase its clients. More favorable scheduling of tee times in golf and match times in tennis – scheduling matches in prime time, on center courts, or as partners with Tiger Woods – may ensure television coverage. Similarly, IMG television crews could make sure that camera angles prominently displayed corporate logos, which, in turn, afforded IMG an advantage in negotiating corporate endorsements for its athletes.[25]

Each of these advantages helped increase the "size of the pie." The net result was that the opportunity cost to an athlete of switching away from IMG was considerably increased, thus reducing the risk of holdup. Moreover, even if some stars were to leave IMG, there were large incentives to return. As senior golfers, for example, they might be invited to play with Tiger Woods in a made-for-TV event or have the courses they designed showcased through an IMG event. Indeed, Jack Nicklaus, Nick Faldo and Nick Price returned to IMG as clients after having left the firm years earlier. IMG's unusual success at retaining and managing star talent led to its founder and CEO, Mark McCormack, being named "the most powerful man in sports" in 2002 – ahead of even the top athletes themselves. IMG itself earned the grudging respect of its competitors, together with such monikers as "I am Greedy," "Incredibly Manipulative Guys," and "I Manage God." As McCormack simply noted, "What we offer are a set of toys that help our athletes, and we have better toys than our competitors."

IMG did not eschew legal defenses, and would often attempt to vigorously defend its non-compete contracts with employees. But, the key point is that by building a vast network of complementary assets and businesses around its clients, IMG simultaneously reduced its vulnerability to personal relationships and holdup by individual sports agents. Indeed, McCormack often observed that "an agent's role was part mentor, advisor, and friend, but an agent remained, ultimately, an employee."

3.4. Strategies that Exploit Weak Property Rights

Control over Complementary Assets

The strategies described in the previous section work by preventing holdup. This section discusses strategies that allow firms to appropriate value by exploiting weak property rights over assets that they invest in. The argument is simple and runs as follows. When surpluses are larger elsewhere, it may be difficult to prevent holdup, but sometimes the firm may profit from employees leaving, competitors copying or customers pirating. Why? The central idea rests, as before, on exploiting complementarities with assets or products that the firm *can* own.

To see this, consider a stylized example. Two products, call them A and B are costless to produce, but the firm has to incur a sunk and fixed up-front cost to "invent" A. A and B are complements in consumption, so that $\mathrm{WTP}(A, B) > \mathrm{WTP}(A, 0) + \mathrm{WTP}(0, B)$.[26] Let the prices of A and B be denoted by p_a and p_b, and demand for each by $D_a(p_a; p_b)$ and $D_b(p_b; p_a)$.

Now, suppose that consumers can access good A at zero cost (by virtue of cheap copies of A, through piracy, or the like). Then, $D_a(p_a; p_b) = 0$, for any $p_a > 0$, and the cost of producing A cannot be recovered. Nevertheless, demand for B would be stimulated if A were available for free. Provided that $p_b > 0$,

$$\pi_b' = D_b(p_b; c') \times p_b > D_b(p_b; p_a) \times p_b$$

for all $p_a > 0$. In other words, B's profits increase when A is cheap and in some cases may be larger than the cost of inventing A. Then there is a straightforward solution to the problem of weak property rights – namely, for a single firm to offer both products A and B.

The key assumption is that there exists a complementary product B over which property rights can be established even though they cannot be over product A. The following discussion offers examples of where this occurs.

To begin, consider pirated software applications. It is known that in some countries, piracy is rampant, but it less known that even in the United States some 25% of all corporate software is purloined, and in other law-abiding Western countries, the figure often approaches 50%. Nevertheless, while stand-alone applications like those used as spreadsheets (e.g. Excel or Lotus) or for word processing (e.g. Word or Word Perfect) can be more easily copied and disseminated, the operating system, which is embedded in the computer microprocessor, is much harder to copy. Of course, if Microsoft were only in the business of developing software applications, it might never do so. But even free software applications cannot be used without the operating system, for which Microsoft can charge a high price. Indeed, as is well known and as our simple model suggests, intense competition and low prices in the software applications market increases demand for the operating system. Thus, Microsoft may *want* to give away software for free – or train employees who then leave and write programs that use the operating system!

The point is that looking to achieve 100% deterrence is not only impossible; it is not even optimal. As a *Wall Street Journal* article on software piracy noted:

> If economic theory has any claim on the real world, Microsoft's pricing should naturally gravitate towards producing an optimum amount of theft. That is, thieves who wouldn't use the product if they had to pay for it, but who might become future customers or who become part of a network of users that makes the software more valuable to legitimate buyers.

Thus free pricing might increase the size of the user network, in turn making the software more valuable to legitimate users. Or, it might induce customers to try a product for free and then end up paying for it in the longer term once they get used to it. These are "complements" in action.

The analogy with free goods suggests a useful lesson. Even if firms cannot retain control of a key asset, they can still capture value by controlling or owning a complementary asset. Consider the impact of biotechnology on pharmaceutical companies since circa 1980. Changes in the technology of drug discovery – from large scale trial and error methods to systematic drug design – made it easy, even efficient, for small biotechnology firms to compete in research with pharmaceutical giants, as the key asset shifted from financial or physical capital to human capital. Indeed, many biotechnology firms trace their roots to these large pharmaceutical companies themselves, as hundreds of scientists left to start or join small companies since the larger ones no longer had an advantage in retaining key talent. Strikingly, however, pharmaceutical giants have earned most profits in the industry over the past two decades, and their average returns on equity have remained amongst the highest in the economy. The reason? Even though pharmaceutical firms no longer controlled a key asset – research talent – they still owned the key complementary assets needed to commercialize drugs: the large distribution facilities and marketing networks.

Complementary assets need not be physical or tangible. Control or ownership of complements that are intangible suffice. One common case is the provision of services. For example, many software firms price their software applications for free, and aim to generate profits by offering complementary products or services that customers value. Netscape, for example, made its browser available for free to most individuals, but charged corporations for the more complex Web server versions of the browser.[27] A recent example is the open source software market where firms like Red Hat and VA Linux specialize in commercializing Linux, an open source UNIX-based operating system that is freely available. Lerner and Tirole (2002) note that:

> Red Hat provides support for Linux-based products, while VA Linux provides hardware products optimized for the Linux environment. (They) exemplify the strategy of commercially providing complementary services and products that are not supplied efficiently by the open source community. (Thus,) while improvements in the open source software are not appropriable, commercial companies can benefit indirectly in a complementary proprietary segment.

The example of pharmaceutical companies illustrates that firms who own key complements can benefit even when key employees leave. Consider the case of McKinsey. At first glance, its hiring and training policy appear puzzling. Every year, it recruits in droves, then forces out large numbers of these same hires after a few years. But, as *The Economist* notes, the vast old-boy network of McKinsey has its own benefits: "The flow of individuals from firms to clients, and back again, is continuous and intensive," with the result that McKinsey alumni would typically be more inclined to "call upon former colleagues to provide services for their new employers."[28] In other words, ownership of the McKinsey name and its associated reputation – the complementary asset to mobile human capital – allows the firm to act as a broker of valuable relationships amongst those who benefit from the name.

Revenue models that rely on complementary assets come in different forms. Consider, for example, various types of online entertainment content – short films, music videos, or cartoons – that are relatively easily pirated. At first glance, it would appear difficult to bundle these types of content with complementary products that consumers would value. But, even when complements do not benefits primary consumers directly, bundling assets and their complements can benefit third parties. For example, television networks bundle products that consumers want (television shows) with products that consumers dislike (advertisements), and the benefits from this bundle are then captured by charging advertisers for the *access* to captive consumers.

As these examples illustrate, adopting strategies that exploit weak property rights may change the boundaries and scope of a firm. Expanding the set of assets that are owned or controlled allows firms to price bundles more accurately, to enhance coordination, or to exploit complementarities across products. But, in addition, diversification and scope expansion works by increasing reliance on the core asset, not by attempting to diversify revenue streams. The premise that underlies strategies that exploit weak property rights is to generate revenues from investments in core assets.

The examples suggest another necessary condition for these strategies to work:, firms must have some market power in the complementary asset market. Excess profits from the complementary asset can then be used to recover investments in assets or products that the firm cannot own. By implication, "carrots" are ineffective when there exists competition in the market for complementary assets.

4. APPLICATION: STRATEGIES TO COPE WITH PIRACY IN THE MUSIC INDUSTRY

Perhaps no industry has seemed more vulnerable to weak property rights than music in recent years. The 1990s saw the development of new technologies for compressing, storing, and playing music over the Internet, which, in turn, spawned websites that facilitated peer-to-peer file sharing on an unimaginable scale. The first such site, Napster, was founded in early 1999, and by the end of the year users could access over 450,000 songs free of charge (Anand & Cantillon, 2003). The popularity of file sharing caused great concern in the music industry because of its potential deleterious impact on CD sales and industry profits.

In this section, we use our framework to explore strategies that may work in the music industry and discard those that probably will not. Our aim is not to make predictions on how the industry will evolve (although we do to some extent), but to use the framework developed in earlier sections to organize the evidence, interpret various events, and shed light on what strategies may work. In the process, we critically examine the popular view that there has been "a fundamental collapse of the music business."[29] We also examine the alleged ineptitude of record companies in devising strategies to cope with piracy. Our analysis suggests reason to doubt both these suppositions.

4.1. Strategies that May Prevent Holdup and Piracy

The Legal and Technological Fix: Limits and Some Opportunities

The recording industry's first reaction was to file a lawsuit against Napster for aiding the infringement of copyrights. In December 1999, the Recording Industry Association of America (RIAA) filed suit on behalf of 18 major record companies (including Bertelsmann, EMI, Sony, Universal, and Warner). After several months of legal wrangling, the Federal Court ruled in February 2001 that Napster could no longer allow its users to swap files of copyrighted music, and the company shut down its site and filed for bankruptcy in June 2002. The record companies' next line of defense was to devise technical solutions that could halt online file-sharing. Examples of such solutions included copy-protected CDs that prevented the copying of files, "digital watermarks" that were invisible markers aimed at tracking file movements and limiting reproduction, and even encryption chips embedded in PCs and other devices that might allow for stronger protection.

As in other sectors, however, neither legal remedies nor anti-piracy technologies were very successful, and by the time music companies succeeded in shutting down Napster, numerous file-sharing sites, such as Gnutella, Kazaa, and Morpheus, had already replaced it. Unlike Napster, though, they did not store files in a central server but offered software that allowed users to search and access files stored on personal computers. Not surprisingly, legal actions became more difficult. In 2001, a Dutch court ruled that Kazaa was not liable for copyright infringements committed by its users, foreshadowing the difficulty that record companies in the United States may face in the courts.[30] This series of events led an industry observer to note:

> [W]hile the entertainment industry may be winning most of its legal battles, it seems to be losing the war . . .. [T]he startling truth [is] that downloading of music, and now films, continues unabated despite the myriad lawsuits.[31]

Neither were technical solutions successful, either because of technical limitations (e.g. some copy-protected CDs could not be played on computers, DVD players, game consoles, and MP3 players) or, more importantly, because they could be circumvented as programmers broke codes regularly. According to a recent account:[32]

> In Silicon Valley, copy protection is seen as folly. Not only do geeks treat code cracking as a contact sport, but the software industry has been trying – and failing – to combat piracy for years . . .. When the [music] industry came up with a supposedly secure format called SDMI, it took Princeton computer-science professor Edward Felten only two weeks to crack it.

More fundamentally, as the *Wall Street Journal* argued:

> A technical fix for piracy was never on the cards. The software code to break DVD encryption has been perfected to the point where it fits on a T-shirt. The music industry's own proudly proclaimed digital watermark was busted in three weeks by four different groups. Now a Dutch cryptographer claims to have found "a fatal flaw" in Intel's design to stop digital broadcasts from being copied and redistributed on the web. That leaves only a business solution.

Apple Computer's CEO Steve Jobs summarized the inherent limits of legal and technological strategies when he argued that "[p]iracy is a behavioral issue, not a technological one."[33]

Yet selective legal and technological strategies may be of some help, as recent actions by record companies suggest. First, they targeted heavy file-sharing users, suing four students from reputed universities – very publicly. Second, realizing that suing each of the millions who occasionally download a song would be fruitless, they targeted firms that either directly or indirectly facilitated piracy. Specifically, they sent "warning letters to major U.S. companies and universities urging them to cut down on file-swapping"[34] by employees or students. Next, they turned their attention to the financiers of firms that developed file-sharing technologies. For example, Universal Music Group sued a leading venture capital firm for its investment in Napster on the grounds that its financial support allegedly contributed to copyright infringement. Last, the RIAA looked to subpoena Internet access providers, in order to obtain the names of customers who had illegally downloaded hundreds of music files. Each of these strategies delegated the task of monitoring illegal downloading and file-sharing to firms that had the ability to do so.

Sticks and Carrots: Can They Work?

Can music companies use sticks or carrots to deal with weak property rights and piracy? Our model suggests some limitations. First, consider sticks. The model presented in Section 2 suggests that sticks work because spillovers enable the "corporation" (here, the studios) to threaten the "researcher" (here, the pirates) with intense market competition if she decides to leave. But, with piracy, competition is not much of a threat because music is available for free anyway! Thus sticks can work only if record companies directly harm pirates.

To some extent, record companies have tried to do just that by engaging in a sort of guerrilla warfare against selected pirates. For example, users who attempt to download the latest songs by rap artist Eminem may obtain empty files or files with bogus tunes that were dumped onto several file-sharing networks and disguised as legitimate tunes. Or these users might be automatically redirected to websites where they could legitimately download the songs (for a fee). Other tactics were more insidious. For example, it was alleged that all five major music record labels contacted several small software companies in California to develop software that might disable the computers of individuals who were suspected of downloading files without permission. This software included viruses, hunter bots, or programs that would freeze the users' Internet connections or freeze their personal computers directly. As recently noted, these tactics

> . . . could be more successful than suing the companies involved since it strikes at the hearts of consumers whose actions on the Internet . . . are measured more by convenience than free or not free. Make it difficult, frustrating, or risky to get the song you want, and, chances are, you'll think twice before venturing onto Morpheus.[35]

Note that record companies are taking advantage of a key "spillover": they manufacture the master copy that is pirated, and so are, in principle, able to introduce devices to harm pirates.

Now consider carrots. In the model, they work because the corporation exploits complementarities, enlarges the pie and offers a better deal to the researcher. The fundamental limitation of carrots in the music industry is that music can be consumed independently, and it is not obvious which other goods or services controlled by the studios may be used to create complementarities that can be accessed only by buying from them.

Music studios have begun to take actions that may be intended to take advantage of complementarities. Thus Vivendi Universal acquired EMusic.com, the first commercial site that offered customers a downloadable music subscription service. Soon after that, other music majors (BMG, Warner, and EMI) launched MusicNet, which licensed to several distribution partners (AOL and Real Networks) both its music content and the technology for streaming, downloading, and burning music. Then, Universal and Sony launched Pressplay, another online subscription-based service. (More recently, technology providers like Apple launched online commercial services with more success. Later, we discuss reasons for the differential success of these different online ventures).

Relatedly, record companies began bundling CDs with other low-cost complementary products or services on which consumers placed substantial value. This made it worthwhile to purchase the CD rather than download it illegally and forego the sweeteners. For example, the band Bon Jovi hoped to establish a serial relationship with fans, offering extras such as preference in buying concert tickets, the possibility to climb on stage, and the possibility to chat with the group online. Similarly, Interscope Records (part of Universal) included a free DVD with the first two million copies of the new Eminem album.[36]

So, how effective can carrots be? As seen, they hold some promise, but a caveat is in order since it is unclear whether these are truly carrots as we have defined the term here. The key issue is whether the sweeteners are either cheaper or more attractive *because* they are sold together with records. Otherwise, the critical reader may observe, record companies may as well charge directly for the sweetener.

4.2. Exploiting Weak Property Rights: Redefining the Boundaries of Music Firms

Several critics of the record companies have argued that they should embrace free music rather than fight it. But this, of course, raises a question: demand for music may increase because it is freely available, but how can record companies capture any value from a free good? The answer is by reaping profits from complements, those activities and goods that become more valuable when music is free. As we will see next, this is perhaps the most effective strategy to cope with piracy, but it may require firms' boundaries to change.

A cursory analysis suggests a broad range of such complements: music concerts, artist merchandise, MP3 players, storage devices, personal computers, high-speed Internet access, and so on. Clearly, producers of complementary products and services might benefit from free music even while producers of music content lose. Early evidence appears to bear this out in striking fashion.[37] For example, in 2001, unit sales of blank CDs (1.1 billion) exceeded sales of pre-recorded CDs (968 million) for the first time. In the same year, there was a 90% rise in the number of computer owners with a CD burner. A third of all PCs have one; 54% of new computers come with one installed. Similarly, in 2001, sales of CD burners rose by 14%, to $684 million. Moreover, the number of portable MP3 players (such as Apple's iPod) more than doubled in the previous two years. And, according to *Fortune*,

> . . . companies like Intel and Microsoft are eager to get a wide variety of movies online because they believe films will drive the large-scale adoption of broadband and help the PC become the center of the home entertainment network.[38]

Indeed, when the RIAA objected to an advertisement by Gateway Computers that appeared to condone CD burning, the company ". . . happily conceded that CD burning was a cash cow." Taken together, the implication is clear. The sales of music and its complementary products and services *may have actually increased*, not declined, in recent years.[39]

All this implies that the main challenge facing record companies, then, *is not necessarily how to increase the music pie, but how to redistribute it*. Indeed, the conflicting objectives between record companies and companies in the "complement" business have been revealed through their actions in recent years – specifically, on whether to fight piracy or to embrace it. Consider, for example, that when several record companies served a subpoena on the Internet service division of Verizon, asking for the names of two customers the RIAA claims downloaded hundreds of music files without authorization, Verizon chose to appeal the ruling:

> [T]he ruling set too low a standard for the circumstances under which it would have to turn over the name of a customer, because a copyright holder wouldn't have to file suit against unidentified peer-to-peer users before obtaining their identities.[40]

It might be tempting to attribute Verizon's stance to its objective of protecting privacy, but the differences between the objectives of the content and technology companies came into sharper relief when, not coincidentally,

> . . . the record labels drew backing from the Motion Picture Association of America, which fears that movies could soon be widely swapped online. For its part, Verizon drew supportive briefs from an array of technology interests, including the U.S. Internet Service Providers Association, which includes EarthLink Inc. and AOL Time Warner Inc.'s America Online unit![41]

Record companies and movie studios recognize these differences in incentives. In his 2002 testimony on a bill before the U.S. Senate that would order the FCC to find a way to stem piracy if the firms in the entertainment and related sectors could not devise effective solutions on their own, Disney CEO Michael Eisner bashed Silicon Valley: "We're dealing with an industry where an unspoken strategy is that the killer app is piracy."[42] Interestingly, the same bill was "anathema to technology leaders like Intel Chairman Andy Grove and Apple Computer CEO Steve Jobs."[43]

But how might record companies establish claims over the new pie? There appear to be at least three ways to do so: (i) acquiring companies that own the complements or being acquired by one; (ii) establishing or renegotiating contracts with such companies either directly[44] or indirectly;[45] and (iii) entering these markets themselves.

Some companies – such as Sony (with Sony Music and Sony Electronics) and AOL Time Warner (with Time Warner Music and Time Warner Cable) – already have an existing stake in both music content and its complementary technologies or products. This results in inter-divisional conflicts or coalitions within the same firm that find themselves on opposite sides in the battle against piracy.[46] Others, such as Bertelsmann, have sought to increase their presence in the complementary parts of the value chain even while their music divisions were combating piracy, leading to inevitable conflicts between music divisions and corporate headquarters. The key point is that, just as the availability of free music has reconfigured the music pie by shifting value from producers of music content to producers of music complements, it will probably reconfigure the scope of operations of record companies, as well. Moreover, because, as we have argued here, any eventual solution will involve some redistribution or reallocation of claims between existing firms over the music pie, the transition is likely to be marked by conflicts.[47] Either way, the music industry will survive simply because, as noted in *Fortune* (May 27, 2002), ". . . unless somebody flinches, both sides stand to lose."[48]

4.3. Ineptitude vs. Incentives

Many have argued that record companies have been inept in coping with piracy. Indeed, their online ventures were, at least in the near term, failures. Within a few weeks of the launch of MusicNet, its CEO declared that the current version of the service was not viable. In diagnosing the reasons for failure, *The Wall Street Journal* noted:

> It isn't hard to see why. [They] created a service that lacked just about everything that makes online music downloads appealing. It didn't allow consumers to keep downloaded songs permanently. It didn't allow them to transfer songs to portable devices or exchange them with friends who weren't signed up to the service. And it charged a monthly fee. Renegade Internet services allow all that and more, for no charge.[49]

Indeed, it was clear from the outset that, when it came to the commercial online music services of the record companies, the "online pricing is way off. The reason: companies look at an online customer as a threat, someone who might stop buying CDs. So they try to set the price at nearly the CD level, even though the costs of online delivery are negligible."[50]

In stark contrast, Apple Computer launched its own commercial online service, iTunes, in April 2003 and allowed users to legally download songs at a price of $0.99 per song. Although, at first glance, the price of downloading ten songs might not appear too different from that of buying a ten-song CD in a store, Apple's program included several attractive features. First, in contrast to the terms offered by record companies on their websites that allowed legal downloads, consumers could now listen to a downloaded song an unlimited number of times. Second, consumers had the right to listen to the song on upto three computers and to legally burn it on an unlimited number of CDs. Third, all five major music labels licensed their content to Apple, resulting in a menu of over 200,000 songs. Finally, the variable pricing model meant that, like other file-sharing sites, consumers did not have to pay an upfront subscription fee. During the first week after launch, there were over one million downloads, or an increase in online music purchases of almost one million dollars.[51]

Indeed, based on their differing records in developing commercial alternatives to free online music, the conventional wisdom is that "... the recording industry has ... been notable mainly for clumsiness, proprietary paranoia, and a condescending attitude towards its customers." In contrast, technology companies have been perceived as forward-thinking, fast-moving, and entrepreneurial.[52] Our analysis suggests a reason for caution because the differences in success may simply reflect differences in *incentives*, not simply in *aptitude* or creative genius. On the one hand, record companies had every incentive *not* to make downloading truly viable because its first-order impact would have been to accelerate the trend towards free music, thereby cannibalizing offline revenues. On the other hand, as a producer of computers, CD burners, and MP3 players, whose revenues were propped up by all forms of music downloading and sharing, Apple had incentives to encourage the trend towards cheap and ubiquitously available music. As an industry observer noted: "In many ways, [Apple's] business model for the iTunes service is very closely aligned to that of the online pirates they're so desperate to destroy."

5. CONCLUSION: THREE LESSONS FOR MANAGERS

We started by arguing that weak property rights over assets and ideas have not deterred firms from investing in them. We conclude with three lessons for managers.

(1) *Coping with weak property rights through the law or the market.* The usual presumption is that the problems caused by weak property rights can be solved by toughening the law and its enforcement. Thus it is often argued that piracy, illegal copying, or rights infringement must imply, inter alia, that the laws to establish and enforce property rights are either not strong enough or not enforced.

But when it comes to appropriating value from investments in certain types of assets, legal means may be inadequate. Weak property rights may just reflect that it is impossible to establish or enforce such laws. At the same time, this does not mean that firms are helpless to recover their investments in such assets. The point is that, in many cases, it may be more productive to search for solutions that rely on the market rather than focusing effort and attention on legal strategies.[53] We have argued that, in these cases, strategies that work through incentives can substitute quite effectively for formal mechanisms that work through fiat.[54]

It is worth noting that firms entering or doing business in many emerging markets face challenges of weak or insecure property rights as well. Consequently, the types of strategies described in this paper are likely to resonate in those markets as well. However, there is one fundamental difference. In the settings that we have described in this paper, weak property rights are more sensibly viewed as a primitive of the contracting environment rather than a consequence of intentional design. In contrast, the differences in strength of property rights between emerging and developed markets are more likely endogenous to the actions of host regimes (and surely over the longer term). Moreover, property rights in those markets are insecure in general, and not only for knowledge and information.

(2) *Type 1 errors: overstating the scope of the problem.* As seen here, numerous firms have designed solutions to cope effectively with weak property rights.

In spite of this, many firms that are challenged by weak property rights often think that they are unable to cope. Why do firms typically overstate the scope of the problem ("type 1 errors") and find it difficult to identify solutions that can circumscribe weak property rights and reduce their vulnerability to holdup? Indeed, can firms avoid this type of error?

One answer is that they have incentives to do so. Overstating the threats is one form of lobbying and can be an effective strategy to rally sympathetic lawmakers. But, too often, firms genuinely believe that they have no other recourse when, in fact, they might. Indeed, we believe that framing the problem too narrowly often tends to limit the search for effective answers. To see this, recall that the music studios' initial (and pessimistic) response to illegal file sharing was to argue that a resulting 5% drop in CD sales would serve to wipe out their entire profits. And correctly so – if they continued to rely on CD sales for the bulk of their profits! As seen, however, effective solutions often require firms to change their existing strategies, their existing activities, and maybe even their existing boundaries. In the future, it would not be surprising for record companies to own or be owned by corporations selling goods or investing in assets that are complements of music.

Recall that an effective strategy for firms to overcome holdup, imitation, or piracy of key assets is to own or control complementary assets (just like firms that intentionally give away a product for free to make money from sales of another). Firms spend R&D dollars on open-source software, knowing that they can extract revenues on related consulting services. Netscape voluntarily priced its browser for free – knowing that this might increase demand for web servers by corporations from whom they could generate revenues. And Microsoft spent millions of dollars to develop a competitive product to Real Network's software for live streaming of audio and video content, then gave it away for free. Why? Because such software applications would only enhance the value of its operating system where it made money. In each case, the firm managed to reduce its reliance on a key asset or product by expanding its scope of activities, so that it could control or own additional complementary assets and products. And, in each case, the argument that weak property rights make it impossible to capture profits from the relevant asset or product would have missed the possibility that a firm may capture rents on complementary assets or products, or elsewhere in the value chain. The point is that looking for solutions using a piecemeal approach can overemphasize the costs of holdup and miss feasible solutions.

(3) *Type 2 errors: believing that weak property rights are not a problem.* In contrast to type 1 errors described above, firms sometimes fall into the opposite trap and underestimate the problem. Consider a simple example. Suppose that one

> observes a given market and sees firms regularly making large investments in assets they cannot legally own and that generate substantial profits. Seeing this, one may incorrectly conclude that either property rights over these assets must be reasonably secure or that weak property rights are not a problem after all. The logical flaw, of course, is to omit that the observed outcomes may reflect effective strategies rather than effective property rights. Consequently, in such markets, one should ask a simple but central question: are weak property rights not a concern per se, or aren't they a concern in *equilibrium*?

Consider the following examples. Book authors rarely switch between publishing houses when it comes time to renegotiate contracts for future book deals. Entrepreneurs who receive funding offers from a particular venture capitalist rarely end up switching financiers to get better terms elsewhere. Athletes who are nurtured from an early age by particular agents or sports marketing companies rarely switch to others later on when they become stars. One might incorrectly conclude that in each case switching costs between firms are high. However, as should be clear from our cases, firms typically design smart strategies to ensure that holdup is mitigated and switching costs are high *in equilibrium*. Consequently, understanding the nature and impact of firms' actions can be central to understanding whether property rights are intrinsically secure or *de facto* adequate.[55]

NOTES

1. See "Copy Cat Culture," *Business and Finance* (2002), 33, Ham and Atkinson (2000), and Langenderfer and Cook (2001).
2. "This is War: Should The Computer Industry Protect Hollywood from Digital Theft?" *Fortune* (2002, 27 May) *145*(11).
3. The rest of this section draws on Anand and Galetovic (2000a).
4. In Anand et al. (2003), we endogenize the extent to which projects within a corporation are kept separate or are integrated, and show that it will be optimal not to provide complete autonomy to project managers.
5. This may be due to several factors: (a) isolating the components of cost of the sources of revenue is simply a harder task when projects are integrated; and (b) the corporation may have an incentive to misrepresent costs and revenues so as to decrease payouts to the researcher. Holmstrom (1989), for example, argues that either "the best intentioned firm does not know capital costs," or, even if it does, it "has control of many levers to make accounting measures less reliable."
6. See Anton and Yao (1994).
7. "Synergize This," by Cynthia Littleton, *Variety* (1999, 23–29 August), 27.
8. A few months later, producer Steven Bochko also filed a lawsuit against Fox for selling the show NYPD Blue (in which Bochko had a profit participation stake) to its own

cable channel, FX, at below-market prices, and argued that "[...] this is just the 1999 version of what they used to call 'creative bookkeeping.' " In "Media Mergers Bottom Line So Far? Lawsuits," by Brian Lowry, *Los Angeles Times* (1999, 17 September).

9. "NBA Lockout Appears Certain," by Eddie Sefko, *Houston Chronicle* (1998, 28 June).

10. Anand and Khanna (2000) argue that because it is often difficult to clearly specify the boundaries and content of some intangible assets, it should also, a posteriori, be difficult to allocate property rights over these assets.

11. For example, Anand and Khanna (2000) note that copyright attaches only to an expression of an idea, not the idea itself, making it considerably easier to invent around these rights. Furthermore, another important feature of copyright law, distinct from patent law, is that it is not held to be infringed if another firm independently creates the article in question.

12. See also Lerner (2002).

13. But see recent analyses by Anton and Yao (2002, 2003) that study how a seller can appropriate value even under these circumstances. They study the role of partial disclosure, and the inclusion of certain protective clauses in contracts with potential buyers.

14. "Adopting Orphans" (1999, 20 February).

15. "Living with Digital Piracy? It's Been Done Before," by Holman Jenkins, *Wall Street Journal* (2002, 22 August).

16. As discussed above, a defining characteristics of "firms" is that there is typically some interaction between employees and across projects, so that the case where assets can be exploited by the firm and not just an employee is likely to be the norm rather than the exception.

17. Brandenburger and Nalebuff (1996) discuss several examples of this general principle, including negotiations between regional utilities and railroads that supply coal, and solicitations of competing bids by a public firm that is a takeover target. Even when a potential entrant has a competitive *disadvantage* relative to incumbents, the threat of competition can result in substantial gains for the entrant. Consider the production of aspartame, for example. Until the expiration of its patents in 1993, Nutrasweet was a monopolist producer of aspartame in the U.S., and extracted substantial rents from its two major buyers, Coca Cola and Pepsi Co. The entry of Holland Sweetener Company (HSC), even as an unbranded, high-cost aspartame producer, was likely to shift the bargaining power substantially. Nutrasweet's value to Coca Cola and Pepsi Co. was now only the *incremental benefit* above generic aspartame. Indeed, as Brandenburger and Nalebuff argue, "[...] what was left was Nutrasweet's brand equity and manufacturing cost advantage." This in turn would imply that HSC should have been able to extract sizeable rents from Coca Cola and Pepsi Co. before entry even though it had higher costs and its product was unbranded. The reason is simple: the pie would have been no smaller without HSC – it just would have been divided up differently. In other words, as the authors elegantly note, HSC was in a very weak position to sell aspartame, but very strong to sell competition.

18. Gallini and Scotchmer (2002) summarize the argument:

> Competitive markets may not be conducive to innovation, for a reason. [...] Inventions are information, and information is a public good. An invention such as a wireless palmtop is a combination of tangible embodiments and an intangible idea, as well as information about how to manufacture it. Typically, both the information and the tangible embodiments are costly to the inventor, but only the tangible components are costly to a rival. Without some sort of protection or reward, the inventor will therefore be at a market disadvantage relative to rivals, and will possibly be dissuaded from investing.

19. Interestingly, Eccles and Crane (1988) note that future payment for current advice is not an industry practice that could be easily changed because the value of advice is not known before it is given.

20. As Anand and Galetovic (2000b) note, "In just over a year, Deutsche Bank lured more than 200 employees of other investment banks, including many senior managers." Within a year, the bank claimed that it was the global market share leader in several product areas, where these market shares were assessed purely on the basis of its bankers' relationships with clients and on the deal volume that they had generated in the previous year!

21. This section is based on International Management Group, by Bharat Anand, Harvard Business School Case Number 702-409.

22. Court documents revealed that agent Jeffrey Schwarz's total compensation from IMG (including salary, bonus, and housing allowance) was slightly over $160,000 in 1997, or less than 10% of IMG's fees from Sampras' endorsements alone. In contrast, note that partners in professional service firms appropriate the entire value created by the firm.

23. In tennis, both Martina Hingis and Anna Kournikova, with whom IMG had established relationships at an early age, left IMG before they had turned 18.

24. In 2002, the average purse for an event was $3.5 million. Since IMG's clients typically win 50% of prize money, IMG would "get back" approximately $175,000 in prize money per event, which afforded them a cost advantage of 5% in bidding for such events.

25. Note the difficulties in contracting at arms-length in each case. For example, forcing event organizers to use its host exemptions for IMG clients in return for the promise of getting Tiger Woods would be illegal because of "tying."

26. Formally, the WTP function is superadditive in its components.

27. As Netscape co-founder Marc Andressen noted: "A lot of companies who are going to use it are going to pay for it, because among other things they want to pay for it. Free software is usually more expensive in the long run for companies to use." In *The Browser Wars, 1994–1998*, by David Yoffie and Mary Kwak, Harvard Business School Case No. 9-798-094.

28. "Spoilt For Choice," *The Economist* (2001, 5 July). The article also notes that "McKinsey's old-boy network is the envy of all. Its role call of chief executives includes Lou Gerstner of IBM, Harvey Golub of American Express, Lukas Muhlemann of Credit Suisse, Jeffrey Skilling of Enron, and Anton van Rossum of Fortis, to name a few. The power of the service firms' old-boy networks increases exponentially . . .. A slight variation on (Metcalfe's Law) can be applied to old-boy networks: their value grows in relation to the square of the number of former employees working for clients, and their elevation in the corporate hierarchy."

29. "This is War: Should the Computer Industry Protect Hollywood from Digital Theft?" *Fortune* (2002, 27 May), *145*(11).

30. "E-Commerce: A Special Report," *Wall Street Journal* (2003, 27 January).

31. *Wall Street Journal* (2003), op. cit.

32. See "Burn, Baby, Burn," *Time* (2002, 20 May), *159*(20).

33. *Time*, op. cit.

34. "Judge Orders Verizon to Name Song-Swapper," *The Wall Street Journal* (2003, 22 January).

35. "Fake Files, User Caps Hit Music Swappers," *The Toronto Star* (2002, 17 June).

36. See "Music Industry Hopes Exclusives Can Blow Pirates Out of the Water," *Wall Street Journal* (2002, 16 September).

37. See "Burned?" *Boston Globe* (2002, 12 May); "Burn, Baby, Burn," *Time* (2002, 20 May).

38. "This is War: Should the Computer Industry Protect Hollywood from Digital Theft?" *Fortune* (2002, 27 May), *145*(11).

39. As the President of New Media for Universal Records notes, "I find it incredibly ironic that some people will spend an extra $1,000 on their hard drives just so they can store more music, but they won't pay for music" ("Burned?" *Boston Globe* (2002, 12 May)).

40. "Judge Orders Verizon to Name Song-Swapper," *The Wall Street Journal* (2003, 22 January).

41. "Judge Orders Verizon to Name Song-Swapper," *Wall Street Journal* (2003, 22 January).

42. See "This is War: Should the Computer Industry Protect Hollywood from Digital Theft?" *Fortune* (2002, 27 May), *145*(11). His comments were echoed in a recent article on piracy that observed that file sharing services "[. . .] are the killer apps of the broadband computing world."

43. *Fortune*, op. cit. The article also cites an op-ed piece in which Grove questioned the logic of regulatory intervention, and the incentives of the entertainment companies in fighting technological change rather than embracing it: "Were the manufacturers of printing presses forced to protect the monks? Was the PC industry forced to protect the mainframe industry? Why is this case any different? . . . The message is clear: the studio owners are dinosaurs. If they can't adapt the to the brave new world the companies like Intel and Apple have ushered in, extinction is what they deserve." In response, one observer simply noted: "Grove has a point. But so do Hollywood executives who accuse their Silicon Valley counterparts of ignoring Internet piracy because it helps them sell gadgets."

44. One example might include renegotiating contracts with MTV and other promotional channels benefit from an increase in the demand for music by listeners, but (as in the case of MTV) obtain music videos for free from the record companies.

45. One example may include lobbying for a tax increase on blank CD sales, whose revenues would flow back to the record companies.

46. Interestingly, the aforementioned Senate bill was endorsed by movie studios from Disney, News Corporation, and Viacom, whereas it was "supported somewhat less enthusiastically by AOL Time Warner CEO Richard Parsons and Sony Corp. of America CEO Howard Stringer." ("This is War: Should the Computer Industry Protect Hollywood from Digital Theft?" *Fortune* (2002, 27 May), *145*(11).

47. Conflicts between owners of content and owners of complements are not recent. Indeed, as one observer noted:

> The entertainment industry has been engaged in a reactive struggle against new technologies for most of the last century, sometimes succeeding in the legal battle but ultimately unable to stop technology's advance. Often surprisingly, the technologies that copyright holders initially sought to suppress were eventually embraced as an integral part of their business. The owner of sheet music copyrights sued the maker of piano rolls used in player pianos. A music producer sued a radio station over its broadcasts when radio threatened the traditional music copyright model. A movie studio sued a fledgling cable television operator for rebroadcasting television signals. More recently, a motion picture company filed a claim against a VCR manufacturer, adding another technological advance to the litany of inventions predicted to

doom intellectual property rights. In each case, content owners eventually embraced the new technology they originally opposed.

Raysman and Brown (2000) note that:

> although the relationship between file sharing and large companies is antagonistic right now, history suggests that this new method of content distribution may be a significant commercial opportunity for the recording industry. [...] The industry needs to look back only to the early 1980s, when Universal City Studios sued Sony for manufacturing videotape recorders, which Universal perceived as a threat to its movie business. The decision initially startled the movie industry, but its effect turned out to be positive: today, revenue from video rentals comprises a substantial share of total movie studio revenues. Similarly, baseball teams initially protested free radio broadcasts of their teams' games, fearing the listening audience would cannibalize ticket sales. Today, revenue from television and radio broadcasting dwarfs that of tickets.

48. See "This is War: Should the Computer Industry Protect Hollywood from Digital Theft?" *Fortune* (2002, 27 May), *145*(11). As the *Wall Street Journal* recently speculated, "eventually, the music companies and consumer electronic companies need to join forces. There's even been whistling that Apple or Microsoft might buy a music company."

49. "Off-Key: The Music Industry is Finally Online, But Few Listen," *The Wall Street Journal* (2002, 7 May).

50. "Digital Media: Don't Clamp Down Too Hard," *Business Week* (2002, 14 October).

51. See "Copy Cat Culture," *Business and Finance* (2002), 33.

52. *The New York Times* ("Downloading Music Over the Internet Without Feeling Like A Criminal," June 7, 2003) noted that "[...] it is ironic that Apple should have introduced the first really successful commercial Internet music service," in contrast to the record industry that had contrived to "alienate its ideal audience – young people who live and breathe music – by being way behind the technological curve and by repeatedly sounding as if its main job was law enforcement rather than selling music."

53. In their survey analysis of protection for intellectual property, Levin et al. (1987) similarly conclude: "Since the impact of legal protection of intellectual property . . . varies widely among industries, focused efforts to solve problems in specific markets would be more prudent than a broad attempt to upgrade protection. There is little point in expending diplomatic capital to compel foreign countries to pass or enforce laws that, *in most industries, would have minimal impact on the competitive process*."

54. Boldrin and Levine (2003) take this logic a step further to make a case against "intellectual monopoly" rights. They argue that even when legal mechanisms are effective, granting monopoly rights to inventors may be *unnecessary* since, in practice, they can often devise ways of recouping the investments in their innovation. Focusing on the benefits of first-mover advantage, they note that "ultimately no academic work can do more than scratch the surface of (this) advantage: it is limited only by human ingenuity."

55. Note that similar problems might arise with surveys or questionnaires as well. Summarizing the results of their survey analysis of managers in 45 industries about the effectiveness of protection afforded by patent laws for innovations, Levin et al. (1987) noted that "a respondent's perception . . . may also be affected by his firm's policies or strategies."

REFERENCES

Alchian, A. (1967). Property rights. *The Concise Encyclopedia of Economics*.

Anand, B., & Attea, K. (2002). International management group. Harvard Business School. Case Number 9-702-409.

Anand, B., & Cantillon, E. (2003). The music industry and the internet. Harvard Business School. Case Number 9-703-513.

Anand, B., & Galetovic, A. (2000a). Weak property rights and holdup in R&D. *Journal of Economics and Management Strategy*, *9*, 615–642.

Anand, B., & Galetovic, A. (2000b). Information, non-excludability and financial market structure. *Journal of Business*, *73*, 357–402.

Anand, B., Galetovic, A., & Stein, A. (2003). Incentives vs. synergies in the market for talent. Work in Progress, CEA and Harvard Business School.

Anand, B., & Khanna, T. (2000). The structure of licensing contracts. *Journal of Industrial Economics*, *47*, 103–135.

Anand, B., Rukstad, M. G., & Paige, C. H. (2000). Capital One financial corp. Harvard Business School Case Number 9-700-124.

Anton, J., & Yao, D. (1994). Expropriations and inventions: Appropriable rents in the absence of property rights. *American Economic Review*, *84*, 190–209.

Anton, J., & Yao, D. (2002). The sale of ideas: Strategic disclosure, property rights, and contracting. *Review of Economic Studies*, *69*, 513–531.

Anton, J., & Yao, D. (2003). Attracting skeptical buyers. Working Paper, Wharton School.

Arrow, K. (1962). Economic welfare and the allocation of resources for innovation. In: R. Nelson (Ed.), *The Rate and Direction of Inventive Activity*. Princeton: Princeton University Press.

Bhattacharyya, S., & Nanda, V. (2000). Client discretion, switching costs, and financial innovation. *Review of Financial Studies*, *13*, 1101–1127.

Bhide, A. (1994). How entrepreneurs craft strategies that work. *Harvard Business Review*, Reprint #94202.

Boldrin, M., & Levine, D. K. (2003). The case against intellectual monopoly. Manuscript in progress.

Brandenburger, A. M., & Nalebuff, B. J. (1996). *Co-opetition: A revolutionary mindset that combines competition and co-operation: The game theory strategy that's changing the game of business*. New York: Doubleday.

Casadesus-Masanell, R., Yoffie, D., & Mattu, S. (2002). Intel Corp. – 1968–2003. Harvard Business School Case Number 9-703-427.

Cohen, W., Nelson, R., & Walsh, J. (2000). Protecting their intellectual assets: Appropriability conditions and why U.S. manufacturing firms patent (or not). Working Paper No. 7552, National Bureau of Economic Research.

Eccles, R., & Crane, D. (1988). *Doing deals: Investment banks at work*. Boston: Harvard Business School Press.

Endlich, L. (2000). *Goldman Sachs: The culture of success*. New York: Touchstone Books.

Gallini, N., & Scotchmer, S. (2002). Intellectual property: When is it the best incentive system? In: A. Jaffe, J. Lerner & S. Stern (Eds), *Innovation Policy and the Economy* (Vol. 2). Cambridge: MIT Press.

Ham, S., & Atkinson, R. (2000). Napster and online piracy: The need to revisit the digital millennium copyright act. *E-Business Law Bulletin*, *1*.

Hart, O., & Moore, J. (1994). A theory of debt based in the inalienability of human capital. *Quarterly Journal of Economics*, *109*, 841–879.

Holmstrom, B. (1989). Agency costs and innovation. *Journal of Economic Behavior and Organization*, *12*, 305–327.

Horstmann, I., Slivinski, A., & MacDonald, G. (1985). Patents as information transfer mechanisms: To patent or (maybe) not to patent. *Journal of Political Economy*, *93*, 837–858.

Langenderfer, J., & Cook, D. L. (2001). Copyright policies and issues raised by A&M Records v. Napster: The shot heard 'round the world' or not with a bang but a whimper? *Journal of Public Policy and Marketing*, *20*.

Lerner, J. (2002). Where does state street lead? A first look at finance patents. *Journal of Finance*, *57*, 901–930.

Lerner, J., & Tirole, J. (2002). Some simple economics of open source. *Journal of Industrial Economics*, *50*, 197–234.

Levin, R. C., Klevorick, R., Nelson, A., & Winter, S. (1987). Appropriating the returns from industrial R&D. *Brookings Papers on Economic Activity*, 783–820.

Merges, R. P., & Nelson, R. R. (1990). On the complex economics of patent scope. *Columbia Law Review*, 90.

O'Donoghue, T., Scotchmer, S., & Thiesse, J. F. (1998). Patent breadth, patent life, and the pace of technological progress. *Journal of Economics and Management Strategy*, *7*, 1–32.

Pakes, A. (1986). Patents as options: Some estimates of the value of holding European patent stocks. *Econometrica*, *54*, 755–784.

Pakes, A., & Schankerman, M. (1984). The rate of obsolescence of knowledge, research gestation lags and the private rate of return to research resources. In: Z. Griliches (Ed.), *Patents, R&D, and Productivity*. Chicago: University of Chicago Press.

Ritter, J., & Chen, H. C. (2000). The seven-percent solution. *Journal of Finance*, *55*, 1105–1131.

Rolfe J., & Troob, P. (2001). *Monkey business: Swinging through the Wall Street jungle*. New York: Warner Books.

Schankerman, M., & Pakes, A. (1986). Estimates of the value of patent rights in European countries during the post-1950 period. *Economic Journal*, *96*, 1052–1077.

Shapiro, C., & Varian, H. R. (1998). *Information rules: A strategic guide to network economy*. Boston: Harvard Business School Press.

Shirky, C. (2001). Where Napster is taking the publishing world. *Harvard Business Review*, Reprint #R0102M.

von Hippel, E. (1988). *The sources of innovation*. New York: Oxford University Press

Wilhelm, W. J., & Downing, J. D. (2001). *Information markets: What businesses can learn from financial innovation*. Boston: Harvard Business School Press.